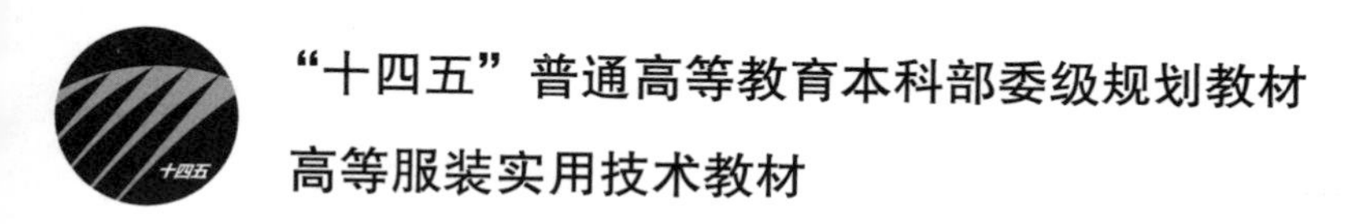

SHIYONG FUZHUANG ZHUANYE YINGYU

# 实用服装专业英语

## （第4版）

张小良　编著

中国纺织出版社有限公司

## 内　容　提　要

本书是“十四五”普通高等教育本科部委级规划教材。编写资料来源于企业各种实用文件，海外服装院校的讲义、书籍、杂志等。本书内容结合目前服装企业的设计、生产、销售与贸易等活动，包括服装面料的选择、设计与纸样、服装工艺、品质检验、服装市场营销、沟通英语和常用商业书信等，在附录中还选编了常见服装款式、服装细节、服装部件、贸易术语、缩略语等相关英语词汇。本书内容丰富而新颖，专业词汇覆盖面广，实用且方便读者学习服装专业英语词汇，有利于提高专业文章的阅读理解能力和对企业文件资料的读译能力。另外，实用图表可供服装企业借鉴使用。

本书适合高等教育本科院校的服装教学，也可供服装企业及广大服装爱好者参考学习。

**图书在版编目（CIP）数据**

实用服装专业英语 / 张小良编著. —4 版. —北京：中国纺织出版社有限公司，2021.11
“十四五”普通高等教育本科部委级规划教材
ISBN 978-7-5180-5866-2

Ⅰ. ①实…　Ⅱ. ①张…　Ⅲ. ①服装工业—英语—高等学校—教材　Ⅳ. ① TS941

中国版本图书馆 CIP 数据核字（2021）第 028814 号

---

责任编辑：宗　静　　特约编辑：石鑫鑫　李淑敏
责任校对：王花妮　　责任印制：王艳丽

---

中国纺织出版社有限公司出版发行
地址：北京市朝阳区百子湾东里 A407 号楼　邮政编码：100124
销售电话：010—67004422　传真：010—87155801
http：//www.c-textilep.com
中国纺织出版社天猫旗舰店
官方微博 http：//weibo.com/2119887771
三河市宏盛印务有限公司印刷　各地新华书店经销
2008 年 2 月第 1 版　2012 年 1 月第 2 版
2017 年 4 月第 3 版　2021 年 11 月第 4 版
开本：787×1092　1/16　印张：20.75
字数：350 千字　定价：59.80 元

---

凡购本书，如有缺页、倒页、脱页，由本社图书营销中心调换

# 第 4 版前言

目前，服装专业英语对服装企业特别是外贸服装企业非常重要。为方便服装专业院校师生及广大服装工作者能更好地学习与掌握真正适用于企业的服装专业英语，我们参考并汇编了企业中的各种文件资料以及国外服装院校的讲义、书籍、杂志等，根据服装企业的设计、生产、销售与贸易等活动过程，编写了《实用服装专业英语》一书。本书可供服装院校师生，服装设计、生产、营销等各个环节的企业员工及广大服装爱好者参考使用。

全书共分七章，最后为附录内容，每章节后附中英专业词汇对照表（其中标“*”的词汇为本章编外相关专业词汇）以及专业词汇练习。另外，全书增加了中文翻译，方便学生与服装爱好者自学。本书作为服装院校专业英语教材，教学时数约为 54 课时。第一章为服装材料的选择，首先介绍了有关纤维、纱线、布料等服装材料的基础知识，包含大量服装材料的专业词汇，以便学生学习与查阅。然后以原材料的采购作为切入点，介绍服装企业的采购职能与工作要求。第二章为设计与纸样，简单介绍服装设计效果图的分类、纸样构造、放码与排料等基本知识，并进一步介绍了时装信息与名词术语，以提高读者对杂志、专业文件的阅读理解能力。第三章为服装生产，简述各种服装类型的制作流程、工序要求与生产方式等内容，并阐述了服装生产工艺中的各种文件系统，从而进一步加强了学生对服装与技术操作词汇的学习。此章重点培养学生对英语文件、制单的翻译与编写能力。第四章为品质检验，主要从 QC（Quality Control）岗位的角度进行编写，内容包括 QC 尺寸检验技能、各种质检文件的应用与编写以及各种服装次品的表达方式。该章节内容对从事 QC 岗位工作的专业人员非常重要。第五章为服装市场营销与采购，简单介绍市场有关部门的职能，以案例形式介绍了休闲装的市场策略以及成衣采购与各种表格文件的应用，内容涉及大量的专业营销与采购词汇。第六章为商务英语，汇编了各种情景下的英语对话，内容涉及营销、办公室电话沟通以及业务洽谈等常用语句。第七章为服装行业商业信函，主要以案例形式分析多种常用服装贸易信件、电子邮件的编写方法与应用技巧。附录则选编了服装款式、服装细节、服装部件、组织结构、职位与头衔、求职应聘申请表、贸易术语、缩略语等，可方便学生的学习与参考。另外，各章节拟定的课时数仅作为参考，可按具体需求进行调节。本教材能顺利出版，除了作者的努力之外，还要感谢那些参与编写与审稿工作的其他教师与专业人士，其中包括北京服装学院的刘利、郑州中原工学院的刘娟、广东五邑大学的江汝南、华南农业大学的范福军、无锡太湖学院的董雪峰、惠州职业经济技术学院的陈文焰、惠州学院的外教 Kelly Runcie 等，以及负责全书插图绘制的惠州学院的侯开慧。他们从最初的资料收集、编著、修改，到最后的定稿、审稿工作，都给予本人很大的帮助，在此向他们表示最诚挚的谢意。

因编著者水平有限，书中内容难免有不妥之处，敬请读者指正。

编著者

2020 年 10 月

# 第 1 版前言

纺织服装工业是我国历史悠久的传统产业，在解决我国“三农”问题、城镇职工就业、增加资金积累、带动第三产业发展以及促进民营企业发展等方面都发挥了重大作用。经过“十五”时期的快速发展，我国纺织服装工业现已形成拥有纤维、纺织、织造、染整、服装等上、中、下游衔接配套的完整产业体系，产能不断扩大，产品结构日趋多样。可以看出，在纺织服装出口贸易中，服装出口处于主导地位。在纺织服装出口总额中，尽管加工贸易出口仍占较高比重，但一般贸易出口呈现出快速增长势头，纺织服装产业的创汇能力与附加值正在提升，这些都推动了我国纺织服装产业的结构升级。

“十一五”期间，我国纺织服装工业将进入后配额时代，尽管目前欧盟、美国对我国纺织服装产品出口增长过快采取了限制措施，导致我国纺织服装产品出口存在一些不确定因素，但国际纺织服装产品市场进一步开放与国内纺织服装产品市场进一步增长是必然的趋势，我国纺织服装工业将面临新的机遇和挑战。目前，我国服装企业多数为中小型企业，而且以乡镇集体企业、民营企业或三资企业为主，因此，在未来的市场竞争中，我国纺织服装企业要提升核心竞争力与可持续发展能力，一方面需要加快经营规模的扩张与经营创新，开展二次创业；另一方面，需要培养或储备大量既懂服装生产工艺，又懂服装生产技术与管理的实用型、技术型、管理型人才。

“服装实用技术教材”系列丛书正是针对服装行业发展的形势及服装企业对人才需求的特点编写而成的，具有实用性和可操作性。该套丛书 2000 年出版以来，深受服装企业及服装职业技术教育院校的欢迎。目前该套丛书结合服装行业发展的实际需求，进行了较大的修订，增加了新的形式与内容，可以作为服装专业的配套教材或在职服装企业经营管理人员及有志于成为服装企业经营管理人员的参考丛书，该套丛书也被正式指定为广东省服装设计与工程专业自考教材。本套丛书由中国纺织出版社组织惠州学院服装系（又称西纺广东服装学院）一批多年从事服装教学工作的教师编写。西纺广东服装学院与香港旭日集团合作办学二十多年，培养了大量服装企业第一线实用型经营管理人才，深受服装企业的欢迎与好评，其新颖的办学模式在珠江三角洲地区产生了广泛的影响，享有较高的声誉，并得到了中国纺织工业协会全国纺织服装教育学会的肯定。我们编写这套丛书，旨在总结西纺广东服装学院合作办学的成果，并通过这套丛书与从事服装教育的广大工作者及从事服装企业经营管理的同仁进行广泛交流，共同促进我国服装行业的发展。

本套丛书包括《成衣基础工艺》、《成衣生产工艺》、《服装纸样设计》、《服装立体裁剪实用教程》、《成衣纸样电脑放码》、《服装设计学》、《服装品质管理》、《实用服装专业英语》、《服装企业督导管理》等十余册，由惠州学院服装系吴铭、刘小红担任编委会主任，参加编写的人员包括刘小红、刘东、杨雪梅、范强、李秀英、陶钧、张小良、万志琴、严燕连、冯麟、陈霞、王秀梅、陈学军、宋惠景、罗琴、徐丽丽、李郁纯等。希望本套教材能受到广大读者的欢迎，不足之处恳请读者批评指正。

编著者

2007年12月

# 第 2 版前言

目前，服装专业英语对服装企业特别是外贸服装企业非常重要。为了方便服装院校师生及广大服装工作者更好地学习与掌握真正适用于企业的服装专业英语，我们参考并汇编了企业中的各种文件资料以及国外服装院校的讲义、书籍、杂志等，根据服装企业的设计、生产、销售与贸易等活动过程，编写了本书。本书可供服装院校师生，服装企业设计、生产、营销等各个环节的员工以及广大服装爱好者参考使用。

全书共分七章，最后为附录内容，每章节后附有中英服装专业词汇对照表（其中有“*”号为本章编外相关专业词汇）以及专业词汇练习。另外，全书在光盘中增加了中文翻译，更方便学生与服装爱好者自学。作为服装院校专业英语教材，教学时数约为 64 学时。第一章面料的选择，首先介绍了有关纤维、纱线、布料等服装材料的基础知识，包含了大量服装材料的专业词汇，以便学生学习与查阅；然后以原材料的采购作为切入点，介绍了服装企业的采购职能与工作要求。第二章设计与纸样，以款式描述形式，简单介绍了服装设计效果图的分类，纸样构造、放码与排料等基本知识，并进一步介绍了时装信息与名词术语，以便提高读者对杂志、专业文件的阅读理解能力。第三章服装工艺，为本书的重点章节，简单概述了各种服装类型的制作流程、工序要求与生产方式等内容，并阐述了服装生产工艺中的各种文件系统，从而进一步加强了学生对服装与技术操作词汇的学习。此章重点培养学生对英语文件、制单的翻译与编写能力。第四章品质检验，主要从 QC (Quality Control) 岗位的角度进行编写，内容包括 QC 尺寸检验技能、各种质检文件的应用与编写以及各种服装次品的表述方式。该章节内容对从事 QC 岗位工作的专业人员非常重要。第五章服装市场营销，简单介绍了市场有关部门的职能，以案例形式介绍了目前国内休闲装的市场策略，内容涉及大量的专业营销词汇。第六章沟通英语，汇编了英语的各种情景对话，内容涉及营销、办公室电话沟通以及业务洽谈等常用英语。第七章常用商业书信，主要以案例形式分析多种常用服装贸易信件、电子邮件的编写方法与应用技巧。附录则选编了常见的服装款式、服装细节、服装部件，组织结构、职位与头衔、求职应聘申请表、缩略语等，可方便学生学习与参考。各章节具体学时数可按实际需求进行调节。

本教材能顺利出版，除了作者努力工作外，还须感谢那些给出编写意见并参与审稿工作的其他教师与专业人士。其中包括北京服装学院的刘莉老师、郑州中原工学院的刘娟老师、广东五邑大学的江汝南老师、加拿大籍外教Sebastien Bousquet老师、Canada Delta College商务英语系主任朱亿老师（加拿大籍）等，还有负责绘制插图的服装设计师林松涛先生。他们从最初的资料收集、编写与翻译、修改，到最后的定稿审稿工作，都给予了本人很大的帮助，在此我向他们表示最诚挚的谢意。

因水平有限，书中内容难免有不妥之处，敬请读者指正。

编著者

2011年10月

# 第3版前言

目前，服装专业英语对服装企业特别是外贸服装企业非常重要。为了方便服装院校师生及广大服装工作者更好地学习与掌握真正适用于企业的服装专业英语，我们参考并汇编了企业中的各种文件资料以及国外服装院校的讲义、书籍、杂志等，根据服装企业的设计、生产、销售与贸易等活动过程，编写了《实用服装专业英语》一书。本书可供服装院校师生，服装设计、生产、营销等各个环节的企业员工以及广大服装爱好者参考使用。

全书共分七章，最后为附录内容，每章节后附有中英专业词汇对照表（其中标“*”的词汇为本章编外相关专业词汇）以及专业词汇练习。另外，全书增加了中文翻译，更方便了学生与服装爱好者自学。本书作为服装院校专业英语教材，教学时数约为54课时。第一章为服装材料的选择，首先介绍了有关纤维、纱线、布料等服装材料的基础知识，包含大量服装材料的专业词汇，以便学生学习与查阅；然后以原材料的采购作为切入点，介绍服装企业的采购职能与工作要求。第二章为设计与纸样，简单介绍了服装设计效果图的分类，纸样构造、放码与排料等基本知识，并进一步介绍了时装信息与名词术语，以便提高读者对杂志、专业文件的阅读理解能力。第三章为服装生产，简述了各种服装类型的制作流程、工序要求与生产方式等内容，并阐述了服装生产工艺中的各种文件系统，从而进一步加强了学生对服装与技术操作词汇的学习。此章重点培养学生对英语文件、制单的翻译与编写能力。第四章为品质检验，主要从QC（Quality Control）岗位的角度进行编写，内容包括QC尺寸检验技能、各种质检文件的应用与编写以及各种服装次品的表达方式。该章节内容对从事QC岗位工作的专业人员非常重要。第五章为服装市场与采购，简单介绍了市场有关部门的职能，以案例形式介绍了休闲装的市场策略以及成衣采购与各种表格文件的应用，内容涉及大量的专业营销与采购词汇。第六章为英语沟通，汇编了英语的各种情景对话，内容涉及营销、办公室电话沟通以及业务洽谈等常用英语。第七章为服装行业商业信函，主要以案例形式分析多种常用服装贸易信件、电子邮件的编写方法与应用技巧。附录则选编了常见服装款式、服装细节、服装部件、组织结构、职位与头衔、求职应聘申请表、缩略语等，可方便学生的学习与参考。另外，各章节拟定的学时数仅作为参考，可按具体需求进行调节。

本教材能顺利出版，除了作者的努力之外，还要感谢那些参与编写与审稿工作的其他教师与专业人士。其中包括北京服装学院的刘利、郑州中原工学院的刘娟、广东五邑大学的江汝南、惠州职业经济技术学院的陈文焰、惠州学院的外教 Kelly Runcie 等，以及负责全书插图绘制的惠州学院的侯开慧。他们从最初的资料收集、编著、修改，到最后的定稿审稿工作，都给予本人很大的帮助，在此向他们表示最诚挚的谢意。

因编著者水平有限，书中内容难免有不妥之处，敬请读者指正。

编著者

2016 年 10 月

# 教学内容及课时安排

| 章 | 专业知识 / 课时 | 节 | 课程内容 |
|---|---|---|---|
| 第一章<br>服装材料的选择 | 服装材料基础知识与面辅料采购跟单<br>（8 课时） | | CHAPTER 1　GARMENT MATERIAL OPTIONS |
| | | 一 | Knowledge of Material　材料知识 |
| | | 二 | Fabric Pretreatment During Production　生产中的面料预处理 |
| | | 三 | Material Purchasing　材料采购 |
| 第二章<br>设计与纸样 | 服装设计与打板技术<br>（8 课时） | | CHAPTER 2　DESIGN AND PATTERN |
| | | 一 | Fashion Design　服装设计 |
| | | 二 | Fashion Trends　流行趋势 |
| | | 三 | Pattern Drawing　纸样绘制 |
| | | 四 | Relative Glossary　相关术语 |
| 第三章<br>服装生产 | 服装生产工艺与流程<br>（8 课时） | | CHAPTER 3　GARMENT MANUFACTURING |
| | | 一 | Sample Manufacturing　样衣制作 |
| | | 二 | Production Sheet　生产制作通知单 |
| | | 三 | Making up Men's Shirt　男式衬衫制作 |
| | | 四 | Pants Construction　裤子结构 |
| | | 五 | Flow Chart of Garment Manufacturing　服装制作流程图 |
| | | 六 | Garment Part Explaining　部件解释 |
| 第四章<br>品质检验 | 服装品质控制与疵点的表达<br>（8 课时） | | CHAPTER 4　QUALITY INSPECTION |
| | | 一 | Quality Standards　品质标准 |
| | | 二 | Inspection Report　检验报告 |
| | | 三 | Expression of Garmen Defects　服装疵点的表述 |
| 第五章<br>服装市场营销与采购 | 服装市场与采购<br>（6 课时） | | CHAPTER 5　APPAREL MARKETING AND MERCHANDISING |
| | | 一 | Introduction　简介 |
| | | 二 | Leisure Wear Market in China　中国的休闲服市场 |
| | | 三 | Case Analyzing— Baleno　案例分析——班尼路 |
| | | 四 | The Apparel Merchandising Cycle　成衣采购循环 |
| | | 五 | Relative Glossary　相关术语 |

续表

| 章 | 专业知识 / 课时 | 节 | 课程内容 |
| --- | --- | --- | --- |
| 第六章<br>商务英语 | 服装企业沟通业务洽谈<br>（4 课时） | | **CHAPTER 6 BUSINESS ENGLISH** |
| | | 一 | Practical English in Chain-Store 连锁店实用英语 |
| | | 二 | Telephone Conversation 电话交谈 |
| | | 三 | Business Demonstration 商务洽谈示范 |
| 第七章<br>服装行业商业信函 | 商业信函与电子邮件<br>（6 课时） | | **CHAPTER 7 BUSINESS LETTER IN CLOTHING INDUSTRY** |
| | | 一 | Introduction 简介 |
| | | 二 | Envelope 信封 |
| | | 三 | The Parts of Business Letters 商业信函的构成 |
| | | 四 | Model Letters 示范信函 |
| | | 五 | E-mail 电子邮件 |
| * 附录Ⅰ ~ Ⅲ<br>款式描述 | 课堂练习与讨论：服装款式描述<br>（4 课时） | | **Appendix Ⅰ ~ Ⅲ STYLE DESCRIPTION** |
| | | 一 | Fashion Style 服装款式 |
| | | 二 | Fashion Details 服装细节 |
| | | 三 | Garment Parts 服装部件 |
| * 附录Ⅳ ~ Ⅷ<br>求职应聘申请表及其他 | 课堂练习与讨论：求职应聘申请表的应用<br>（2 课时） | | **Appendix Ⅳ ~ Ⅷ APPLICATION FORM AND OTHERS** |
| | | 一 | Organization Chart 组织结构 |
| | | 二 | Position or Title 职位与头衔 |
| | | 三 | Application Form 求职应聘申请表 |
| | | 四 | Trading Terms 贸易术语 |
| | | 五 | Abbreviation of Terms 缩略语 |

**注** 各院校可根据自身的教学特点和教学计划对课时数进行调整。

# 目　录

CHAPTER 1

# GARMENT MATERIAL OPTIONS 服装材料的选择

**课题名称：**GARMENT MATERIAL OPTIONS 服装材料的选择

**课题内容：**Knowledge of Material 材料知识

Fabric Pretreatment During Production 生产中的面料预处理

Material Purchasing 材料采购

**课题时间：**8课时

**教学目的：**让学生进一步了解服装材料的基础知识，重点掌握并熟记各种纤维、纱线、面料以及辅料等相关术语与名称。

**教学方式：**结合PPT与音频多媒体课件，以教师课堂讲述为主，学生可结合专业知识适当参与讨论学习。

**教学要求：**1. 明确服装用纤维、纱线、面料以及辅料的基本分类。

2. 了解特殊面料的种类与工艺处理技巧。

3. 熟悉各种面料采购文件的形式与应用。

**课前（后）准备：**结合专业知识，课前预习课文内容。课后熟读课文主要部分，并熟记各种纤维、纱线、面料以及辅料等专业名称。

# CHAPTER 1

# GARMENT MATERIAL OPTIONS 服装材料的选择

## 1.1 Knowledge of Material 材料知识

Fibres, yarns, fabrics and accessories are the basic requirement of any textile, apparel or related industry. It is very important to select the appropriate fabric according to customers' requirements. So the manufacturers select fibres, yarns and fabrics according to the product they manufacture.

### 1.1.1 Textile Fibre 纺织纤维

Textile is applied to woven and knit fabric, such as threads, cords, ropes, braids, laces, embroideries, nets, and clothes made by weaving, knitting, felting, bonding, tufting. ① And the different kinds of textiles possess characteristics that make them useful in clothing applications. Textiles are still the major component of the clothes we wear and of many furnishings in our homes and offices. Fibres are the most basic raw material for the textile industry. And fibres are divided into various categories on the basis of different criteria. The most common categorization is that of natural fibres and man-made fibres. Fibres are also categorized as filament and staple fibres. Fibres in the form of strands are called filaments, example being nylon filament yarn. Very short fibres are known as staple fibres. The figure 1-1 shows a common classification of textile fibres.

①纺织业生产机织与针织产品，如通过机织、针织、制毯、毡合和起毛等方式制成的缝线、绳线、绳索、织带、花边、绣品、网布和衣服等。

Originally, all textiles were made from natural fibre such as cotton, wool, mohair, linen, ramie, and vicuna. All of these were available only as staple fibres that had to be spun into yarns before they were to cloth. ② And natural fibres are derived from plants, animals, or minerals, and include cotton, flax, wool, or silk, etc. Manufactured fibres are chemically produced. Each fibre has characteristics that make them suitable for various uses. Fibre blends are combinations of two or more different fibres. Usually the fibre present in highest percentage dominates the fabric, but a successful blend will exhibit the desirable qualities of all.

②最初，所有纺织材料都由天然纤维制作而成，如棉、羊毛、马海毛、亚麻、苎麻和小羊驼毛等。在织成布料前，只有将这些短纤维纺成纱线后才能有效应用。

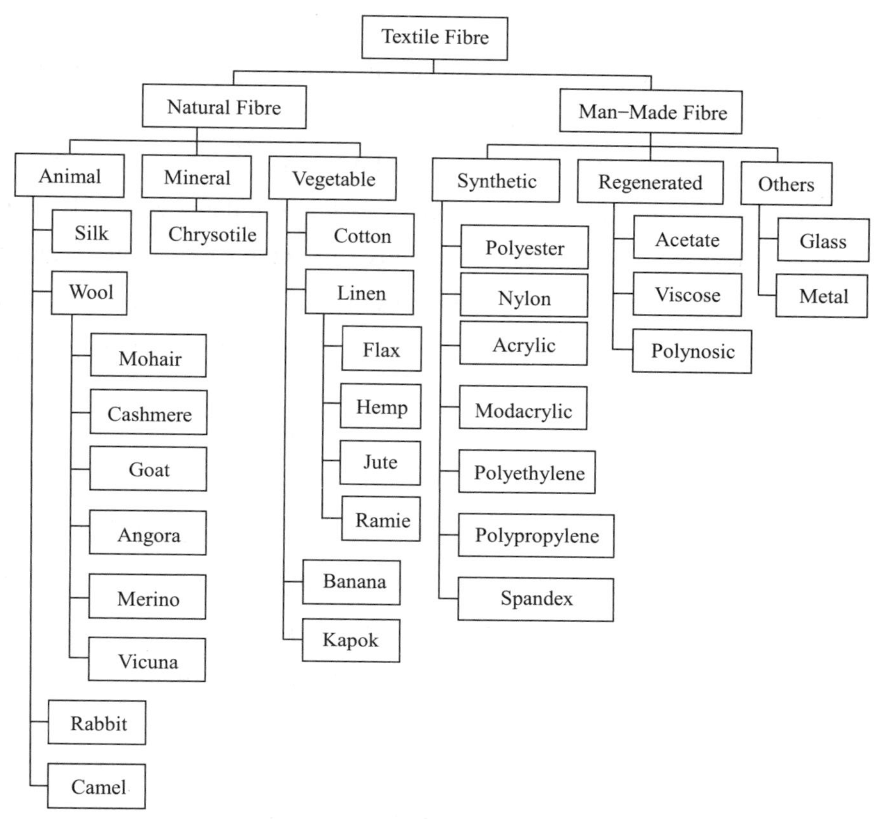


**Figure 1-1 Classification of Textile Fibre**

Silk was the first monofilament material, and for years scientists were obsessed with trying to make 'artificial' silk. Some new textiles possess qualities that make them stain-resistant, flameproof, and even stiff. Innovation in textile technology continues and more unusual products will almost surely emerge. These developments laid the foundation for the scientific principles that led the Du Pont Company to invent nylon and polyester fibres and yarns. Artificial fibres are produced from natural products, such as cellulose and proteins. These fabrics have a silk appearance, shape very well and are therefore ideal in the production of drapery. It can be used for lingerie, suits, blouses and lining.

Synthetic fibres are made entirely from chemicals. Synthetic fibres are usually stronger than either natural or regenerated fibres. Synthetic fibres and the regenerated acetate fibre are thermoplastic, they are softened by heat. Therefore manufacturers can shape these fibres at high temperatures, adding such features as pleats and creases. Synthetic fibres will melt if touched with a hot iron. The most widely used kinds of synthetic fibres are nylon (polyamide), polyester, acrylic and polypropylene.

The most important animal fibres are wool, mohair, cashmere. It is used in suitings, coatings, dress-goods, knitwear, rugs. Highest quality mohair is expensive, and it is a difficult fibre to spin. Cashmere is the under-coat hair of the Cashmere goat. Cashmere goats grow two coats, a fine soft under-coat and a coarse outer coat of long hair. ③ Cashmere is beautifully soft and lustrous, it has a slightly slippery handle and is used in high quality knitwear, coatings and suitings. Fabrics made from 100% cashmere are expensive, therefore mixtures with wool are quite common.

③开司米山羊长了两层皮毛，包括幼细而柔软的内层与粗糙的长毛外层（此处“coat”指皮毛层）。

### 1.1.2 Yarn 纱线

In general, yarn may be defined as a linear assembled stage of fibres or filaments, which is formed into a continuous strand having textile-like characteristics. The textile-like characteristics refer to good tensile strength and high flexibility. Many non-textile materials can be designed to have similar strength in continuous-strand form.

There are four basic staple yarn manufacturing systems that have become well standardized. These staple yarn systems are the carded cotton, the combed cotton, the woolen, and the worsted. The carded and combed cotton systems were developed to convert short and long cotton and cotton-like fibres into yarn. The woolen and worsted systems are developed to convert short and long wool and wool-like fibres into yarn.

Before the advent of man-made fibres, silk was the only continuous-filament yarn available. The desired frequencies and directions of twist were added the single yarn, and subsequently multiple yarns.

If the filament is to be processed in a staple yarn system, thousands of fibres are brought together into a twist-less linear assemblage known as tow for subsequent crimping and cutting operation. ④ One of the advantages of man-made fibres is the possibility to exercise over each step of the production process.

④如果长纤丝是通过短纱线方法进行制作的，那么要用上千根纤维一起加少量捻度合并成粗纤维，以配合随后的皱缩与剪切工序。

A high-bulk yarn is a staple or continuous-filament yarn that has a normal extensibility with having a high degree of fullness. These yarns retain their bulkiness under both relaxed and stressed conditions. Great covering power with little weight is possible in fabrics composed of high-bulk yarns.

### 1.1.3 Fabric 织物

According to the garments purposes and demands manufacturer will use, fabric can be chosen. There are many exciting fabrics in shops, with different weights, styles, and price ranges. The quality (durability, warmth, structure, surface, bulkiness) of fabrics depends upon which raw materials are used to make threads, how they are spun, woven and finished (shiny, furry, waterproof, etc). ⑤

⑤布料的质量（耐用性、保暖、构造、外观、厚度）依赖于原材料所用的纱线种类、纺纱及织造和后处理的方法（光面、起毛、防水等后处理）。

Most garments are woven fabrics made of wool or wool blend fabrics, which are divided into two categories —woolen or worsted. Many people can distinguish woolen fabrics by their heavier weight & handling from worsteds by their lightness & smoothness, but what they may not realize is that the two terms refer to the yarns from which the cloth has been woven. ⑥ Some fabrics, such as blazers, are woven with a worsted warp and woolen weft. The difference between them reflects their suitability for different uses —woolen for jackets and overcoats, worsteds for suits and trousers.

⑥很多人可以通过羊毛面料较重的质感、手感和精纺面料较轻的质感、滑糯的手感将两者进行区分。但他们不知道的是，这两个词其实指的织制面料所用的纱线。

The weaving fabric includes plain, twill and satin, the classic types include cotton, linen and man-made as well as woolen and worsted, etc. The best-known terms for woven fabrics are as follows:

(1) Calico: the name is applied to any plain cotton. Used for "trying out" patterns prior to cutting out in the deserved fabric.

(2) Towel cloth: the surface of fabric, tufts or loops formed by extra threads are woven into the fabric and these are either cut or left as loops. ⑦

⑦织物表面的毛丛或线圈是由织入面料的纱线形成的。这些纱线或线经修剪而成毛丛，未修剪而成线圈。

(3) Bedford cord: strong cotton or wool cloth with a definite rib in the warp. Unlike cord, the rib is not formed by a pile but is made by extra warp "stuffing" threads.

(4) Corduroy: a cotton or cotton blend fabric with a pile of ribs in the warp, known as wale.

(5) Duck: strong, closely woven cotton with a smooth surface.

(6) Flannel: a woolen or worsted twill fabric with a soft handle and surface achieved by milling. Best known as gray, it originated from wale.

(7) Gabardine: a fine twill fabric, usually worsted, cotton or linen, with a diagonal appearance.

(8) Herringbone: a weave giving a zig-zag effect resembling the backbone of a fish, achieved by alternating the direction of the twill.

(9) Hopsack: a construction based on the plain weave in which two or more

warp and weft threads weave as one, forming a basket-effect.

(10) Panama: a lightweight plain weave suiting fabric.

(11) Serge: robust twill weave fabric, usually wool and plain dyed.

(12) Tweed: sturdy woolen cloth, originally woven in southern Scotland but now applied to wide range of fabrics, characteristics by natural colors in mixtures.

(13) Twill: description of any fabric woven in twill weaves (such as denim, gabardine and serge). Twills can easily be recognized by the diagonal appearance of the surface of the fabric. ⑧ The cloth is traditionally cut so that in the finished garment the diagonal effect runs from bottom left up to top right.

⑧对机织斜纹布的描述（如牛仔布、华达呢、哔叽呢）。斜纹布因其斜纹外观很容易辨认。

(14) Venetian: a satin weave fabric (smooth surface) with a lustrous look woven with a worsted warp and a woolen weft.

The knitting fabric divides into two basic knitting forms of warp-knitting & weft-knitting. Knitted fabric is made by intermeshing loops, and would consist of any kind of fibres, yarns, stitches or patterns for apparel, home-furniture and industrial end-uses. ⑨ Knitted fabric is stretched more than a woven fabric. The weft knitting has remained popular for garment sections making, and the fabric is used as a cut and sew basic. The common knitted fabrics are as follows:

⑨针织面料由线圈的相互交织而成，成衣、家居、产业用的针织布一般由各种纤维、纱线、针法与样式组成。

(1) Weft knit: knitting fabric where a yarn forms loops across the width of the fabric, it can be either hand-made or machine processed.

(2) Warp Knitting: knitting fabric in loops of yarn running vertically, such as tricot knits and raschel knits.

(3) Plain/jersey: semi-circular needle loops shown in the front, and V-shaped loops shown in the back. And smooth side is the front, while the back is rough. It curls easily at both ends, and unravels readily from both ends. End uses are underwear, shirts, dresses, sweaters, stockings and T-shirts.

(4) Rib/ Double jersey: the 1 × 1 rib is the simplest rib fabric in which the structure includes alternate wale of plain and reversed plain stitches. Examples of the other typical constructions of ribs are 2 × 2 rib and 3 × 3 rib. A broken stitch will cause laddering, and will not curl at both ends and good stretch ability in widthwise direction. End uses are cuffs, collars, waistbands, sweaters and other garments.

(5) Purl: this weft knitted fabric has better extensibility in lengthwise direction, both sides of the fabric have similar appearance to the reverse side of plain knitting, and the fabric does not curl at both ends. End uses are children's wear and sweaters.

（6）Interlock：can only be unraveled from the last end，without tendency to ladder and curl，and smooth on both sides. End uses are ladies dresses，blouses，T-shirts，sweaters，outerwear，sport-wear and underwear.

（7）Terry knit：a broken stitch will be running，more flexible and more absorbent than woven terry cloth. End uses are robes，beachwear and other fashion apparel.

（8）Velour knit：better draped ability，soft handle and suede-like surface. And end uses are men's shirts and women's dresses.

（9）Pile knit：better draped ability，can be laundered and cold tumble-dried，and fur-like surface. End uses are fur fabrics and rugs.

The main competitor to weave and knit in fabric production is the rise of the "non-woven". These fabrics are directly converted from fibre to fabric，without the usual stages of yarn production. There are clear differences in general terms between woven，knitted and non-woven.

### 1.1.4 Garment Trimmings 服装辅料

There are many important supporting accessories（named as trimmings）applied in garment manufacturing，such as lining，interlining，adhesive tape，stay，shoulder pads，etc. In fact，these components take a very important role in the construction of garments，they tend to enhance the appearance and improve the handling of sewing operators. [10]

⑩事实上，这些部件在服装结构上扮演了非常重要的角色，它们美化服装外观，方便车工的操作。

（1）Lining：cloth shaped to cover the inside of the garment. It can improve the garment quality，give a better shape retention and provide more comfort for the wearer.

（2）Interlining：an inner lining between garment and proper lining to give body shape or extra warmth. There are several types of interlining，such as fusible interlining，non-fusible interlining or sew-in interlining，and may be classified into woven interlining，knit interlining，non-woven interlining or paper interlining.

（3）Tape：a narrow strip of material applied to a garment part for either a functional or a decorative purpose. Tapes can be applied to many parts of a garment，depending on its usage and its design. For example，if the tape is for hanging，it will be placed at the top edge of a skirt or underarm seam/ shoulder seam of dress，but if it is a decorative tape，it will be placed on the garment according to the required design.

（4）Stay：a material applied to garments for strengthening purpose. There are various kinds of stays made from different materials for different applications, such as collar stay, supporting stay and other special stays.

（5）Shoulder pads: the most popular padding to give firmness to the armhole. In fact, the application of shoulder pads always gives a garment good quality and a luxurious feeling.

Besides above supporting accessories of manufacturing, there are many other accessories, such as buttons, threads, zippers, elastics, belts, rivets, magic tape or Velcro, eye-lets, care labels, size labels, brand labels, and other packing accessories.

Otherwise, the fashion accessory is a part that used to achieve a specific outlook of the wearer, such as handbags, umbrellas, wallets, boots and shoes, cravats, ties, hats, belts and suspenders, gloves, jewelry, watches, shawls, scarves, socks, and stockings.

## *Words and Expressions*

flexibility [ˌfleksə'biliti] 适应性，灵活性 *
textile ['tekstail] 纺织的，纺织品
woven fabric['wəuvən 'fæbrik] 机织布
fibre ['faibə] 纤维
filament ['filəmənt] 长纤丝
staple yarn ['steipl jɑːn] 短纤丝
interlacing [ˌintə(ː)'leisiŋ] 交织
felting ['feltiŋ] 制毡法
rope [rəup] 绳索
braid [breid] 饰带
tufting ['tʌftiŋ] 栽绒
natural fibre ['nætʃərəl 'faibə] 天然纤维
mohair ['məuhɛə] 马海毛
cashmere [kæʃ'miə] 山羊绒，开司米毛
linen ['linin] 亚麻布
flax [flæks] 亚麻，麻布
textured yarn ['tekstʃəd jɑːn] 膨松纱，变形纱
glass fibre [glɑːs 'faibə] 玻璃纤维
metal fibre ['metl 'faibə] 金属纤维
nylon ['nailən] 尼龙
cotton ['kɔtn] 棉
viscose rayon ['viskəus 'reiɔn] 黏胶纤维
polyester ['pɔliestə] 聚酯纤维，涤纶
synthetic [sin'θetic] 合成纤维
acetate ['æsiˌteit] 醋酯纤维
man-made fibre [mæn meid 'faibə] 化学纤维
regenerated fibre [ri'dʒenərit 'faibə] 再生纤维
mineral ['minərəl] 矿物质
silk [silk] 丝
wool [wul] 羊毛
towel ['tauəl] 毛巾
suiting ['sjuːtiŋ] 西装料
coating ['kəutiŋ] 外衣料
dress-goods [dres-gudz] 服装产品
knitwear ['nitˌwɛə] 针织服装
rug [rʌg] 厚毯，小地毯
yarn [jɑːn] 纱线
acrylic [ə'krilik] 腈纶
print fabric [print 'fæbrik] 印花布
artificial [ɑːti'fiʃəl] 人造的
blend fibre [blend 'faibə] 混纺纤维
property ['prɔpəti] 特性，性质
woolen ['wulin] 粗纺毛

strand [strænd] 绳，股纱
worsted ['wustid] 精纺毛
carded cotton [kɑ:d 'katn] 粗梳棉
combed cotton [kəumd 'katn] 精梳棉
density ['densiti] 密度
advent ['ædvənt] 到来，出现
multiple yarn ['mʌltipl jɑ:n] 多股纱
end-use [end -ju:s] 最后用途
bulk yarn [bʌlk jɑ:n] 膨体纱
shrinkage ['ʃrinkidʒ] 缩水
loop formation [lu:p fɔ:'meiʃən] 线圈形成
spun [spʌn] 捻成丝状的，纺
entrapping [in'træpiŋ] 使陷入
counts [kaunts] 棉纱支数
spinning ['spiniŋ] 纺纱
spun yarn [spʌn jɑ:n] 细纱，短纤纱
camel hair ['kæməl heə] 骆驼毛
rabbit hair ['ræbit heə] 兔毛
shiny ['ʃaini] 发亮的（广东话：起镜面）
furry ['fə:ri] 毛皮制品
waterproof ['wɔ:təpru:f] 防水的
warp-knitting [wɔ:p-'nitiŋ] 经编针织
weft-knitting [weft -'nitiŋ] 纬编针织
durability [ˌdjuərə'biliti] 耐用的
non-woven ['nɔ'-wəuvən] 非织造布
blanket ['blæŋkit] 毛毯
suits [sju:t] 套装
coating ['kəutiŋ] 大衣
diagonal [dai'ægənl] 斜纹的
herringbone ['heriŋbəun] 人字形
zig-zag ['zig-zæg] 人字纹
hopsack ['hɔpˌsæk] 方平织物
plain weave [plein wi:v] 平纹组织
panama [ˌpænə'mɑ:] 巴拿马薄呢
serge [sə:dʒ] 哔叽
robust [rə'bʌst] 粗壮的
tweed [twi:d] 粗花呢
sturdy ['stə:di] 强健的
Scotland ['skɔtlənd] 苏格兰
Denim ['denim] 牛仔布，粗斜纹棉布
traditionally [trə'diʃən(ə)li] 传统的
venetian [vi'ni:ʃən] 直贡呢，威尼斯缩绒呢
sock [sɔk] 短袜
stocking ['stɔkiŋ] 长袜
lustrous ['lʌstrəs] 有光泽的
luxurious [lʌg'zjuəriəs] 华丽的
outerwear ['aʊtəweə(r)] 外套
distinguish [dis'tiŋgwiʃ] 区别，分类
blazering ['bleizə] 鲜艳的运动上衣
robe [rəub] 长袍
jacket ['dʒækit] 夹克
overcoat ['əuvəkəut] 外套
trousers ['traʊzəz] 西裤
Bedford cord [bedfɔ:d kɔ:d] 经条灯芯绒
stuffing ['stʌfiŋ] 填充料
wale [weil] 纵行线圈
duck [dʌk] 帆布
flannel ['flænl] 法兰绒
twill [twil] 斜纹布
gabardine ['gæbədi:n] 斜纹呢
velour knit [və'luə nit] 丝绒针织布
interlock [ˌintə'lɔk] 双面针织
jersey ['dʒɜ:zi] 平纹针织
looped fabric [lu:pt 'fæbrik] 起圈布
pile fabric [pail 'fæbrik] 起毛织布
purl [pə:l] 双反面针织布
raschel [rɑ:'ʃel, rə] 拉舍尔经编织物
tricot ['trikəu] 特里科经编织物
rug [rʌg] 地毯
stripe [straip] 条纹布
corduroy ['kɔ:dərɔi] 灯芯绒
sweater ['swetə] 毛线衫，羊毛衫
loop ['teri lu:p] 线圈，毛圈
terry knit ['teri nit] 厚绒针织布
rib [rib] 罗纹织物
double jersey ['dʌbl 'dʒɜ:zi] 双面平纹
lining ['lainiŋ] 里料
interlining ['intə'lainiŋ] 里衬
adhesive tape [əd'hi:siv teip] 黏性带条
stay [stei] 拉条（广东话：扁带条）
shoulder pad ['ʃəuldə pæd] 垫肩
component [kəm'pəunənt] 部件

shape retention [ʃeip ri'tenʃən] 定形
fusible interlining ['fju:zəbl 'intə'lainiŋ] 黏合衬，热熔衬
non-fusible interlining ['nɔn-'fju:zəbl 'intə'lainiŋ] 非黏合衬
sew-in interlining [sju:-in 'intə'lainiŋ] 车缝里衬（广东话：生里衬）
paper interlining ['peipə 'intə'lainiŋ] 纸衬
tape [teip] 牵条，带条
decorative tape ['dekərətiv teip] 装饰带
underarm seam ['ʌndərɑ:m si:m] 腋下缝
arm-hole [ɑ:m-həul] 袖窿
button ['bʌtn] 纽扣
thread [θred] 缝纫线
zipper ['zipə] 拉链
elastic [i'læstik] 松紧带
belt [belt] 皮带
rivet ['rivit] 铆钉（广东话：撞钉）
magic tape/ Velcro ['mædʒik teip / velcre] 魔术贴
eye-let [ɑi-let] 孔眼，眼孔
care label [keə 'leibl] 洗涤标
size label [saiz 'leibl] 尺码标
brand label [brænd 'leibl] 主商标
packing accessory ['pækiŋ æk'sesəri] 包装辅料
warp float [wɔ:p fləut] 经向跳花（广东话：浮经）
hat [hæt] 草帽
handbag ['hændbæg] 手提袋 *
hemp [hemp] 大麻
jute [dʒu:t] 黄麻
mono-filament ['mɔnəu-'filəmənt] 单股长丝 *
multi-filament [,mʌlti-'filəmənt] 多股长丝
carpet ['kɑ:pit] 地毯
poplin ['pɔplin] 纺，毛葛
sucker ['sʌkə] 泡状布
velveteen ['velvi'ti:n] 仿天鹅绒
velvet ['velvit] 天鹅绒
bleaching ['bli:tʃiŋ] 漂白
brocade [brə'keid] 织锦
Aberdeen [,æbə'di:n] 阿巴丁布
canvas ['kænvəs] 马尾衬布，帆布 *
crepe [kreip] 绉布，绉绸
chiffon ['ʃifɔn] 薄纱，雪纺
fur [fə:] 皮裘
gray cloth/ calico [grei klɔ:θ / 'kælikəu] 平布，白布（广东话：胚布）*
melton ['meltən] 麦尔登呢 *
taffeta ['tæfitə] 塔府绸 *
towel cloth ['tauəl klɔ:θ] 毛巾布
waterproof fabric ['wɔ:təpru:f 'fæbrik] 防水布 *
selvedge ['selvidʒ] 布边 *
chambray ['ʃæmbrei] 有条纹或格子花纹的布 *
crepe de-chine [kreip di:-tʃain] 双绉布 *
dobby ['dɔbi] 小提花织物 *
sateen [sæ'ti:n] 纬向缎纹（广东话：纬向色丁）*
satin ['sætin] 经向缎纹（广东话：经向色丁）
mildew resistant finish ['mildju: ri'zistənt 'finiʃ] 防霉整理 *
moth resistant finish [mɔθ ri'zistənt 'finiʃ] 防虫整理 *
flameproof fabric ['fleımpru:f 'fæbrik] 防火织物 *
perspiration resistant finish [,pə:spə'reiʃən ri'zistənt 'finiʃ] 防汗整理 *
shrink-resistant / preshrunk finish [ʃriŋk ri'zistənt / 'pri:'ʃrʌŋk 'finiʃ] 防缩整理 *
antistatic finish [,ænti'stætik 'finiʃ] 防静电整理 *
atmospheric fading resistant finish [ætməs'ferik 'feidiŋ ri'zistənt 'finiʃ] 防褪色整理 *
crease & wrinkly resistant finish [kri:s & 'riŋkli ri'zistənt 'finiʃ] 防皱整理 *
resin finish ['rezin 'finiʃ] 树脂整理 *
antiseptic finish [,ænti'septik 'finiʃ] 防蛀整理 *
glazed finish [gleizd 'finiʃ] 轧光整理 *

bleeding ['bli:diŋ] 洗水后褪色渗化 *
battik ['bætik] 蜡染 *
opaque finish [əu'peik 'finiʃ] 防透光整理 *
odorless & perfumed finish [ˈəudəlis 'pə:fju:m,d 'finiʃ] 防臭整理 *
ramie [ˈræmi] 苎麻纤维
vicuna [vi'kju:nə ] 小羊驼
lingerie [,lænʒə'ri:] 妇女贴身内衣
polyamide [pɔli'æmaid] 聚酰胺
polypropylene [,pɔli'prəupili:n] 丙纶，聚丙烯
angora [æŋ'gɔ:rə] 安哥拉山羊毛
polyethylene [,pɔli'eθili:n] 聚乙烯
spandex [spændeks] 氨纶、弹性纤维
merino [mə'ri:nəu] 美利奴羊毛
modacrylic [,mɔdə'krilik] 变性腈纶
Polynosic [,pɔli'nɔsik] 富纤
umbrella [ʌm'brelə] 雨伞
boots [boots] 靴子
cravat [krə'væt] 围巾
belt [belt] 腰带
suspender [sə'spendə] 吊裤带，吊袜带
glove [glʌv] 手套
shawl [ʃɔ:l] 披肩

## *Exercises*

1. Translate the following terms into Chinese.

（1）textile
（2）viscose
（3）woven fabric
（4）polyester
（5）synthetic
（6）filament
（7）acetate
（8）triacetate
（9）felting
（10）regenerated fibre
（11）rope
（12）braid
（13）natural fibre
（14）mohair
（15）cashmere
（16）linen
（17）flax
（18）towel
（19）rayon
（20）suiting
（21）nylon
（22）coating
（23）knitwear
（24）rug
（25）double end
（26）acrylic
（27）warp float
（28）artificial fibre
（29）woolen
（30）taffeta
（31）loop
（32）spinning
（33）fur
（34）multiple yarn
（35）silk
（36）lining
（37）wool
（38）fabric
（39）cotton
（40）shell fabric
（41）interlining
（42）furry
（43）waterproof
（44）warp-knitting
（45）weft-knitting
（46）worsted
（47）non-woven
（48）blazer
（49）jacket
（50）trousers
（51）loop pile
（52）duck
（53）blanket
（54）flannel
（55）suit
（56）twill
（57）gabardine
（58）herringbone
（59）hopsack
（60）serge
（61）tweed
（62）denim
（63）satin
（64）leather
（65）stripe
（66）corduroy
（67）sweat-shirt
（68）sweater
（69）crepe
（70）chiffon
（71）care label
（72）size label

(73) brand label
(74) overcoat
(75) spandex
(76) merino
(77) modacrylic
(78) umbrella
(79) boots
(80) cravat
(81) belt
(82) suspender
(83) glove
(84) shawl

2. List 5 types of natural fibres and 4 types of man-made fibres.
   Example: Natural Fibre: Cotton
   Man-Made Fibre: Polyester
3. List 5 kinds of woven fabrics and 5 kinds of knitted fabrics.
   Example: Woven Fabric: Denim
   Example: Knitted Fabric: Jersey
4. List 10 kinds of garment accessories.
   Example: Button

## 1.2 Fabric Pretreatment During Production 生产中的面料预处理

Besides understanding the knowledge of material, manufacturer must understand the characteristics of certain materials after purchasing them. For example, it is very important to handle the shell fabric in the garment workshop or in finished operation, because adopting a wrong operation to handle and finish the problem fabric may cause the poor quality products, such as transparent fabric, shrinkage fabric, pile fabric, knit fabric, leather, loose or tight fabric, thermoplastic fabric.

### 1.2.1 Fabric Shrinkage 面料缩水

The quality of a particular piece of fabric determines whether it needs to be washed or dry-cleaned. Apparel producer should ask for the washing/ cleaning instructions when they purchase the fabric. As a rule, fabrics should be put through a shrinking process before cutting and making the garment. On the other hand, fabric that contains a great deal of sizing becomes soft, especially if rinsed in fabric softener or in a vinegar solution. ① All-cotton knits and sweat-shirt fabrics should always be washed before they're sewn, as these can shrink up to 10%.

①另一方面，面料中的大量浆料可以被软化，特别是布料在柔软剂或醋酸溶剂中洗涤后。

Always remember to turn the fabric inside out when washing, especially in a machine, as the color can lose its richness. If washing operators plan to dry the garment in a clothes dryer, the fabric should probably be dried in it before cutting and sewing. Combines fabrics or colors in a garment, the fabric should be laundered separately to avoid colors from running together and different rates of shrinkage. When a garment is made up, it should be treated as most delicate fabric. Most fabrics shrink only the first time they are washed, and mostly in the length. An option with pants is to be basted in the hem, and remove them before the first laundering. If manufacturer don't want to wash the fabric before making the garment, you can make a laundry sample of fabric.

### 1.2.2 Handling Skills 处理技术

There are now many fabrics used in garment manufacturing. Manufacturers need further knowledge, equipment and handling skills than those required to make simple cotton or woolen garments. The design, cutting and construction of the garment must allow for the quality fabric to be used. For successive sewing, the stitch formation made with the need of thread, needle, stitch type and feeding mechanism of the machine must be compatible with the structure and fibre content of the fabric. ② Pressing and finishing techniques must cooperate with the fabric's initial characteristics and finishing.

②为使缝纫协调，缝纫线、车针、线迹种类和缝纫机送布机构应与面料结构和纤维成分协调一致。

### 1.2.3 Consideration of Fabric Characteristics 面料性能的考虑

The designer and pattern cutter's responsibility when using problem fabric is to use the quality of the fabric to enhance the design effect, but to be aware of the construction problems whenever possible, or with the knowledge that the skills and equipment necessary to produce the garment are at an adequate standard as possible.③ Some basic problems must be considered when manufacturing problem fabrics are stretch, seam pucker, damage to fabric structure, transparency, inflexibility, thermoplastic, etc. There are various fabric constructions which can stretch, for example, knitted, specially constructed woven fabrics, bias-cut woven fabrics, etc. The amount of stretch will depend on the specific fabric, some will stretch in one direction only, others in all directions, therefore the design, cutting, and manufacturing techniques must vary accordingly. Seam pucker affects many fabric structures which are caused by a variety of reasons, such as seam twist, tight stitching, distortion of the fabric structure.

③使用有问题的面料时，设计师和样板师的工作重点是利用面料的特性提升设计效果，但应意识到随时可能出现的结构问题，或具备操作技能和设备方面的知识，以便尽可能地生产出符合标准的服装。

Most fabrics can be spoilt if misused during manufacturing or wearing. These include tightly woven fabrics, loosely woven fabrics, pile fabrics, etc. ④ Care must be taken so that damages are not obvious during manufacturing but apparent on the garment. The styles and constructions of transparent or semi-transparent garments

④很多面料在制作与穿着过程中可能会因使用不当而被损坏。这些面料包括组织紧密型面料、组织疏松型面料、起绒面料等。

must avoid any obstruction to this feature. Consideration must be given to edge-finish, seam, interlining & lining, etc. Materials such as leather and plastic can be inflexible to handle, patterns of a garment in such material is much more than softer materials, which is easy to gather. Leather must be dry-cleaned through a special process. A garment trimmed with small pieces of leather can be hand washed. Checks, stripes and other special printed fabrics require matching, which can cause stitching difficulties because of seam twisting.

## *Words and Expressions*

handling [ˈhændliŋ] 处理（广东话：执手）
characteristic [kæriktəˈristik] 特性
shell fabric [ʃel ˈfæbrik] 面料
workshop [ˈwəːkʃɔp] 工厂，车间
dry-clean [drɑi kliːn] 干洗
washing instructions [ˈwɔʃiŋ inˈstrʌkʃənz] 洗涤说明
shrink [ʃriŋk] 缩水
outer clothes [ˈɑutə kləuðz] 外套
sizing [ˈsaiziŋ] 上浆
rinsed [rins] 洗涤，漂洗
softener [ˈsɔfnə] 柔软剂
vinegar [ˈvinigə] 醋酸
solution [səˈljuːʃən] 解决，溶解
sweat-shirt [swet-ʃəːt] 针织套头衫式衬衫
laundry [ˈlɔːndri] 洗衣
all-cotton knits [ɔːl ˈkɔtn nits] 全棉针织品
clothes dryer [kləuðz ˈdraiə] 干衣机
cutting [ˈkʌtiŋ] 裁剪
sewing [ˈsəuiŋ] 缝制
delicate [ˈdelikit] 精致的
sample [ˈsæmpl] 样板
manufacturer [mænjuˈfæktʃərə] 制造商
equipment [iˈkwipmənt] 设备
design [diˈzɑin] 设计
construction [kənˈstrʌkʃən] 结构
stitch formation [stitʃ fɔːˈmeiʃən] 线迹的形成
thread [θred] 缝纫线
needle [ˈniːdl] 缝针
feeding mechanism [ˈfiːdiŋ ˈmekənizəm] 送布机构
problem fabric [ˈprɔbləm ˈfæbrik] 问题布，疵布
seam pucker [siːm ˈpʌkə] 缝迹起皱
transparency [trænsˈpɛərənsi] 透明的
inflexibility [inˌfleksəˈbiliti] 不灵活
thermoplastic [ˌθəːməˈplæstik] 热塑性塑料
knitted [ˈnitid] 针织
woven fabric [ˈwəuvən ˈfæbrik] 机织面料
bias cut [ˈbaiəs kʌt] 斜裁
specific [spiˈsifik] 明确的
seam twist [siːm twist] 缝迹扭曲
distortion [disˈtɔːʃən] 歪斜，扭曲
technique [tekˈniːk] 技能
spoilt [spɔilt] 破坏，损坏
obstruction [əbˈstrʌkʃən] 障碍物
edge-finishes [edʒ - ˈfiniʃ] 边脚处理
interlining [ˈintəˈlainiŋ] 里衬
lining [ˈlɑiniŋ] 里料（广东话：里布）
leather [ˈleðə] 皮革
print fabric [print ˈfæbrik] 印花布
synthetic [sinˈθetic] 合成的
needle heating [ˈniːdl ˈhiːtiŋ]（摩擦产生）针热
pressing [ˈpresiŋ] 熨烫
trimmed [trimd] 修整，装饰

baste [beist] 假缝，粗缝
beachwear [ˈbiːtʃweə(r)] 沙滩装
pants [pænts] 裤子

## Exercises

1. List 5 kinds of problem fabrics.
   Example：Transparent fabric
2. Translate the following terms into Chinese.

(1) fabric characteristic
(2) shell fabric
(3) dry-clean
(4) cleaning instruction
(5) shrinkage
(6) sweat-shirt
(7) sweater
(8) laundry
(9) all-cotton knits
(10) cutting
(11) sample
(12) manufacturer
(13) cotton & woolen
(14) designer
(15) stitch
(16) thread
(17) seam pucker
(18) knitted
(19) woven fabric
(20) bias cut
(21) seam twist
(22) edge-finishes
(23) interlining
(24) lining
(25) leather
(26) beachwear
(27) printed fabric
(28) seam
(29) synthetic
(30) pressing
(31) baste
(32) pants
(33) pressing shiny

## 1.3 Material Purchasing 材料采购

### 1.3.1 Purchasing Function 采购功能

An important activity of the purchasing department is to monitor the performance of suppliers and track new orders, also to ensure timely delivery of the proper quantity and quality of goods or services at reasonable prices. ① There are many activities in purchasing.

(1) The evaluation of alternative suppliers, and the use of a small but effective number of these suppliers for the more important goods and materials. ②

(2) To provide the chance to buy effective materials with competitive prices.

(3) The rating or assessment of suppliers on delivery, service and quality, as well as price.

(4) The enhancement of the company reputation in the eyes of suppliers and competitors.

(5) Have an ability to select new materials by constant communication with the marketing department.

①采购部的一项重要任务是监督供应商及跟踪新订单的进展，确保对方及时以合理的价格交付规定的数量和质量的商品或服务。

②对不同供应商进行评估，较为重要的产品和原材料应选用少量但效率高的供应商。

（6）The maintenance of adequate stock levels throughout the company, in conjunction with other departments.

It will be evident from these activities that the purchasing function could have an effect on the profitability of the company.

### 1.3.2 Purchase Order & Contract 采购订单与合同

The range of contracting services offered in apparel producing center spans all steps of production and design.[3] Whether a small or moderate-sized firm will be successful is determined by how much investment capital is available, the talent of the originator, and the age of the business. The vendor has the authority to ship materials to firms, and the purchasing department makes the purchase order out. The purchase order should contain the following information:

③在成衣生产部门，合同上的服务范围包括生产和设计在内的所有环节。

（1）Order date.

（2）Order No. & style No.

（3）Delivery date.

（4）Firm's name.

（5）Full description on color, style, quantity and specification.

（6）Prices and terms.

（7）Where and how to be shipped along with packing instructions.

5 copies may be used with the following distributions.

（1）Purchasing Dept.: for keeping record and booking.

（2）Vendor Dept.: for the arrangement of the materials as the purchase order's specification.

（3）Finance Dept.: for accountability.

（4）Shipping Dept.: for applying the letter of credit if purchase order is needed.

（5）Sales Dept.: for reference and checking.

Advantages of this form are easy to follow up and complaint or claim according to the purchase orders specification. But if details are not clearly shown on purchase order, company will sometimes get loss（Table 1-1, Table 1-2）.

**Table 1-1 Purchase Order**

**PURCHASE ORDER**

NO. 0688

| FABRIC SWATCH |
|---|

PURCHASE NO. TK-21

DATE : 2020-3-25

STYLE NO.: JN-012

DESCRIPTION / QUANTITY: BLEND WITH 30% WOOL & 70% POLYESTER, PLAIN TWEED, CONSTRUCTION 76 × 68/48S × 48S, BLACK COLOR, WIDTH 60INCHES. TOTAL ABOUT 550 METERS, FINISHING IS FURRY.

UNIT PRICE: US$16/METER INCLUDING LOCAL DELIVERY.

PACKING: PACKAGE IN POLYBAG.

SHIPMENT / DELIVERY: DELIVERY BEFORE 25TH JUNE 2020.

REMARK: RECEIVED ISPECTION BY SGS

BUYER'S SIGNATURE
CONFIRMED & ACCEPTED BY:ALICE LIU

SELLER'S SIGNATURE
FOR: DK ZHANG

ATTENTION: SHIPPING QUANTITY +/-5% IS ACCEPTABLE.

**Table 1-2 Purchase Contract**

**PURCHASE CONTRACT**

REF. NO.: MSC81009 DATE: 2020. 2.28

CONTRACT BEING MADE THIS DAY BETWEEN POLO ENTERPRISE LIMITED ( CALLED THE BUYER ) AND RYKIEL TRADERS CO.

ADDRESS: # 6 YIN ON ST. KOWLOOG HK.

( CALLED THE SELLER, WHERE THE BUYER AGREES TO BUY AND SELLER AGREES TO SELL THE FOLLOWING MERCHANDISE. SUBJECT TO THE TERMS AND CONDITIONS AS SPECIFIED HEREUNDER. )

DESCRIPTION OF GOODS: 100% COTTON CHAMBRAY, SULFUR DYED, BLACK COLOUR.
CONSTRUCTION: 72 × 42 / 10S × 10S.
WIDTH: 48 INCHES.
FINISH: SIZING & STARCHED.

QUANTITY: TOTAL ABOUT 1550 YARDS. DIFFERENCE OF ABOUT 5% IS ALLOWED.

PRICE: HK$ 1200 PER YARD INCLUDING LOCAL DELIVERY.

AMOUNT: HK$ 18600. ( HK DOLLARS EIGHTEEN THOUSAND SIX HUNDRED. EXACT AMOUNT WILL BE BASED ON ACTUAL QUANTITY DELIVERED. )

PACKING: ROLL ON TUBE AND IN POLYBAGS.

PAYMENT: BY 30 DAYS CHEQUE AGAINST DELIVERY.

SHIPMENT/ DELIVERY: PROMPT DELIVERY BY NOT LATER THAN 30TH MAY 2020.

OTHER CLAUSES: QUALITY: AS PER SAMPLE SWATCH SUPPLIED.
ORIGIN: HONG KONG.

**Continued**

<table>
<tr><td colspan="2">PLEASE RETURN THE SIGNED COPY WITHIN 3 DAYS.</td></tr>
<tr><td>ACCEPTED AND SIGNED BY BUYER<br>FOR AND ON BEHALF OF POLO<br>INTERNATIONAL LTD.<br><br>________________________<br>AUTHORIZED SIGNATURE</td><td>SIGNATURE OF SELLER<br>RYKIEL TRADERS CO.<br><br><br>________________________<br>AUTHORIZED SIGNATURE</td></tr>
</table>

### 1.3.3 Material Warehousing 原材料存储

Under the purchasing department, there is a stock-keeping section to ensure the operation processes would not run out of any needed items, never have more items in hand than needed, and do not pay excessive money because of small purchases.[④] In general, the following are those important functions of the stock keeping:

④在采购部门下设有库房管理部，其运作是保证成衣生产中所需材料既不会用尽，也不会囤积，从而不会因小额采购而花多余的钱。

（1）To check all incoming and outgoing materials.

（2）To provide service information for stock management.

（3）To keep stock at an economic level.

（4）To provide enough stock to prevent production from stopping when suppliers do not deliver on time.

（5）To allocate storage space for incoming materials, and distribute materials to other departments when needed.

（6）To ensure materials usage with available financing capital.

（7）To plan accurate materials provision.

Regards to the above activities, the responsibility of stock-keeper is to minimize the cost associated with the inventory level.

### 1.3.4 Application of Control Chart 控制表的应用

Control Chart may include “Material Control Chart” and “Material Purchasing Sheet” etc.（Table 1-3, Table 1-4）, and the purchasing department provides these data. It collects all sales orders to determine the total raw materials and accessories needed per style and color. It can enable the purchasing department to order the correct amount of raw materials and accessories needed. A suggested distribution of 4 copies is:

（1）Purchasing Dept.: keeping records and estimation of materials used by collating with this chart.

（2）Sales Dept.: for reference on usage, re-checking on the quantity and for calculation of costs.

（3）Accounting Dept.: for reference and checking on bills.

（4）Production Dept.: for reference and checking on consumption. It is good for checking on consumption and calculation of costs.

**Table 1-3 Material Control Chart**

**FABRIC CONTROL CHART**

BUY BY: ________ APPROVED BY: ________ DATE: ________

| ORDER NO. | | | TOTAL | |
|---|---|---|---|---|
| STYLE NO. | | | | |
| QUANTITY | | | | |
| YARD × WIDTH | | | | |
| FACTORY | | | | |
| PRICE | | | | |
| COLOUR | | | | RECEIVING QTY. ON DATE: |

**Table 1-4 Material Purchasing Sheet**

**MATERIAL PURCHASING SHEET**

INITIAL SAMPLE: ______ APPROVAL SAMPLE: √ BULK-PRODUCTION: √

VENDOR: TOMMY DATE: 2020-4-5 ORDER NO.: T-05248

| ITEM | MODEL NO. | SIZE | CLR. | SUPPLIER | DATE | SWATCH | QTY. |
|---|---|---|---|---|---|---|---|
| TQ-08 | M8745 | S-XXL | W-01 | DALIN | 4-25 | — | 100Y |
| TQ-09 | M8746 | S-XL | B-021 | DALIN | 4-30 | — | 120Y |
| — | | | | | | | |
| — | | | | | | | |

APPLICANT: MAY 2020-4-5 MATERIAL DEPT.: BADY 2020-4-6 SAMPLE DEPT.: DK LIU 2020-4-6

## Words and Expressions

evaluation [i,vælju'eiʃən] 评定，估价
alternative [ɔ:l'tə:nətiv] 选择性的
assessment [ə'sesmənt] 估定，评定
reputation [repju(:)'teiʃən] 名誉，信誉
enhancement [in'hɑ:nsmənt] 增加，提高
conjunction with [kən'dʒʌŋkʃən wið] 联接，结合
yard/ yardage [jɑ:d / 'jɑ:didʒ] 码数
supplier [sə'plaiə] 供应商
letter of credit/ L/C ['letə ɔv, 'kredit] 信用证
shipping department ['ʃipiŋ di'pɑ:tmənt] 船务部门
financial control [fɑi'nænʃəl, kən'trəul] 财政控制
facility [fə'siliti] 设备，工具
consolidate [kən'sɔlideit] 结合，合并
apparel firm [ə'pærəl fə:m] 服装公司
goods/ articles [gudz / 'ɑ:tiklz] 货品
approve [ə'pru:v] 核准
acceptable [ək'septəbl] 可接受的
investment [in'vestmənt] 投资
capital ['kæpitl] 资金
talent ['tælənt] 人才，才干
instruction [in'strʌkʃən] 指令
salesperson [seilz'pə:sn] 销售员
complaint [kəm'pleint] 投诉
advantage [əd'vɑ:ntidʒ] 优点
fabric swatch ['fæbrik swɔtʃ] 布样
delivery/ shipment date [di'livəri / 'ʃipmənt 'deit] 交付期，船期
shipping quantity ['ʃipiŋ 'kwɔntiti] 出货数量
enterprise ['entəprɑiz] 企业
cotton chambray ['kɔtn 'ʃæmbrei] 条纹棉布，钱布雷棉布
sulfur dyed ['sʌlfə dɑid] 硫化染色
construction [kən'strʌkʃən] 构造，结构
sizing & starched ['saiziŋ & stɑ:tʃ] 上浆加硬的
sample swatch ['sæmpl swɔtʃ] 样板
authorized signature ['ɔ:θəraizd 'signitʃə] 认可的签名
performance [pə'fɔ:məns] 性能
competitive [kəm'petitiv] 有竞争力的
marketing ['mɑ:kitiŋ] 市场
communication [kə'mju:nikeiʃən] 沟通
profitability [,prɔfitə'biliti] 收益性，利益率
purchase order ['pə:tʃəs 'ɔ:də] 采购订单
purchase contract ['pə:tʃəs 'kɔntrækt,] 采购合同
manufacturing process [,mænju'fæktʃəriŋ 'prəuses] 制作程序，生产过程
specification [,spesifi'keiʃən] 规格
packing instructions ['pækiŋ in'strʌkʃənz] 包装指示
commission [kə'miʃən] 佣金
style [stail] 款式
warehouse ['wɛəhaus] 仓库
vendor ['vendɔ:] 客商
unit price ['ju:nɪt prɑis] 单价
roll on tube [rəul ɔn 'tju:b] 卷装
poly-bag [poli-bæg] 包装袋，塑料袋
payment ['peimənt] 付款方式
prompt [prɔmpt] 迅速的
stock-keeping [stɔk-'ki:piŋ] 仓库保管，库房管理
economic level [i:kə'nɔmik 'lev(ə)l] 经济水平
inventory ['invəntri] 存货
control chart [kən'trəul tʃɑ:t] 控制图
consumption [kən'sʌmpʃən] 消费，消耗
associated with [ə'səuʃieitid wið] 联合
collate [kɔ'leit] 对照，核对
initial sample [i'niʃəl 'sæmpl] 初样
approval sample [ə'pru:vəl 'sæmpl] 核准样
bulk-production [bʌlk prə'dʌkʃən] 大货生产
applicant ['æplikənt] 申请人

## *Exercises*

1. List content points of purchase order.

   Example: Order date.

2. Translate following Terms into Chinese.

(1) yardage
(2) supplier
(3) letter of credit
(4) shipping
(5) financial control
(6) apparel firm
(7) goods
(8) approved
(9) fabric swatch
(10) delivery date
(11) enterprise
(12) sizing & starched
(13) sample swatch
(14) signature
(15) marketing
(16) Purchase Order
(17) specification
(18) style
(19) warehouse
(20) vendor
(21) unit price
(22) poly-bag
(23) payment
(24) inventory
(25) initial sample
(26) approval sample
(27) bulk-production
(28) applicant
(29) manufacturing
(30) material
(31) accessory
(32) shell fabric
(33) chambray

# 译文

# 第一章　服装材料的选择

## 1.1　材料知识

纤维、纱线、织物和辅料是任何纺织、服装或相关行业的基本需求。根据客户要求选择适当的织物是非常重要的。因此，制造商根据他们制造的产品选择纤维、纱线和织物。

### 1.1.1　纺织纤维

纺织业生产机织与针织产品，如通过机织、针织、制毯、毡合和起毛等方式制成的缝线、绳线、绳索、织带、花边、绣品、网布和衣服等。这些不同纺织品的特性在服装的应用中得到有效使用。纺织品仍然是人们穿着的服装和家里及办公室布置的主要组成。纤维是纺织行业最基本的原料。基于不同的标准，纤维可分为不同类别。最常见的分类是天然纤维和化学纤维。纤维也可分为长纤丝与短纤维，线丝状的纤维叫长纤丝，如尼龙纱线。非常短的纤维叫短纤维。如图 1-1 所示为可纺纤维的普通分类。

最初，所有纺织品都由天然纤维制作而成，如棉、羊毛、马海毛、亚麻、苎麻和小羊驼毛等。在织成布料前，只有当这些短纤维纺成纱线后才能有效应用。另外，天然纤维包括的棉、亚麻、毛和丝等都来源于植物、动物或矿物质。人造纤维是经过化学加工而成。各种纤维各具特色以使它们用途广泛。混纺纤维是由两种或两种以上不同的纤维组合而成。通常占混纺纤维百分比最多的纤维成为这纤维的主导纤维，但一种成功的混纺纤维往往拥有各纤维的优良质量。

丝曾经是第一种单丝材料，多年来科学家一度着迷于尝试制造“人造”丝。一些新型纺织品有抗污、防火、甚至硬挺等质量属性。纺织技术不断创新，肯定有更多的特殊产品出现。这些发展奠定了科学理论基础，并成为杜邦公司研制出尼龙、涤纶和纱线的科学理论。再生纤维由天然纤维生产而成，如纤维素和蛋白质。这些面料有真丝的外观，定型很好，是理想的面料产品。可用作妇女贴身内衣、套装、女装衬衫和衬里。

合成纤维完全由化学制品制成。合成纤维通常比天然纤维或再生纤维有更

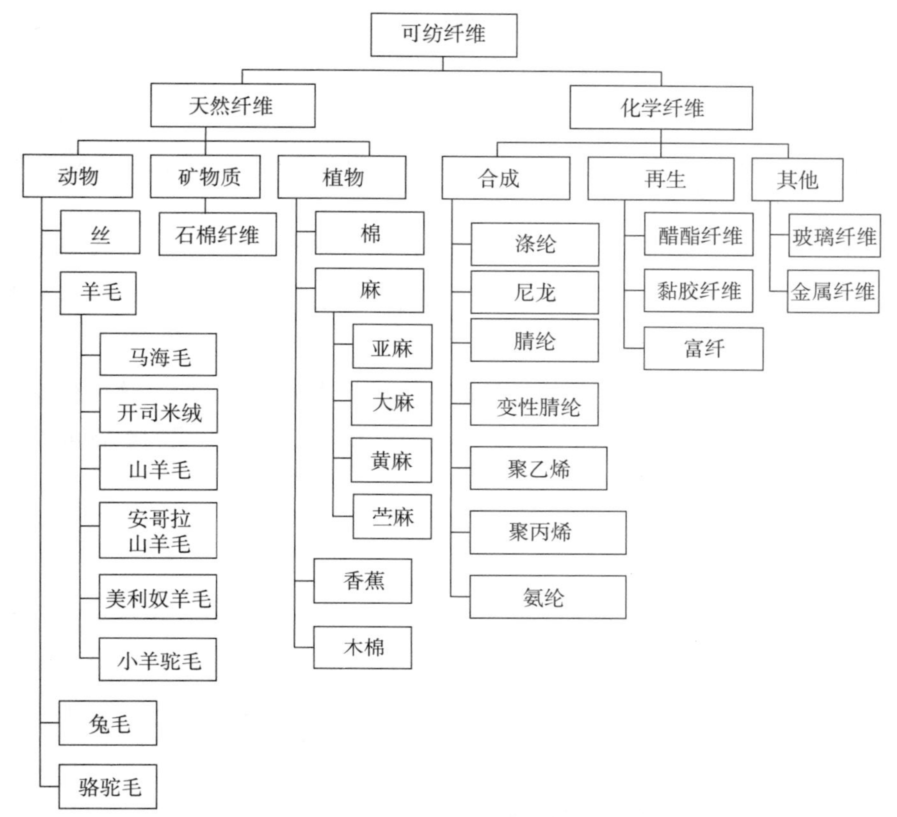


**图 1-1 可纺纤维分类**

高的强度。合成纤维与醋酯纤维均为热塑性纤维，它们在热力下会变软，制造商可以在高温下将该纤维定型，如加上褶裥和绉褶特征。如果熨斗太热，合成纤维将会熔掉。最常用的合成纤维是尼龙（聚酰胺）、涤纶、腈纶和丙纶。

最重要的动物纤维是羊毛、马海毛、山羊绒等。通常用于西装、外衣、长裙产品、针织服装、小地毯等。高质量的马海毛价格昂贵，而且纺纱难。山羊绒是开司米山羊的最底层毛绒。开司米山羊长了两层皮毛，包括幼细而柔软的内层与粗糙的长毛外层（此处“coat”指皮毛层）。山羊绒漂亮柔软、有光泽，有光滑的手感，通常被用于制作高档针织服装、外套和西装产品。100% 的纯羊绒面料昂贵，因此与羊毛混纺较为普遍。

### 1.1.2 纱线

纱线一般被定义为把纤维或长纤维丝集合成连续的股线，使其具有可纺特性。可纺特性是指具有良好的延伸性和高度灵活性。很多非纺织类材料可以设计成有类似强度的连续股线。

四种基本的短纤维丝制造方法已实现了标准化。短纤维纱线方法有粗疏棉、精梳棉、粗纺毛和精纺毛。粗疏棉和精梳棉方法是把短或长的棉或类似棉的纤维纺成纱线。粗纺毛和精纺毛方法是把短的或长的毛或类似毛的纤维纺成纱线。

在化学纤维发明之前，真丝是唯一的连续纤维纱线。加上适当的捻度和捻向后形成单股纱线，然后再纺成多股纱。

如果长纤丝是通过短纱线方法进行制作的，那么要用上千根纤维一起加少量捻度合并成粗纤维，以配合随后的皱缩与剪切工序。人造纤维其中一个优点是在产品的每个生产过程都可被利用。

高膨体纱是由具有高度蓬松延长性的短丝或连续长丝构成。这种纱线在自然状态或受压状态下都会保持它们的膨松度。由高膨体纱织成的织物可以做到重量轻，覆盖能力强。

### 1.1.3 织物

根据服装的用途和要求，制造商可以选择不同面料。商店里有很多令人眼花缭乱的面料，它们有不同的重量、风格和价格。面料的质量（耐用性、保暖、构造、表面效果、厚度）依赖于原材料所用的纱线种类、纺纱及织造和后处理的方法（光面、起毛、防水等后处理）。

大多数服装由毛纺或混纺毛的机织面料制成，毛料分为两大种类——粗纺毛和精纺毛。很多人可以通过羊毛面料较重的质感、手感和精纺面料较轻的质感、滑糯的手感将两者进行区分。但他们不知道的是，这两个词其实指的是织制面料所用的纱线。例如，一些颜色鲜艳的运动上衣面料，是采用精纺毛经纱和粗纺毛纬纱制成的。不同的是它们适用于不同用途——粗纺毛适用于夹克和外套，而精纺毛适用于西装和西裤。

机织面料包括平纹、斜纹和缎纹，典型的机织面料有棉、麻、人造纤维、粗纺毛和精纺毛。较常用的机织面料如下：

（1）白布：名称适用于任意一种平纹棉织物。用作裁剪之前的试样。

（2）毛巾布：织物表面的毛丛或线圈是由织入面料的纱线形成的。这些纱线经修剪而成毛丛，未修剪而成线圈。

（3）经条灯芯绒：在经纱上有明显纹状的棉布或毛料。这种纹状不像绳索，不是由毛丛而成，而是由额外的经纱填充而成。

（4）灯芯绒：在经纱上有纹状的坑纹棉或混纺棉面料。

（5）帆布：表面平滑、牢固而紧密编织的棉面料。

（6）法兰绒：通过表面磨毛后，有滑糯手感的粗纺毛或精纺毛的斜纹面料。最出名的是源于凸条纹的灰色织物。

（7）斜纹呢：一种斜纹外观的纤细面料，通常是一些精纺毛、棉或麻料。

（8）人字纹织物：一种通过改变斜纹方向而成的织物，与鱼背骨的人字形效果相似。

（9）方平织物：以两条或两条以上的经纱和纬纱作为一条纱线进行编织，形成箩筐效果的平纹组织结构。

（10）巴拿马薄呢：一种轻质的平纹组织西装面料。

（11）哔叽呢：粗犷的斜纹机织面料，通常是毛料和平纹色织布。

（12）粗花呢：结实的粗纺毛织物，最初织造于苏格兰南方地区，现在广泛用于各种织物，混织后的自然颜色成为其特点。

（13）斜纹布：对机织斜纹布的描述（如牛仔布、华达呢和哔叽呢）。斜纹布因其斜纹外观很容易被辨认。在成衣制作中，该布料一般采用从左下角向右上方的斜纹裁剪的效果。

（14）直贡呢：一种用精纺毛为经纱、粗纺毛为纬纱的经向缎纹机织面料，表面光泽度强。

针织面料分为经编针织和纬编针织。针织布料以线圈相互交织而成，成衣、家居、产业用的针织布一般由各种纤维、纱线、针法与样式组成。针织面料比机织面料有更好的伸缩性。纬编仍是成衣制造中流行的方法，这面料可用于基本的裁剪和缝纫。常用的针织面料如下：

（1）纬编针织料：纱线沿着纬向成圈的织物，它们既可手织又可机织。

（2）经编针织料：纱线沿纵向成圈的针织织物，如特里科经编织物和拉舍尔经编织物。

（3）纬平针组织：针编半圆弧显于正面，沉降 V 形弧藏于反面；正面平滑，反面粗糙。两端易卷边和散脱。可用作内衣、衬衫、连衣裙、毛衣、长袜和 T 恤衫。

（4）罗纹组织：1×1 组织是最简单的罗纹织物，它由一个正面线圈纵行和一个反面线圈纵行的交替结构。其他特别的罗纹结构的例子有 2×2 罗纹和 3×3 罗纹。漏针会引起纵向散脱，罗纹两端不会卷边，而且有较好的横向伸展性。可用作袖口、领子、腰带、毛衣和其他服装。

（5）双反面组织：这种纬平组织有更好的纵向延展性，织物正反面像纬平组织的反面，而且两端不卷边。可用作儿童装和毛衣。

（6）双罗纹组织：可从最后一行散脱，但不会延纵向散脱和卷边，双面光滑。可用作女装连衣裙、女衬衫、T 恤衫、毛衣、外套和内衣。

（7）毛圈针织布：比机织厚绒织物更具弹性和吸水性。可用作长袍、沙滩装和其他时尚服饰。

（8）针织天鹅绒：有更好的覆盖能力、柔软手感和有仿鹿皮的表面。可用作男装衬衫和女装裙。

（9）毛绒针织布：有更好的覆盖能力，可洗涤和低温甩干，有毛料类似的表面。可用作皮毛织物和小毛毯。

机织和针织面料的主要竞争者是无纺布。这些面料直接由纤维转换成织物，无须经过纺纱的生产阶段。一般来说，机织、针织和无纺布之间是有明显差别的。

### 1.1.4 服装辅料

很多重要的辅料应用于服装的制造过程，如里料、里衬、黏性带条、拉条、肩垫等。事实上，这些部件在服装结构上扮演了非常重要的角色，它们美化服装外观，方便车工的操作。

（1）里料：用于覆盖服装内部的布料。它可提高服装的质量，给出更好的定型效果，使穿着者有更好的舒适性。

（2）里衬：一种在服装面料和里料之间的内里，可以定型和保暖。里衬可分为机织衬、针织衬、无纺衬和纸衬几种类型，如热熔衬、非黏合衬和车缝里衬。

（3）牵条：一种应用在服装部件上作为功能性或装饰性用途的窄条物料。牵条可根据它的用途和设计应用于服装的多个部位。例如，用于悬挂作用的牵条可缝制于裙子的顶端、侧缝或肩缝处，但一条装饰的牵条会缝在设计所要求的服装部位上。

（4）拉条：一种用于加固作用的服装物料。根据不同的应用，可选择不同的材料制成各种拉条，如领子拉条、支撑拉条和其他特殊拉条。

（5）肩垫：一种最流行的垫块可用于加固袖窿部位。事实上，肩垫的应用可使服装的质量提高，给人华丽的感觉。

除了以上生产上用辅料外，还有很多其他种类，如纽扣、缝线、拉链、松紧带、皮带、铆钉、魔术贴、眼孔、洗涤标、尺码标、主商标和其他包装辅料等。

此外，时尚配饰也是一项用于达到穿着者特定外观形象的部件，如手袋、雨伞、钱包、鞋靴、围巾、领带、帽子、腰带、吊裤带、手套、珠宝、手表、披肩、围巾和袜子等。

## 1.2 生产中的面料预处理

除了掌握面料的知识外，生产商一定要了解所买物料的特性。例如，在缝纫车间或在后整工序环节，处理好面料是非常重要的，因为采用错误的工序方法处理和整理问题布料可能产生劣质产品，如透明面料、收缩面料、起毛面料、针织面料、皮革、结构松散或紧密型面料、热塑性面料等。

### 1.2.1 面料缩水

一块特殊布料的质量取决于它是否要水洗或干洗。成衣生产商可在购买面料时索取洗涤指南。面料在裁剪和服装制作之前通常需要经过预缩处理。另一方面，面料中的大量浆料可以被软化，特别是布料在柔软剂或醋酸溶剂中洗涤后。因全棉针织和运动衫面料的缩水率可达 10% 以上，所以在缝制前需经过水洗预缩。

切记面料在洗涤时要里朝外，特别是机洗时，否则鲜艳的颜色会变黯淡。如果水洗操作人员打算用干衣机甩干面料，那么它需要充分甩干才能裁剪和缝制。一件服装有不同种类和颜色的面料，应该分开洗涤，避免颜色互渗以及产生不同的缩水。当缝制完一件服装时，应作为高质面料小心处理。多数面料通常会在第一次洗涤时缩水，且在纵向缩水。在第一次洗涤前，可选择拆除裤子下摆的针粗缝。如果生产厂家不想在制作衣服前水洗布料，可以制作一件面料的洗涤样板。

### 1.2.2 处理技术

目前有很多面料应用于服装生产上。与制造简单的棉或毛衣相比，制造商需要更多的知识、设备和处理技术。衣服的设计、裁剪和制造需使用高质面料。为使缝纫协调，缝纫线、车针、线迹种类和缝纫机送布系统应与面料的组织结构和纤维成分协调一致。压烫和整理技术需要与面料的原本特性和后整理相一致。

### 1.2.3 面料性能的考虑

使用有问题的面料时，设计师和样板师的工作重点是利用面料的特性提升设计效果，但应意识到随时可能出现的结构问题，或具备操作技能和设备方面的知识，以便尽可能地生产出符合标准的服装。采用问题面料进行生产时，必须考虑一些基本问题，如伸长、缝迹起皱、织物组织的损坏、透明度、缺乏弹性、热可塑性等。很多面料的结构可以伸长，如针织、特殊结构的机织面料、斜纹机织面料等。伸长量取决于具体的面料，有一些在一个方向伸长，其他的在各个方向伸长，所以设计、裁剪和制造的技术需要相应的变化。由于多种原因，缝迹起皱影响很多面料结构，如缝迹扭曲、线迹太紧、面料结构扭曲等。

很多面料在制作与穿着过程中可能会因使用不当而被损坏。这些面料包括组织紧密型面料、组织疏松型面料、起绒面料等。注意避免一些在缝制过程中不明显但会显露在服装上的损坏。透明或半透明服装的款式和结构需避免障碍物影响透明特性。必须考虑边脚处理、缝迹、衬布和里衬等处理方法。像皮革和塑料等面料处理起来不灵活，这种面料的服装纸样比柔软面料更多，更容易缩褶。皮革要经过特别过程来干洗。一件用小片皮革装饰的衣服可以用手洗。格子、条纹面料和其他特殊印花面料需对条对格处理，易引起缝迹扭曲而造成缝纫困难。

## 1.3 材料采购

### 1.3.1 采购功能

采购部的一项重要任务是监督供应商及跟踪新订单的进度，确保对方及时以合理的价格交付规定数量和质量的商品或服务。采购活动包括如下过程：

（1）对不同供应商进行评估，较为重要的产品和原材料应选用少量但效率高的供应商。

（2）以有竞争力的价格提供机会购买有效的物料。

（3）对供应商的交货期、服务、质量和价格进行分级和评估。

（4）提高公司在供应商和竞争者心中的声誉。

（5）通过与市场部的不断交流而获得选择新材料的机会。

（6）通过与其他部门的联系，确保全公司有足够的材料存货水平。

从以上活动可清楚看到采购对公司的利润会产生影响。

### 1.3.2 采购订单与合同

成衣生产部门，合同上的服务范围包括了生产和设计的所有环节。一个中小企业能否成功决定于有多少投资的资金、创始人的能力以及经营的时效。客商有权把原材料付运到公司，采购部门需要制订出采购订单。采购订单应包括以下信息：

（1）落单日期。

（2）订单号和款号。

（3）送货日期。

（4）公司名称。

（5）颜色、款式、质量和规格的详细描述。

（6）价格和交易条件。

（7）装船地点、船运方式和包装说明。

五份复印件需要送往以下部门：

（1）采购部门：用于存档和登记。

（2）客户部门：作为采购订单的规格说明，以便安排原材料采购。

（3）财务部门：用于会计明细。

（4）船务部门：如有需要可与采购订单一起申请信用证。

（5）销售部门：用于参考和检查。

这类表格的好处是根据采购订单的规格说明，便于跟进、投诉和索赔工作。但如果采购单的细则不清晰，公司可能受到损失（表 1-1、表 1-2）。

**表 1-1　采购订单**

| 采购订单 | |
|---|---|
| | 文件编号：0688 |
| 布板 | 采购订单编号：TK-21<br>日期：2020-3-25 |
| 款式编号：JN-012 | |
| 描述 / 数量：30% 羊毛 & 70% 涤纶混纺，平纹绒面布组织，76 × 68/48S × 48S，黑色 WIDTH，布幅 152 ㎝，总数量大约 550 m，后整为毛面效果。 | |
| 单价：US$16/m，包括本地运输费用。 | |
| 包装：折叠装入透明包装袋。 | |
| 付运：在 2020 年 6 月 25 日前交付。 | |
| 备注：接受 SGS 检测 | |
| 买家签名确认：ALICE LIU | 卖家签名确认：DK ZHANG |
| 注：交货数量 +/-5% 可以被接受。 | |

表 1-2　采购合同

| 采购合同 | |
|---|---|
| 参考编号：MSC81009 | 日期：2020-2-28 |
| 本合同由 POLO 企业有限公司（称为买家）与 RYKIEL 贸易公司之间签订。<br>地址：中国香港九龙远安街 6 号（称为卖家，买卖双方同意以下商品的交易。主要贸易方式与条款规定如下。） | |
| 产品描述：100% 棉条纹格子布、硫化染色、黑色。<br>结构：72×42/ 10 英支 ×10 英支。<br>布幅：48 英寸。<br>后整理：上浆处理。 | |
| 数量：总数量大约 1550 码。大约 5% 的出入可以被接受。<br>价格：每码 1200 港币，包括本地运输费用。<br>总金额：18600 港币（实际金额根据实际付货数量决定）。<br>包装：卷装入透明包装袋。<br>付款方式：付货 30 天内支票支付。<br>交货：在 2020 年 5 月 30 日前尽快交货。<br>其他条款：质量：必须与提供的样品一致。<br>原产地：中国香港。 | |
| 请在 3 天内将复印件签名后交回 | |
| 买家签名确认<br>POLO 企业有限公司<br><br>______________<br>有效签名 | 卖家签名确认<br>RYKIEL 贸易公司<br><br>______________<br>有效签名 |

### 1.3.3 原材料存储

在采购部门下设有库房管理部，其运作是保证成衣生产中所需材料既不会用尽，也不会囤积，从而不会因小额采购而花多余的钱。一般情况下，存货有以下几点重要作用：

（1）核对所有的入库和出库物料。

（2）向仓库管理部门提供服务信息。

（3）使库存量保持在经济的水平。

（4）提供足够的库存量，防止因供应商的延迟送货而造成生产停滞。

（5）为入库物料预留储存空间，并把所需要的物料运送到其他部门。

（6）确保有足够的财务资金供物料采购使用。

（7）有准确的物料供应计划。

注意以上活动，管仓员的责任是结合存货水平尽量降低成本。

### 1.3.4 控制表的应用

控制表包括物料控制表和物料采购单等（表 1-3、表 1-4），采购部提供有关数据。控制表收集了全部销售订单，以便决定每种款式和颜色的服装所需的原材料和辅料用量，使采购部门购买准确数量的原材料和辅料。建议把控制表的四份复印件送到：

（1）采购部门：做记录并借此表格估计已用物料情况。

（2）销售部门：用于使用情况的参考、质量的再次核对和成本的计算。

（3）会计部门：用于账单的参考和核对。

（4）产品部门：用于消耗量的参考和核对，也用于成本的计算。

**表 1-3 物料控制表**

| 布料控制表<br>卖家：______ 核准：______ 日期：______ | | | | |
|---|---|---|---|---|
| 订单编号 | | | 总数量 | |
| 款号 | | | | |
| 数量 | | | | |
| 码长 × 布幅 | | | | |
| 工厂 | | | | |
| 价钱 | | | | |
| 颜色 | | | | 该天收货数量： |

**表 1-4 物料采购单**

| 物料采购单<br>初板：______ 核准板：√ 大货生产：√<br>客商：TOMMY 日期：2020-4-5 订单编号：T-05248 | | | | | | | |
|---|---|---|---|---|---|---|---|
| 物料种类 | 样式编号 | 尺码 | 颜色 | 供应商 | 日期 | 样板 | 数量 |
| TQ-08 | M8745 | S-XXL | W-01 | DALIN | 4-25 | — | 100Y |
| TQ-09 | M8746 | S-XL | B-021 | DALIN | 4-30 | — | 120Y |
| — | | | | | | | |
| — | | | | | | | |
| 申请人：MAY 2020-4-5 物料部：BADY 2020-4-6 样板部：DK LIU 2020-4-6 | | | | | | | |

## 本章小结

- 总体上介绍了服装用纤维、纱线、面料以及辅料的基本分类，并列举了服装用料可选用的常见种类与名称。
- 通过对特殊面料预处理的简单描述，介绍了主要特殊面料在缝制与生产中需注意的事项，并描述了面料预处理的相关专业词汇。
- 从服装材料采购职能入手，导入采购订单与合同、面辅料控制表等服装材料采购与跟单工作涉及的相关文件，从而引出各种样板名称等有关专业词汇在各文件表格中的应用。
- 此章节重点加强学生对纤维、纱线、面料以及辅料等专业术语的学习与记忆。

**CHAPTER 2**

# DESIGN AND PATTERN　设计与纸样

**课题名称：** DESIGN AND PATTERN　设计与纸样

**课题内容：** Fashion Design　服装设计

Fashion Trends　流行趋势

Pattern Drawing　纸样绘制

Relative Glossary　相关术语

**课题时间：** 8 课时

**教学目的：** 让学生了解服装设计效果图的分类与款式描述、时装信息文章的解读以及尺码测量与纸样构造等基本知识。掌握不同服装款式的基本描述、时装或纸样术语的解释。重点掌握并熟记各种服装款式、部件名称、尺寸测量部位等相关术语与名称。

**教学方式：** 结合 PPT 与音频多媒体课件，以教师课堂讲述为主，学生结合案例分析讨论与学习相关款式描述。

**教学要求：** 1. 了解服装设计效果图分类与描述。

2. 熟悉时装信息文章的解读与理解。

3. 明确尺寸测量、纸样制作基本方法的应用。

4. 熟悉各相关术语的解释与不同款式的描述。

**课前（后）准备：** 结合专业知识，课前预习课文内容。课后熟读课文主要部分，学会款式描述的技巧与方法，并熟记各服装款式、部件名称、尺寸测量部位等相关术语与名称。

# CHAPTER 2

# DESIGN AND PATTERN
# 设计与纸样

## 2.1 Fashion Design 服装设计

①时装设计是指对一种风格进行独特的或个性化的处理。一种风格能通过多种设计来传达。

For most people，the word “fashion” means clothes. But people may ask the question，“What clothes are in fashion?” And they use the adjective “fashionable” in the same way: “She was wearing a fashionable coat.” “His shirt was really a fashionable color.” But of course there are fashions in many things，not only in clothes. There are fashions in holidays, in restaurants, in films and books. And the fashion design is a particular or individual interpretation treatment of a style. A style may be expressed in a great many designs, all different, yet all related because they are in the same style.[①] “Fashion Sense” consists of the ability to tell which clothing and accessories look good and which don’t. Since the entire notion of fashion depends on subjectivity, the fashion designers create their collections by combining their own images with fashion trends. The fashion illustration or design sketch shows the effect of wearing (Figure 2-1).

**Figure 2-1 Fashion Sketch**

### 2.1.1 Famous Fashion week 著名时装周

（1）Paris Fashion Week: It is a famous fashion week held semi-annually in

Paris, France, the Spring/ Summer and the Autumn/Winter events held each year. In addition to ready-to-wear shows, there are men's and haute couture shows. ② Dates are determined by the French Fashion Federation（Figure 2-2）.

②除了成衣秀之外，还有男士和高级定制时装秀。

**Figure 2-2 Traditional Styles**

（2）Milan Fashion Week: It is a clothing trade show held semi-annually in Milan, Italy. The Spring/Summer event held in February - March of each year and the Autumn /Winter event held in September - October of each year. The most important fashion show is dedicated to Women's and Men's wear fashion in the events（Figure 2-3）. ③

③在活动期间，最重要的时装表演是致力于女装和男装的时尚。

**Figure 2-3 Fashion Show**

④ 2010年春季，伦敦时装周成为第一个全面拥抱数字媒体的重要时装周，它为所有在T台上展示自己时装的设计师提供了在互联网上直播节目的机会。

（3）London Fashion Week: It is an apparel trade show held in London, England twice each year, in February and September. It is organized by the British Fashion Council. In spring 2010, London Fashion Week became the first major fashion week to fully embrace digital media when it offered all designers who were showing their collections on the catwalk to broadcast their shows live on the Internet（Figure 2-4）. ④

**Figure 2-4　Shows on Catwalk**

（4）New York Fashion Week: It is one of four major fashion weeks held around the world, and it is held in February and September of each year in New York City. Sometimes, the dress styles are the collection of all the characteristics, just like the four different styles of women's clothing in the film *Sexy and the City*（Figure 2-5）.

**Figure 2-5　Different Styles**

### 2.1.2　Fashion Illustration　时装图解

Fashion sketches express the designer's concept and support the creation and marketing of the collection. Before sample making up or apparel production, the fashion sketch should be draw into production sketch with CAD, and it is a clear accurately drawn sketch of a garment that shows the important cutting details such as pockets, darts, seams, top-stitching, proportions, as well as other special design details.

(1) Women's Blouse: It is a soft and loose garment for the upper body, which is worn by women together with skirts or slacks. ⑤ In today's fashion, such as the neckline and the collar, length, cut, details and decoration, together with the material produce a particular style. It is typically gathered at the waist or hips so that it hangs loosely over the wearer's body (Figure 2-6).

⑤女衬衫是一种柔软宽松的上衣，可以搭配短裙或裤子一起穿着。

(2) Uniform: This style of shirt is designed to compliment of the uniform status, or if the shirt is to be worn without a jacket then attention must be given to the practical aspects of the wearer's occupation. Such as police, emergency services, security guards, etc. And some other officials, workplaces and schools also wear uniforms in their duties.

(3) Work Clothing: The shirt is designed with the first consideration given to the kind of work of the wearer, especially some works that involve manual labour. The work clothing provide durability and safety, the fashion and style is a secondary priority, but must not be neglected (Figure 2-7).

**Figure 2-6 Women's Blouse**

**Figure 2-7 Occupation Styles**

(4) Optional Pockets: Pockets are both functional and decorative, and come in a variety of styles, and there are three general manufacturing principles, patch pocket, set-in pocket and seam pocket. And the patch is an extra shell fabric piece lay on top of the garment directly. Patch pockets adorn men's, women's, and children garment. The fabric piece is stitched on three sides, with top open. They can be square, rectangular, pointed, curved, or other special shapes, and may be decorated with top-stitching, lace or braid trims. Fashion designers often create variations

⑥时装设计师经常在明贴袋上创造各种变化。例如，采用装饰面线、拉链、悬垂效果、装饰贴边、纽扣搭配和袋盖等设计元素。

on patch pockets, using such elements as decorative stitching, zippers, draping, welts, button-down closures, and flaps, etc⑥ (Figure 2-8).

(5) A-Line: In 1955, Christian Dior produced a line, which ran from fair narrow shoulder and bust to a wide hemline at low calf length, giving the impression of the shape of a capital letter "A". The cross line of the "A" usually came at or below the knee, such as flounce of pleats ⑦ (Figure 2-9).

⑦在1955年，克里斯汀·迪奥设计了一款新风貌时装，在肩和胸部较窄，小腿较宽。给人留下大写字母"A"的外形印象。"A"字形的横切线一般在膝围或以下的位置，就好像荷叶边裙褶。

Figure 2-8 Optional Pockets　　Figure 2-9 Fashionable Styles

(6) Princess Line: A princess line or princess dress describes a woman's fitted dress or other garment cut in long panels without any seams at the waistline.

(7) Halter Neckline: A neckline with a strap or built up bodice in front, continuing round to the back of the body. It is sleeveless, maybe just single strap around neck or over shoulder.

(8) Breeches: short trousers used for riding. It is normally closed and fastened about the leg, along its open seams at varied lengths, and to the knee, by either buttons or by a draw-string, or by one or more straps and buckle or brooches.

(9) Jacket: It is a basic item of outerwear, and it is a mid-stomach-length garment for the upper body. Sometimes, it refers particularly to the tailored types in men's clothing, e.g. men's tailored jacket. In a broad sense it includes the denim or jumper as well, and is worn by men, women and children. The fabric can vary with solid colors, plaids, strips or checks (Figure 2-10).

(10) Skirt: The skirt is clothing worn from the waist down and which is the basic element of women's clothing matching jacket with a skirt makes a suit, or costume, such as women's tailored suit with skirt. Many styles of skirts are subject to changes in fashion, such as jeans skirt. Individual styles are distinguished by length, width, silhouette, cut and details, etc (Figure 2-11).

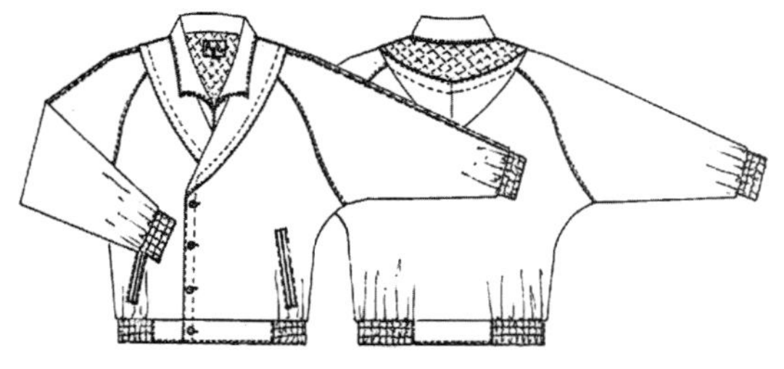

**Figure 2-10 Men's Jacket**

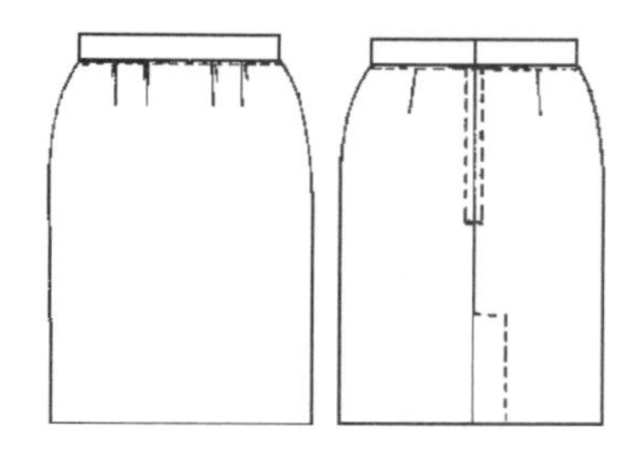

**Figure 2-11 Tailored Skirt**

(11) Jeans: The term jeans are used to describe the vary close fitting trousers, e.g. five pockets jeans. The most famous jeans being known as Levis are made by the large trousers makers in the world.[⑧] The first jeans were made in 1850 by Levi Straus for gold prospectors in USA. Now, Jeans have now become a fashionable garment for both men and women as leisure and casual wear, and they come in various fits, including skinny, tapered, slim, straight, boot cut, narrow bottom, bell bottom, low waist and flare, etc (Figure 2-12).

⑧牛仔常指不同合体的裤子，如五袋款牛仔裤。最著名的牛仔裤就是利维斯牛仔裤，由世界上著名裤业制造公司制造。

(12) Sport Wear: a style should be made of the tops & bottoms for different sports. It will be found that practical consideration about colors and designs of fabric also play a large part in the designing of sports shirts (Figure 2-13).

**Figure 2-12 Cargo-Pocket Jeans**

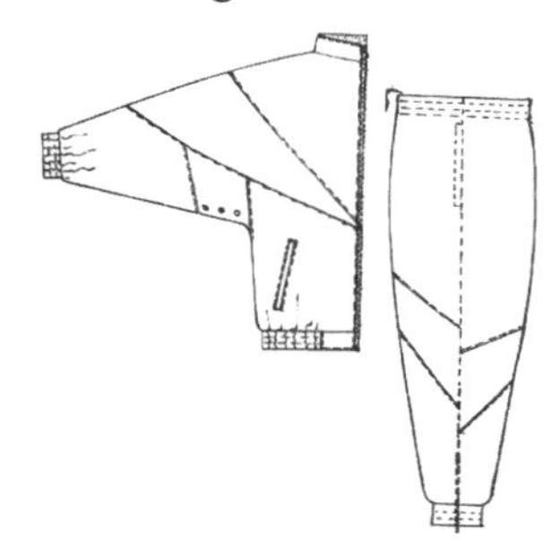

**Figure 2-13 Sports Wear**

(13) Mandarin Dress: It is a Chinese Mandarin Dress, named as Qipao Dress, with a narrow upstanding collar meeting at the front of the neck[⑨]. The original Qipao was wide and loose, and it covered most of the woman's body, revealing only the head, hands, and the tips of the toes. With time though, the Qipao was tailored to become more form fitting and revealing (Figure 2-14).

⑨旗袍是一款中式长裙。旗袍有一窄而直立的衣领，领尾在前领圈处相扣。

(14) Undergarment: It is an item of clothing worn beneath outer clothes, usually in direct contact with the skin, to shape the body. Undergarment is known by a number of terms, such as Unders, Underwear, Underclothes and Underclothing, etc. Women's undergarments collectively are called lingerie too. If made of suitable

material or textile, some undergarments can serve as nightwear or swimsuits（Figure 2-15）.

**Figure 2-14 Mandarin Dress**　　**Figure 2-15 Underwear**

## *Words and Expressions*

ready-to-wear [redi-tu-wiə] 成衣
haute couture [əut ku'tju:r] 高级女装定制
French Fashion Federation [frentʃ 'fæʃən ˌfedə'reiʃən] 法国时装联盟
sleeveless [ˊsli:vlɪs] 无袖的
British Fashion Council ['britiʃ 'fæʃən 'kaunsil] 英国时装委员会
catwalk ['kætwɔ:k] T台
fashion sense ['fæʃən sens] 时尚感觉
notion ['nəuʃən] 观念，意见，概念
illustrate ['iləstreit] 图示，图解
fashion trend ['fæʃən trend] 流行趋势
fashion sketch ['fæʃən sketʃ] 时装画
design sketch [di'zɑin sketʃ] 设计图，效果图
casual shirt ['kæʒjuəl ʃə:t] 便装衬衫
trousers ['trauzəz] 西裤
collar ['kɔlə] 领子
sleeve [sli:v] 袖子
pocket ['pɔkit] 口袋
style [stail] 款式
front opening [frʌnt 'əupəniŋ] 前片开襟
bottom ['bɔtəm] 下摆
eyelet holes ['ailit həulz] 孔眼
lacing ['leisiŋ] 花边
button-holes ['bʌtn həulz] 纽扣孔（广东话：纽门）
zipper fastener ['zipə 'fɑ:snə] 拉链系结物
formal shirt ['fɔ:məl ʃə:t] 礼服衬衫
pleating [pli:tiŋ] 褶裥（广东话：活褶）
frilling [ˊfriliŋ] 皱边 *
fly front [flɑi frʌnt] 暗门襟
double cuff ['dʌbl 'kʌf] 双层袖克夫
sweat shirt [swet ʃə:t] 运动衬衫
pullover ['puləuvə(r)] 无领无扣衫，套衫
terry cloth ['teri klɔ:θ] 毛圈织物 *
fleece-lined cotton [fli:s- lɑind 'kɔtn] 厚绒棉布 *
plaid [plæd] 格子布
stripes [straip] 条纹布，带条
checks [tʃeks] 大格布
uniform ['ju:nifɔ:m] 制服
compliment ['kɔmplimənt] 恭维，称赞
status ['steitəs] 地位，状况
aspect ['æspekt] 局面，容貌
jacket ['dʒækit] 夹克
occupation [ɔkju'peiʃən] 职业
work clothing [wə:k 'kləuðiŋ] 工作服
consideration [kənsidə'reiʃən] 考虑，要点
neglected [ni'glektit] 忽略的
business shirt ['biznis ʃə:t] 职业衬衫

smart appearance [smɑːt ə'piərəns] 潇洒的外貌
town-wear [ˈtaun-weə(r)] 上街装，外出服
dinner jacket ['dinə 'dʒækit] 晚礼服
tuxedo [tʌk'siːdəu] 男士无尾晚礼服，塔士多
braid [breid] 织锦
tail-coat [teil-kəut] 燕尾服
waist [weist] 腰围，腰
braces [breisiz] 吊带，背带
side straps [said stræp] 侧吊带 *
elastic [i'læstik] 松紧带
tab [tæb] 襻带
metal slide ['metl slaid] 金属滑片
side seam [said siːm] 侧缝
hips [hips] 臀围（广东话：坐围）
cash pocket [kæʃ 'pɔkit] 口袋，表袋
waistband ['weistbænd] 腰头
current fashion ['kʌrənt 'fæʃən] 流行服装
un-cuffed bottom[ʌn-'kʌfd 'bɔtəm] 不翻边裤脚
plain bottom [plein 'bɔtəm] 平折下摆
slacks [slækz] 宽松长裤
leisure occasions ['leʒəə'keiʒənz] 休闲场合
patch pocket [pætʃ 'pɔkit] 明贴袋
side pocket [said 'pɔkit] 侧口袋
hip pocket [hip 'pɔkit] 后袋
up-to-date trends [ʌp-tuː-'deit trend] 最新潮流的
pants [pænts] 休闲裤
sports wear [spɔːtswεə] 运动服
ski-pants[skiː-pænts] 滑雪裤
A-Line[ə'lain] A 字形
Christian Dior ['kristiən diɔ] 克里斯汀 · 迪奥 ( 法国时装设计师 )
shoulder ['ʃəuldə] 肩宽
bust [bʌst] 胸围
hemline ['hemlain] 下摆
calf [kɑːf] 小腿
knee [niː] 膝围
flounce [flauns] 衣裙上的荷边装饰
princess line [prin'ses lɑin] 公主线
halter neckline ['hɔːltə 'neklain] 吊带领圈
breeches ['briːtʃiz] 马裤
production sketch [prə'dʌkʃən sketʃ] 生产图
pattern-cutting ['pætən - 'kʌtiŋ] 纸样裁剪
dart [dɑːt] 省道（广东话：死褶）
top stitching ['tɔp-stitʃ] 面缝线迹( 广东话：缉面线）
leisure shirt ['leʒə ʃəːt] 休闲衬衫
yoke [jəuk] 育克，过肩（广东话：担干，机头）
center back ['sentə bæk] 后中
western shirt ['westən ʃəːt] 西部衬衫
tapered shirt ['teipəd ʃəːt] 紧身衬衫
pointed collar ['pɔintid 'kɔlə] 三尖领
flaps [flæps] 袋盖
fancy snap ['fænsi snæp] 装饰按钮
placket ['plækit] 开口，开襟（广东话：明筒）
Tee-Shirt [tiː-ʃəːt] T 恤衫
plain [plein] 平纹 ( 布 )
printed ['printed] 印花（布）
motifs [məu'tiːf] 意念，要素
underwear ['ʌndəwεə] 内衣
close fitting [kləuz, 'fitiŋ] 紧身的
Levis Jeans ['liːvaiz dʒeinz] 李维斯牛仔裤
prospectors [prɔ'spektə(r)] 采矿者
cargo-pocket jeans ['kɑːgəu 'pɔkit dʒeinz] 大袋牛仔裤
Bermuda shorts [bə(ː)'mjuːdə ʃɔːts] 百慕大短裤
engaged [in'geidʒd] 忙碌的，预定的
mini shorts ['mini ʃɔːts] 超短裤
Mandarin Dress/ Qipao Dress ['mændərin dres] 旗袍
upstanding collar [ʌp'stændiŋ 'kɔlə] 立领
lingerie [ˌlænʒə'riː] 妇女贴身内衣
grommet ['grɔmit] 金属眼孔
national costume ['næʃənl 'kɔstjuːm] 民族服装
frog pocket [frɔg 'pɔkit] 风琴袋
crossed pocket [krɔst 'pɔkit] 横开口袋

tight fit [tait fit] 紧身
denim ['denim] 粗斜纹棉布，牛仔服装
adorn [ə'dɔːn] 装饰
gathered skirt['gæðəd skəːt] 缩褶裙
solid color ['sɔlid 'kʌlə] 单色，纯色
durability [ˌdjuərə'biliti] 耐久性
princess dress ['prinses dres] 公主装
panel ['pænl] 衣片
seam [siːm] 缝骨
waistline ['weistlain] 腰围线
fasten [fasten] 系结物，扣紧
buckle ['bʌkl] 皮带扣
brooch [bruːtʃ] 胸针
jeans skirt [dʒiːns skəːt] 牛仔裙
skinny ['skini] 极瘦的
tapered ['teipəd] 锥形的
slim [slim] 苗条的
boot cut [buːt kʌt ] 喇叭口
narrow bottom ['nærəu 'bɔtəm] 窄脚
bell bottom [bel 'bɔtəm] 喇叭脚
low waist [ləu weist] 低腰
flare [flɛə] 宽摆的
revealing [rɪˋviːlɪŋ] 暴露的，袒胸露肩的
undergarment /underclothing ['ʌndəˌgɑːmənt] / ['ʌndəkləuðiŋ] 内衣
outer clothes ['autə kləuðz] 外套
unders/ underwear ['ʌndəs] / ['ʌndəwɛə] 内衣
nightwear ['naɪtweə] 睡衣
swimsuit ['swɪmsuːt] 游泳衣

## *Exercises*

1. Translate the following terms into Chinese.

(1) bottoms
(2) eyelet hole
(3) lacing
(4) zipper
(5) fastener
(6) pleating
(7) frilling
(8) fly front
(9) single cuff
(10) pullover
(11) terry cloth
(12) cotton
(13) knitted wear
(14) plaid
(15) striped
(16) fashion
(17) casual wear
(18) trousers
(19) collar
(20) sleeve
(21) patch pocket
(22) bottom
(23) jacket
(24) working shirt
(25) town-suit
(26) dinner jacket
(27) coat
(28) waist
(29) brace
(30) strap
(31) elastic
(32) tab
(33) side seam
(34) cash pocket
(35) waistband
(36) current fashion
(37) plain bottom
(38) slacks
(39) side pocket
(40) hip pocket
(41) pants
(42) sports wear
(43) ski-pants
(44) A-Line
(45) shoulder
(46) hemline
(47) flounce
(48) princess line
(49) breeches
(50) underwear
(51) close fitting
(52) Mandarin Dress
(53) lingerie
(54) tight fit
(55) grommet
(56) national costume
(57) pattern
(58) top-stitching
(59) leisure wear
(60) yoke

(61) center back
(62) pointed collar
(63) pocket flap
(64) snap
(65) placket
(66) Tee-shirt
(67) plain
(68) pleat
(69) dart

2. Describe this style in a 200 words paragraph (Including Front View & Back View).

**Five-pockets Jeans**

3. Draw the production sketch according to the following description.
A casual shirt should be designed with the collars, short or long sleeves, and the point pockets are usual features in this style of shirt. The front opening can be from top to bottom with buttons and button-holes, or zipper fastener.

## 2.2 Fashion Trends 流行趋势

### 2.2.1 1980's Style 20 世纪 80 年代的时装风格

Fashion was highly diverse and differentiated during the Eighties. In women's wear the inspiration was the active, self-aware women. Styles were both classic as well as casual. Feminine fashion was persisted, however, with refined, seductive and extravagant styles. ① The casual, elegant style, with masculine shape and details retained for suits and coats. Eveningwear was feminine and elegant, softly flowing or figure-fitting with a tight, or flared or full skirt. ② Skirt lengths were varied from knee length to ankle length according to the style. The miniskirt had returned. The waist may be raised or lowed. Sleeves were cut generously for freedom of movement. The shoulders were often emphasized, and the combination of very different lengths and widths gave rise to new proportions. ③

In men's wear, the traditional suit was elegant with a classic cut, and comfort-

①女装灵感来自活跃、自我意识强的女性。既有经典优雅的风格，也有休闲实用的风格。女装追求精制、富有魅力、奢华的风格。

②晚装很女性化、优雅，有的柔软飘逸，有的紧身贴体，配以紧身的或展开的或蓬起的裙子。

③肩部常常是被重视的，不同长度和宽度的组合形成了新的比例。

④男士晚装是宴会装的生动组合。休闲装采用轻薄面料和休闲款式，但要求有复杂的功能性细节。

able. The waist was slightly fitted, shoulders were lightly padded, and lapels were not very broad. Quilted and light shell coats were popular. Evening men's wear was a lively mixture of party fashion. In casual clothing, light materials, a casual cut and elaborate functional details were required. ④ Jeans wear will be very colorful, if the designer's predictions are not wrong. Somebody said it is now common to have jeans in brown, green, red, gray and white. Other more fashionable versions in bright orange and purple will reach the chain-stores soon.

⑤带着一点幽默感的紧身夹克衫，穿在连有条带的露背裙或紧身高腰裙上。

### 2.2.2 Designer's Collection 设计师系列

Pastel pinks and baby blues are the main colors in which to beat the Karl Lagerfeld's way. Suits are in plain or tweed, wool crepe or soft leathers. The silhouette is traditional. Chanel, Box jackets, new long skirts and cool silk blouses make up a total look that is elegant all the way. Jackets are closed-fitting with a touch of humor, worn over backless dresses with straps or figure-clinging high-waisted skirts. ⑤ In pinks, yellows and blues, they liven up even the most traditional workday suit.

As for the skirts, Karl Lagerfeld played with hemlines going from as short as even to just above the knee, to just above the ankle. Where does it stop ? Anywhere on the leg seems to be the answer, with Chanel showing thigh- cut slits on long skirts and minis everywhere, and the evening-wear with ruffles and flounces.

### 2.2.3 Summer Silhouette 夏季廓型

Soft fabrics, gray colors and smooth silhouettes herald the return of prettier clothes after the rugged influence of winter. A simple wool crepe chemise style dress, or cool cotton and linen frock, completed with a pretty hat, handbag or a pair of flat shoes, provides a comfortable feeling in the city heat.

⑥当鲜绿色、粉色或黄色的漂亮裙子加上一点迷人的黄昏色彩到各款夏天衣服上时，会给性感的莎笼裙和实用的猎装增添更轻松的风格，既可穿着于商务场合，又可穿着于休闲场合。

Business wear is given a fresh and appealing look in soft shades combined with rich cream or navy. Jackets are longer with a soft shoulder line, matching slim pants and skirts. Bright colors, rich prints, and cool fabrics are paired with trendy jeans and leggings for fun holiday dressing. Over-sized shirts tied at the waist accentuate a slim waistline. Added to this more relaxed look is sexy sarong skirts and practical safari jackets, which can be business as well as leisure, while bright dresses in sharp shades of green, pink or yellow add a touch of evening glamour to this versatile summer wardrobe. ⑥

### 2.2.4 Jeans' 2015 Spring/Summer Denim Collection 2015 春夏牛仔系列

Jeans Spring/Summer 2015 collection focuses on casual city style with special cuts. Natural fabrics such as linen and cotton, pure or mixed, play a central role. Outlines are relaxed, yet tailored and slim. Ready-to-wear field jackets with accordion pockets, shoulder flaps and casual seaming are real attention. Casual blousons in high-tech nylons definitely is from sporting inspiration, in textured, washed and bleached effects. ⑦ Classic jackets are comfortable yet retain all fitted pattern, with quality garment-dyed color ways ensuring a smart effect. Trousers again focus on slim styles in linen mixes, plus cotton, that's finished like denim to create special washed effects. Seersucker striped fabrics are also important. ⑧ The collection introduces new variations on denim, from loose-fit styles and raw denim to bleached washes.

⑦ 具有高科技成分的尼龙休闲衫来自运动灵感，在质地上具有水洗与漂白效果。

⑧裤子再次以紧身款式为焦点，布料采用加棉混纺麻布，经过后整理同样可以创造出牛仔布的特殊洗水效果。采用泡泡纱的条纹面料也是非常重要的。

## *Words and Expressions*

diverse [dai'və:s] 变化多的
inspiration [ˌinspə'reiʃən] 灵感
feminine ['feminin] 娇柔的，女性的
persist [pə(:)'sist] 坚持，持续
refine [ri'fain] 精制，使文雅高尚
seductive [si'dʌktiv] 诱人的
extravagant [iks'trævəgənt] 奢侈的，浪费的
masculine ['mɑ:skjulin] 男性的，男子气概的
figure-fitting ['figə 'fitiŋ] 紧身贴体
miniskirt ['miniˌskə:t] 迷你短裙，超短裙
elaborate [i'læbərət] 精心制作的
prediction [pri'dikʃən] 预言，预报
pure silk [pjuə silk] 真丝 *
linen ['linin] 亚麻布
single-breasted ['siŋgl-'brestid] 单排纽扣的 *
waistcoat ['weistkəut] 背心，马甲
polo-shirt ['pəuləu-ʃə:t] 开领短袖衬衫，马球衫 *
casual wears ['kæʒjuəl wiəz] 轻便装 *
shorts [ʃɔ:ts] 短裤 *
knit-wear [nit -wiə] 针织服装 *
pants [pænts] 休闲裤 *
hood [hud] 帽子 *
quilted [kwiltid] 绗缝 *
acrylic [ə'krilik] 腈纶 *
polyester [' pɔliestə] 涤纶，聚酯纤维 *
leisure wear ['leʒə wiə] 休闲服
hemline ['hemlain] 底边，贴边 *
jeans wear [dʒeinz wiə] 牛仔服
chain-store [tʃein - stɔ:] 连锁店
Chanel Look [ʃə'nel luk] 香奈儿风格
Karl Lagerfeld [kɑ:l 'lɑ:gəfeld] 卡尔·拉格斐(法国时装设计师)
tweed [twi:d] 粗花呢
crepe [kreip] 绉纱，绉绸
leather ['leðə] 皮革
box jackets [bɔks 'dʒækits] 箱型夹克
elegant ['eligənt] 文雅的，端庄的
closed-fitting [kləuzd 'fitiŋ] 贴身的
humor ['hju:mə] 滑稽，幽默，幻想
backless dress ['bæklis dres] 露背长裙
straps [stræp] 带条

figure-clinging ['figəd- 'kliŋiŋ] 紧身的
thigh-cut slits [θai-kʌt slit] 大腿开衩
minis ['miniz] 迷你型
evening-wear ['i:vniŋ - wiə] 晚礼服
ruffles ['rʌfls] 皱褶
flounces [flauns] 衣裙上的荷边装饰
dressing ['dresiŋ] 穿衣，装饰
chemise [ʃi'mi:z] 女式无袖衬衫
frocks [frɔks] 上衣，外衣，工装，僧衣
hat [hæt] 帽子
handbag ['hændbæg] 手提包
appealing [ə'pi:liŋ] 吸引人的
creamy ['kri:mi] 奶油色，淡黄色
shoulder line ['ʃəuldə lɑin] 肩线
slim pants [slim pænts] 紧身裤
rich prints [ritʃ prints] 华丽印花布
trendy jeans ['trendi dʒeinz] 新潮牛仔服
legging ['legiŋ] 袜统，绑腿，裹腿
over-sized ['əuvə - 'saizd] 特大型的
accentuate [æk'sentjueit] 强调
slim waistline [slim 'weistlain] 细腰的，修腰的
sarong skirts ['sɑ:rɔŋ skə:ts] 莎笼裙，褶皱裙
safari jackets [sə'fɑ:ri 'dʒækits] 旅行夹克，猎装
glamour ['glæmə] 魅惑，迷人的美
wardrobe ['wɔ:drəub] 衣柜，服装
ready-to-wear ['redi-tu:-wiə] 成衣
accordion pockets [ə'kɔ:djən 'pɔkit] 风琴袋
blouson ['blu:sɔn] 甲克衫，松紧带束腰的女衫
checked fabric [tʃekt 'fæbrik] 格子布 *
evening gowns set ['i:vniŋ 'gɑunz set] 晚礼服 *
see- through [si:-θru:] 透明装 *
fishnet ['fiʃnet] 网眼布 *
balloon sleeves [bə'lu:n sli:vz] 灯笼袖，泡泡袖 *
bustier ['bʌsti] 无吊带紧身褡 *
tulle [tju:l] 薄纱 *
faille [feil] 罗缎，菲尔绸 *
satin ['sætin] 缎子，缎纹 *
swimsuit ['swimsju:t] 泳衣 *
swimming trunks ['swimiŋ trʌŋks] 泳裤 *
butterfly sleeve ['bʌtəflɑi sli:v] 蝴蝶袖 *
overcoat ['əuvəkəut] 大衣 *
athletic shirts [æθ'letik ʃə:t] 运动衬衫 *
raincoat ['reinkəut] 雨衣 *
stage costume [steidʒ 'kɔstju:m] 表演服 *
women's wear [wu:mens wiə] 女装 *
student's wear ['stju:dənts wiə] 学生装 *

## *Exercises*

1. Translate the following terms into Chinese.

（1）culottes
（2）batwing sleeve
（3）armhole
（4）raglan sleeve
（5）men's wear
（6）pure silk
（7）linen
（8）cotton
（9）wool
（10）Hawaii collar
（11）double breasted
（12）jabot
（13）Kimono sleeve
（14）modeling
（15）underarm
（16）cape
（17）apron
（18）overskirt
（19）blouse
（20）sleeveless
（21）waistcoat
（22）blazers
（23）polo-shirt
（24）casual wear
（25）trousers
（26）knit-wear
（27）jacket
（28）pants
（29）hood
（30）acrylic

(31) polyester
(32) leisure wear
(33) business wear
(34) hemlines
(35) jeans wear
(36) chain-store
(37) crepe
(38) slim waistline
(39) wardrobe
(40) balloon sleeves
(41) bustier
(42) tulle
(43) leather
(44) closed-fitting
(45) backless dress
(46) evening-wear
(47) ruffle
(48) frocks
(49) satin
(50) swimsuit
(51) overcoat
(52) athletic shirts
(53) raincoat
(54) stage costume
(55) chemise
(56) shoulder line
(57) slim pants
(58) over-sized shirts
(59) embroidery
(60) evening dress

2. Design a style with a production sketch, and then describe it with a paragraph of about 200 words.
3. Write a 10 sentences description for the following style (Men's Jacket).
   Example: It is a 100% cotton men's jacket in gray color.

**Men's Jacket**

## 2.3 Pattern Drawing 纸样绘制

### 2.3.1 Basic Method 基本方法

There are two main methods to develop garment pattern, they are pattern drafting and pattern draping.

(1) Pattern Drafting: a system of pattern drawing that depends on a figure measurement to finish the paper pattern, it refers to flat pattern. In the flat pattern design process, a basic pattern is developed to fit an individual body or a standard dress form, and the basic pattern already has a designated shape and ease allowance. [①] Drafting method is used more for staple ready-to-wear items than for fashionable apparel, and

①平面纸样设计过程中，先开发基于个体或标准服装号型的基本纸样，并且基本纸样已经初具廓型和放松量。

the tailored garments are most successfully designed from flat pattern drafting.

（2）Pattern Draping: It is a three-dimensional piece of fabric draped around a padded dummy or figure conforming to its shape, creating a three-dimension fabric pattern, so it is the modeling pattern too. And then the fabric pattern is transferred to paper to be used for corrections and creating a final pattern. The modeling can provide a clear view of fabric drape and overall design effect of the finish garment before the garment components are cut and sewn together. ② During draping, basic ease can be pinned into pleats, gather, tucks, etc., which are later released as the pattern is developed. Modeling techniques work best with jersey fabrics and generous amounts of soft materials. It is also used to work fabric on bias. ③

②立体裁剪能提供一个清晰的面料悬垂效果，可以在服装部件裁剪与缝制之前获得服装的整体设计效果。

③立体裁剪技术最适合针织面料与各种柔软面料。它也常采用斜纹面料。

### 2.3.2 Measuring Method 尺寸测量方法

Before pattern drawing, it must be taken the correct measurement, and whether it's measuring body measurement or garment measurement, it always uses a soft flexible tape measure. All specifications are measured in cm or inches. When measure the garment measurement, garment should be measured on a flat, smooth surface, large enough to accommodate the entire garment. Being careful not to stretch a knit garment, smooth or pat until garment is wrinkle free. Now take clothing measurement as an example to illustrate the measurement method.

（1）Tops Measurement.

（a）Neck Girth: undo all closures and lay the collar flat so that the inside of garment is facing you. And measure from the center of the button to the farthest end of the buttonhole.④

（b）Chest/ Bust: measure the straight width between points 2.5 cm down of the bottom of armholes. Then double the measured width as the measurement.⑤ Garment with box pleat must to be measured with the pleat opened.

（c）Waist: measure from side seam to side seam at the narrowest point. Then double the measured width as the measurement. Garment with center box pleat must to be measured with the pleat opened.

（d）Bottom: measure straight across from one side seam end point to the other. And double the measured width as the measurement. Garment with pleats must to be measured with the pleats are fully extended

（e）Front Length: turn the collar up, measure perpendicularly from the highest point of the shoulder line to the hemline.⑥

④敞开开口并放平领子，使服装里侧向自己。从纽扣中位到最远处扣眼的末端测量。

⑤袖窿底端下2.5 cm，测量两点之间的垂直宽度。以两倍量度的宽度作为胸围尺寸。

⑥将衣领向上翻起，从肩线的最高点垂直向下测量到底边线的距离。

( f ) Shoulder width: measure the straight across distance from the right shoulder point to left shoulder point.

( g ) Full Back Length: measure straight from the center back neck down to the hemline.

( h ) Across Back: the width measurement is taken horizontally between the left side and right side back armhole joint point, also named as "X-Back".

( i ) Across Front: the width measurement is taken horizontally between the joints of both armholes.

( j ) Sleeve length ( Long sleeve ) : measure straight from the center back to the top cuff end.

( k ) Sleeve length ( Short sleeve ) : measure straight from the top of the armhole to the end of the sleeve.

( l ) Nape to Waist: measure down the center of the garment from neck seam to waist seam.

( m ) Armhole Depth: measure diagonally from the top of an armhole to the bottom of the same armhole.

( n ) Cuff: measure straight across the bottom of the sleeve fold to fold. For extended measurement, measure as above with elastic or knit fabric fully extended.

( o ) Elbow Girth: fold the sleeve so that the top of the cuff in line with the top of the armhole, measure the width of the fold. And double the measured width as the measurement.

( 2 ) Bottoms Measurement.

( a ) Waist: measure circumference through middle of the waistband. But for the elastic waistband, must stretch the gathers open when take the measurement.

( b ) Seat/ Hips: to open the pleats according to the specified on size chart, and eliminate all wrinkles at the seat-parts. Measure 10.2cm ( 4 inches ) up from crotch along front rise seam and mark that point, and then measure straight across from side-seam to side- seam at right angle. ⑦

⑦根据尺码表的规定将褶裥展开，并抚平臀围位的皱褶。在裤裆点上10.2cm（4英寸）的位置，以直角方式从一边侧缝垂直横量到另一边侧缝。

( c ) Upper Hips: takes the across horizontal measurement midway between the waist and hip level and parallel to the ground.

( d ) Thigh: lay the pants creased ( side view ) , measure at inside of the pants leg with 5cm ( 2 inches ) down from crotch seam, then straight across.

( e ) Knee: lay the pants creased ( side view ) , measure at inside of the pants leg and 30.5cm ( 12 inches ) down from crotch seam, then straight across.

(f) Bottom: lay the pants creased (side view), measure at inside of the pants leg and from edge to edge.

(g) Front rise: slightly pull garment so the rise is in a straight line, measure up from crotch seam along front rise seam to the top of waistband (include waistband).

(h) Back rise: same as front rise but measure along back rise seam.

(i) Out Leg: take from the top of the waistband along side seam to the bottom edge. e.g. trousers length. It is also named as "Out Seam".

(j) In Leg: take from the middle of the crotch seam to the bottom of the opening of the pants. It is also named as "In Seam".

(k) Skirt Length: measure the length between waistline and the hemline of a skirt.

### 2.3.3 Basic Block Pattern for Bodice 上身基本纸样

(1) Required measurements: When making blocks, the body measurements are necessary, which include height, bust, shoulder width, nape to waist, across back, etc. e.g. woman's bodice basic pattern.

⑧对于上身基本纸样，一般采用平面纸样方法以胸围尺寸的半围原理进行绘制（半围＝胸围/2）。

(2) Instruction for drafting: In the bodice drafting, always adopt scale principle with half bust girth measurement (Scale=1/2 Girth Measurement) in the flat pattern.⑧ Now show the following detailed instruction (Figure 2-16 Basic Block Pattern).

1－0: 1/8 height less 4cm.

2－0: Nape to waist measurement.

3－0: Half of nape to waist plus 3.5cm.

4: The point is the midway between point 0 and 1.

5－0: 3cm for all sizes, and square out all these points.

6－0: 1/8 scale plus 1.6cm.

7－6: 1/16 scale less 1cm.

8－4: 1/2 X-back measurement plus 1.3cm and square to point 9 and 10.

9: The meeting point is in the shoulder line.

10: The meeting point is in the armhole level.

11－10: 2.2cm for all sizes, and joins point 7 and 9 to locate point 12.

12: The meeting point is in the across back line.

13－12: 3cm for all sizes, and using point 13 as pivot point, and sweeps arc from point 7 to right side.

14– 11: On this step, locates the shoulder measurement plus 0.6cm.

15– 9: 3/8 Scale less 4.7cm, and square to locate point 16.

16: The meeting point is in the extended position of the shoulder line.

17– 3: Half bust measurement plus 5cm, and squares to locate point 18 and 19, and making point 18.

18: The meeting point with 1.3cm below waist construction line.

19: The meeting point is in the front neck position.

20– 19: 3/16 Scale less 6cm.

21– 20: 1/8 Scale plus 0.6cm.

22– 20: 1/8 Scale plus 0.6cm.

23– 17: 3/16 Scale plus 1.5cm, and using point 23 as pivot point, and sweeps arc from point 22.

24– 16: 2.5cm for all sizes.

25– 24: 4cm for all sizes.

26– 25: Equal to measurement of 11– 14 with the shoulder measurement plus 0.6cm, and have the arc from point 22 to 26, and then shaping bust dart with 1.3cm above point 23.[9]

⑨等于肩长 11–14 加 0.6cm 的尺寸，并以点 23 上 1.3cm 作为褶点，从点 22 至点 26 的距离画弧形成胸省。

27– 1: 1/2 Scale plus 1.5cm, and square to locate point 28, and then dropping 0.6cm below waist level.

28: The meeting point in the waistline and side seam.

29– 28: Point 30 from 28 is each 2.7cm for all sizes, and then shaping side seam.

30: The point must be overlapped when sewing side seams together.

31– 2: 1/8 Scale plus 5.5cm, and then shaping back waist dart.

32– 31: 4cm for all sizes, both points must be dropped 0.3cm below waistline.

33– 1: 1/8 Scale plus 5.5cm, and then shaping back dart.

34– 18: 3/16 Scale less 5.5cm and below waistline 1.5cm, and then pivoting at point 23 sweep arc from point 34.

35– 34: On this arc is 5cm for all sizes, and then shaping front waist dart from point 34 and 35 to 1.3cm below point 23.

When shaping waistline, must close the waist dart and side seam firstly. And check the scye shape at the shoulder ends, also back neck and shoulder seam, etc.

（3）Application of Basic Block Pattern: Pattern is by using the body measurement to trace out a basic pattern or original pattern, and all the dimensions, curves

⑩ 纸样是采用身体尺寸绘制一套基本或原版纸样，并且通过移褶处理，将所有尺寸、曲线和省道等细节都集合进这套纸样里。

and darts are gathered into this pattern through dart manipulation.[10] And that is to say, there are already darts in the basic pattern, and the darts are usually divided into waist dart and bust dart for application.

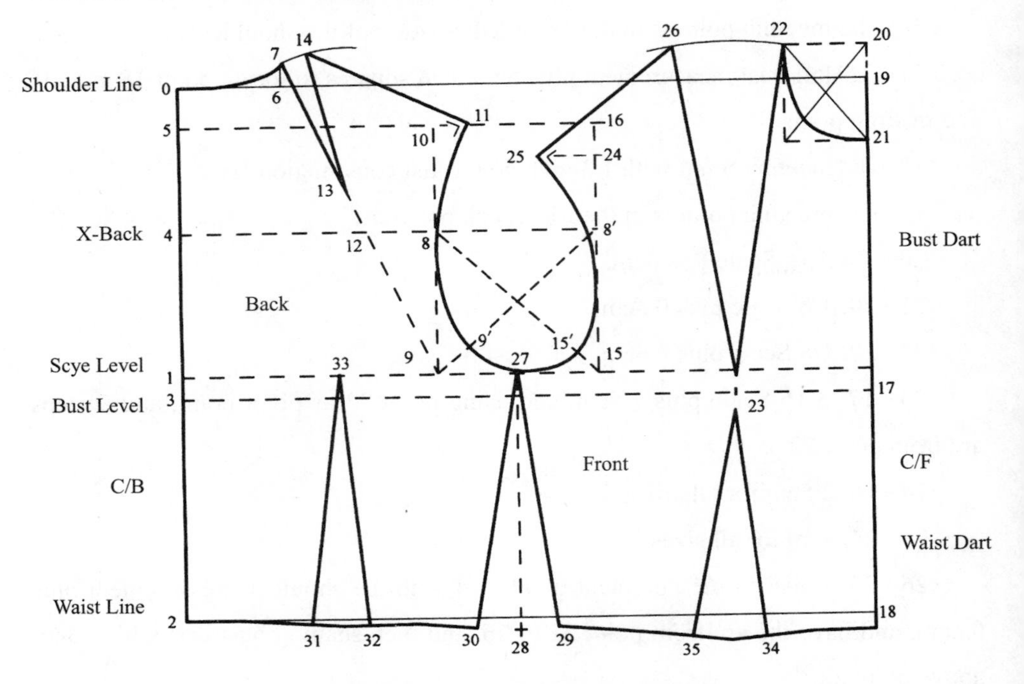


**Figure 2-16 Basic Block Pattern**

## *Words and Expressions*

pattern drafting ['pætən 'drɑ:ftiŋ] 平面裁剪
designate ['dezigneit] 指定，指明
pattern draping['pætən 'dreipiŋ] 立体裁剪
conform [kən'fɔ:m] 使一致，符合
three-dimensional [θri:-'dimenʃənəl] 三维的，立体的
flat pattern [flæt 'pætən] 平面纸样
drafting ['drɑ:ftiŋ] 绘制
modeling ['mɔdliŋ] 立体裁剪
fullness ['fulnis] 丰满的，丰满性
ease [i:z] 松量，松份（广东话：容位）
ease allowance [i:z ə'lauəns] 放松量值
logical approach ['lɔdʒikəl ə'prəutʃ] 合理方法
jersey ['dʒə:zi] 平纹针织布
bias cut ['baiəs kʌt] 斜纹裁剪
ready-to-wear ['redi- tu:-wiə] 成衣
body measurements ['bɔdi 'meʒəmənts] 身体尺寸
outline ['aut-lain] 轮廓线
fashionable ['fæʃənəbl] 流行的，时髦的
draping ['dreipiŋ] 悬垂
pleat [pli:t] 褶裥（广东话：活褶）
gather ['gæðə] 碎裥，碎褶
tuck [tʌk] 裥
girth measurement [gə:θ 'meʒəmənt] 围度尺寸
vertical measurement ['və:tikəl 'meʒəmənt] 垂直尺寸
measurement taking ['meʒəmənt 'teikiŋ] 尺寸测量
bodice ['bɔdis] 紧身衣服

tops [tɔpz] 上装
bottoms ['bɔtəmz] 下装
center back length ['sentə bæk leŋθ] 后中长
center back ['sentə bæk] 后中
center front ['sentə frʌnt] 前中
nape to waist [neip tuː weist] 背长（广东话：腰直）
lay garment flat [lei 'gɑːmənt flæt] 摆平服装
measuring tape ['meʒəriŋ teip] 软尺
shoulder seam ['ʃəuldə siːm] 肩缝
belt [belt] 皮带
slider buckles ['slaidə 'bʌkl] 皮带扣
fold to fold [fəuld tuː fəuld] 折线到折线
free of wrinkles [friː ɔv, riŋkl] 消除皱褶
contour of body ['kɔntuə ɔv, 'bɔdi] 体型
sleeve length [sliːv leŋθ] 袖长
half waist [hɑːf weist] 半腰围
hood width [hud widθ] 帽宽
hood height[hud hɑit] 帽高
upper arm ['ʌpə ɑːm] 上臂围
neck [nek] 领圈
neck seam [nek siːm] 领圈缝口
neck scoop [nek skuːp] 领凹
chest/ bust [tʃest bʌst] 胸围
size chart [saiz tʃɑːt] 尺码表
alternate [ɔːl'təːnit] 不同的
circumference [sə'kʌmfərəns] 围度尺寸
sleeve cap [sliːv kæp] 袖山，袖头
sleeve opening [sliːv 'əupəniŋ] 袖口
collar length ['kɔlə leŋθ] 领长
elastic [i'læstik] 松紧带
knit [nit] 针织
side-seam [said siːm] 侧缝
relax [ri'læks] 放松
back armhole [bæk 'ɑːmhəul] 后袖窿
align [ə'lain] 成一直线
eliminate [i'limineit] 消除
button-hole ['bʌtn həul] 扣眼（广东话：纽门）
knee [niː] 膝围
front rise [frʌnt raiz] 前裆（广东话：前浪）
back rise [bæk raiz] 后裆（广东话：后浪）
slacks [slæks] 休闲裤
upper hips ['ʌpə hips] 上臀围（广东话：上坐围）
seat/ hips [siːt / hips] 臀围（广东话：坐围）
thigh [θai] 大腿围，股围
waistband ['weistbænd] 腰头，裤腰
head girth [hed gəːθ] 头围
waist [weist] 腰围
wrist [rist] 腕围
shoulder width ['ʃəuldə widθ] 肩宽
nape [neip] 后颈
cross back/ X-back [krɔs bæk] 后背宽
cross front/ chest width [krɔs frʌnt / tʃest widθ] 前胸宽
bust point to bust point [bʌst pɔint tuː bʌst pɔint] 胸距
shoulder to waist line ['ʃəuldə tuː weist lain] 肩到腰线距离
shoulder to bust point ['ʃəuldə tuː bʌst pɔint] 肩到胸点距离
full back length [ful bæk leŋθ] 全长
elbow ['elbəu] 手肘围
elbow length ['elbəu leŋθ] 手肘长
crotch point [krɔtʃ pɔint] 裆点
crotch depth / body rise [krɔtʃ depθ / 'bɔdi raiz] 上裆，直裆立裆（广东话：直浪）
skirt length [skəːt leŋθ] 裙长
hem line ['hemlain] 下摆线，底边，贴边
preference ['prefərəns] 偏爱，优先选择
out leg [ɑut leg] 外长
in leg [in leg] 内长
out seam [ɑut siːm] 外缝，外长
in seam [in siːm] 内缝，内长
scale principle [skeil 'prinsəpl] 半围原理
instruction [in'strʌkʃən] 指示
square out [skwɛə ɑut] 成直角地
bust dart [bʌst dɑːt] 胸省
side seam [said siːm] 侧缝
armhole / scye ['ɑːmhəul sai] 袖窿（广东

话：夹圈）
sweep [swiːp] 下摆围，扫过
arc [ɑːk] 弧
shoulder ends [ˈʃəuldə endz] 肩端
back neck [bæk nek] 后领圈
basic block pattern [ˈbeisik blɔk ˈpætən] 基本纸样
production pattern [prəˈdʌkʃən ˈpætən] 生产纸样
original pattern [əˈridʒənəl ˈpætən] 原板纸样
manipulation [məˌnipjuˈleiʃən] 处理，操作
dart pivoting [dɑːt ˈpivətiŋ] 移褶处理

## Exercises

1. Translate the following terms into Chinese.

（1）pleat
（2）gather
（3）tuck
（4）girth measurement
（5）vertical measurement
（6）head girth
（7）measuring tape
（8）dimension
（9）modeling
（10）ready-to-wear
（11）body measurement
（12）garment
（13）apparel
（14）neck
（15）bust
（16）waist
（17）hips
（18）thigh
（19）knee
（20）X-back
（21）nape to waist
（22）waistline
（23）full back length
（24）bust dart
（25）side seam
（26）armhole
（27）shoulder seam
（28）sweep
（29）hips line
（30）body rise
（31）out leg
（32）crotch
（33）in seam

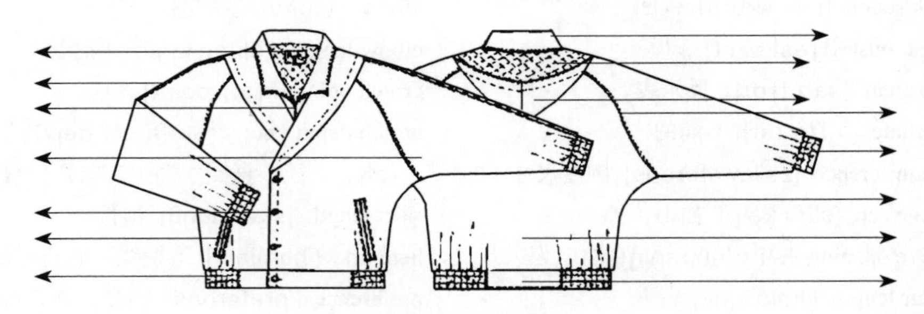

**Men's Jacket**

2. List the body measurement types, and give 4 examples for each different measurement type.
3. Point out the following indications（Men's Jacket）.
4. Explain the following body measurement taking methods.

（1）Waist （2）Shoulder （3）Bust
（4）Sleeve Length （5）Hips （6）In-seam

## 2.4 Relative Glossary 相关术语

### 2.4.1 Design Term 设计词汇

（1）Abbe Cape：cape worn by an abbot originally. It is a sleeveless outer garment or short cloak，which drapes the wearer’s back，arms and chest，and fastens at the neck.

（2）Apron：an overskirt tying at the back. It is an outer protective garment that covers the front of the body. It may be worn for hygienic reasons as well as in order to protect clothing from wear and tear，or else due to a symbolic meaning.

（3）Bell Skirt：bell shaped skirt，bells out form a narrow fitted bodice. At present，the hemline of skirts can vary from short to floor-length and can vary according to cultural conceptions and the wearer’s personal taste，which can be influenced by such factors as fashion and social context.

（4）Bolero：short jacket reaching to waist. The bolero is appropriate for wearing in slow-tempo Latin music and its associated dance origninally.

（5）Circular Skirt：circular piece of material with a hole cut out of center for the waist.

（6）Chic：french word meaning stylish.

（7）Culottes：a skirt divided and usually flared from hips. Originally culottes were normally closed and fastened about the leg，to the knee，by buttons，a strap and buckle，or a draw-string，and now culottes are an item of clothing and popular for ladies.

（8）Dirndl：a circular cut skirt gathered or pleated around the waist，with a waistband and falls below the knee. A dirndl is a type of traditional worn in Germany，Austria，etc. The loosely dirndl based on the country-inspired fashion.

（9）Peplum：short hip frill at waist that is usually attached to a fitted jacket，bodice，blouse or dress.

（10）Bishop Sleeve：a long sleeve with fullness below the elbow，and must be gathered into a cuff. It is a type of shirt made as a loose-fitting blouse，usually decorated with large frills on the front and on the cuffs.

（11）Batwing Sleeve：a long sleeve with fullness at the armhole. This style is cut very large at the armhole and extends to almost the waist，tapering at the wrist. It’s usually not a “set-in” sleeve，but part of the garment.

（12）Convertible Collar：a straight，pointed collar，usually on a shirt or blouse, which can be worn open down as a shawl collar or closed up as a stand collar, and it is similar to Hawaii Collar.

（13）Cowl Neckline: a soft draped neckline, the front of the bodice is cut on the bias of the material and soft folds are formed.

（14）Decollete: It is a very low cut bodice, the upper part of a woman's torso, comprising her neck, shoulders, back and chest, that is exposed by the neckline of her clothing. Decollete is most commonly applied to low-cut necklines of ball gowns, evening gowns, leotards, lingerie and swimsuits, etc.

（15）Dolman Sleeve: one-piece sleeve with deep cut scye, in fact, the armscye may extend to the waistline, in which case there will be no underarm seam in the blouse, and some of these scye are made round or oval and others square at the base.[①] Originally, dolman sleeve is referred to a long and loose garment with narrow sleeves and an opening in the front.

①袖窿很低的一片袖，事实上，袖窿可以延伸到腰线，有些情形在罩衣上是没有侧缝骨的，一些袖窿是圆弧或椭圆形的，也有的底部是方形。

（16）Double-Breasted: a garment wrapping over in the front with a double row of buttons or fastenings. In most modern double-breasted coats, one column of buttons is decorative, while the other functional. By contrast, a single-breasted has a narrow overlap and only one column of buttons.

（17）Empire Line: It is refered a high waistline of women's dress, in which has a fitted bodice ending just below the bust, giving a high-waisted appearance, and a gathered skirt which is long and loosely fitting.

（18）Eton Collar: a stiff white turned down collar as worn by boys at ETON.

（19）Fitting: trying on the garment in order to make adjustments to fit and style.

（20）High-way-men Collar: a high, large, turned down collar, sometimes rising above the ears.

（21）Jabot: a neck frill or ruffles decorating the front of a shirt. It has evolved into a decorative clothing accessory consisting of lace or other fabric falling from the throat, suspended from or attached to a neckband or collar; or simply pinned at the throat.

（22）Kimono Sleeve: sleeve cut all in one with the garment, usually with a seam reach to the back part, and similar to that on traditional Chinese robes.

（23）Line: the silhouette of a garment that makes it look fashionable or unfashionable, and silhouette is the outline of a garment.

（24）Modeling: the art of draping material on a figure to arrive at the shape of the garment.

（25）Magyar Sleeve：sleeve is cut in one with the garment with gussets let into the underarm.

（26）Raglan Sleeve：sleeve with the armhole line extending to the neck so that the shoulder section is joined to the sleeve crown without a seam.

（27）Roll Collar：any collar that softly rolls back and is not pressed flat，as opposed to a collar with a pressed crease at the fold.

（28）Set in Sleeve：a sleeve with a crown fitting into the scye.

（29）Sample：the designing model，e.g. prototype.

（30）Vee-Neck：a neckline cut to a " V " in the front，the depth of the " V " can vary.

### 2.4.2 Pattern Term 纸样词汇

（1）Armhole：also known as armscye or scye.

（2）Basic Pattern/ Block Pattern：a pattern that reflects the shape and posture of the figure without the inclusion of style features. ② The cardboard master forms style that can be worked out.

②一套反映人体体型和形态，但不包含款式特点的纸样。

（3）Bias：angle at which is cut in respect to the selvedge of the material. Weft at an angle of 45° to the warp.

（4）Bust Line：used for block construction and is the line on which the bust measure is taken. The bust point is found on this line，all front darts must be radiated from bust point.

（5）Crease Line/ Fold Line：the line on collar and reverse/ lapel folds back on.

（6）Crutch/ Crotch：angle formed by parting the legs. And the crutch depth is the distance from the center front waist point to the angle formed by parting the legs.

（7）Crown/ Sleeve Head：top part of the sleeve.

（8）Dart Pivoting：pivoting a dart from one site to another，and the point of dart used for dart manipulation.

（9）Dart Site：place on the bodice where the dart originates，e.g. shoulder，bust，waist. This site is optional and depends on fashion and style required.

（10）Draft：the outline drawing of a pattern，which has to be traced-off to obtain the finished pattern.

（11）Ease：for example the sleeve head measurement is longer than the armhole，and the "difference" is called ease. Ease is most important for woven garments cut on the straight or cross grain，allowing little or no stretch.

（12）Grading：producing a range of patterns for different sizes from a master pattern.

（13）Grain：the direction of thread in the fabric，and it referred to straight，cross and bias grain line for the woven fabric.

（14）Marker：a sheet of paper on which the component parts of a pattern are marked in preparation for placing on a lay，prior to cutting.

（15）Off Grain：the term used to describe the distortion of a woven fabric when warp and weft are not at the right angle.

（16）Pattern Cutter：one who converts ideas of fashion or style into the form of a pattern for producing clothes.

（17）Size Chart：a chart of measurements on which block pattern is based.

（18）Seam Allowance：It is the area between the edge and the stitching line on two or more pieces of material being stitched together，and sometimes called Inlay or Turnings. Such as 1cm SA in neck line，armscye or facing seams for sewing.

（19）Toile：a pattern usually in calico of a model garment. The word "toile" can refer to the fabric itself，a test garment sewn from the same material.

（20）Turn-up：used as a feature on trousers cuffs or skirts.

（21）Tolerance：the amount by which the garment measurement exceeds the body measurement.

（22）Dress Stand：a padded model or dummy of the human body for modeling garments.

## *Words and Expressions*

abbe cape ['æbei keip] 修道院长披风
abbot ['æbət] 男修道院院长，大寺院主持
apron ['eiprən] 围裙
overskirt ['əuvəskə:t] 上裙，外裙
apparel [ə'pærəl] 服装，装饰（广东话：成衣）
blouse [blɑuz] 女装罩衣，女衬衫
figure ['figəd] 身段，姿态
bell skirt [bel skə:t] 钟型裙
Bolero [bə'lɛərəu] 短上衣
circular skirt ['sə:kjulə skə:t] 圆台裙
chic [ʃi(:)k] 流行的
cloak [kləuk] 斗篷
sleeveless [ˋsli:vlis] 无袖的
dirndl ['də:ndl] 紧身腰裥服装
gusset ['gʌsit] 三角形插料，衣袖插片
peplum ['pepləm] 褶襞短裙，裙腰剪接片
culottes [kju(:)'lɔts] 裙裤
flared [flɛəd] 喇叭的，宽摆的
bishop sleeve ['biʃəp sli:v] 主教袖
batwing sleeve ['bætwiŋ sli:v] 蝙蝠袖
Hawaii collar [hɑ:'waii: 'kɔlə] 夏威夷领
convertible collar [kən'və:təbl 'kɔlə] 开襟领
cowl neckline [kaulˋneklain] 垂褶领圈
decollete [dɪ'kɔltei] 露胸式衣服

double –breasted ['dʌbl - 'brestid] 双排扣的
empire line ['empaiə lain] 帝国式腰线
Eton collar ['i:tn 'kɔlə] 伊顿式阔翻领
fitting ['fitiŋ] 试身
high-way-men collar [hai - wei- men 'kɔlə] 男装高领
jabot [ʒæ'bəut] 胸前装饰花边
neck frill [nek fril] 领圈装饰花边
kimono sleeve [ki'məunəu sli:v] 和服袖
silhouette [ˌsilu(:)'et] 轮廓
modeling ['mɔdliŋ] 立体裁剪
Magyar sleeve ['mægja: sli:v] 马扎尔袖
underarm ['ʌndəra:m] 衣袖内侧，腋下部（广东话：袖底骨）
raglan sleeve ['ræglən sli:v] 牛角袖，拉格兰袖
sleeve crown [sli:v kraun] 袖头
roll collar[rəul 'kɔlə] 翻领，青果领
set in sleeve [set in sli:v] 普通袖,圆袖(广东话：夹圈袖）
sample ['sæmpl] 样板
prototype ['prəutətaip] 原型板
Vee-neck [vi:- nek] V 形领
armhole/ armscye / scye ['a:mhəul / a:msai / sai] 袖窿（广东话：夹圈）
balance line ['bæləns lain] 平衡线
basic pattern/ block pattern ['beisik 'pætən / blɔk 'pætən] 基本纸样
reflect [ri'flekt] 反映
posture ['pɔstʃə] 姿势，体态
feature ['fi:tʃə] 特征
cardboard ['ka:dbɔ:d] 纸板
bias ['baiəs] 斜纹
selvedge ['selvidʒ] 布边
weft [weft] 纬纱
warp [wɔ:p] 经纱
bust line [bʌst lain] 胸围线
bust point [bʌst pɔint] 胸高点
radiate ['reidieit] 辐射，射线的
crease line / fold line [kri:s lain / fəuld lain] 折线
collar ['kɔlə] 领子
rever/ lapel [rev / lə'pel] 襟贴，挂面
crutch/ crotch [krʌtʃ / krɔtʃ] 裤裆位
crown/ sleeve head [kraun / sli:v hed] 袖头
dart pivoting [da:t 'pivətiŋ] 移褶
dart site [da:t sait] 省位
outline ['aut-lain] 轮廓
traced-off [treis - ɔ:f] 复制
grain ['grein] 布纹，织物纹路
thread [θred] 缝线，纱线
prior to ['praiə tu:] 在前，居先
off grain [ɔ:f 'grein] 布纹不直
distortion [dis'tɔ:ʃən] 变形，扭曲
pattern cutter ['pætən 'kʌtə] 打板师
manipulation [məˌnipju'leiʃən] 处理
size chart [saiz tʃa:t] 尺码表
seam allowance/ turnings [si:m ə'lauəns / 'tə:niŋz] 缝份（广东话：子口）
toile [twa:l] 样衣
calico ['kælikəu] 白棉布（广东话：胚布）
turn-up hem ['tə:n-ʌp hem] 向上翻，折下摆
dress stand[dres stænd] 人体模型，胸架
padded stand [pædid stænd] 泡沫人体模型
hygienic [haɪ'dʒi:nɪk] 卫生的
symbolic [sim'bɔlik] 象征性
hemline ['hemlain] 下摆线
social context ['səuʃəl 'kɔntekst] 社会背景
slow-tempo [sləu 'tempəu] 慢速
draw-string [drɔ: striŋ] 拉绳
strap [stræp] 带条
buckle ['bʌkl] 皮带扣
bodice ['bɔdis] 紧身胸衣
set-in sleeve [set ɪn sli:v] 袖窿袖
shawl collar [ʃɔ:l 'kɔlə] 燕子领
stand collar [stænd 'kɔlə] 立领
ball gown [bɔ:l gaun] 舞会礼服
evening gown ['i:vniŋ gaun] 晚礼服

leotard ['li:əˌtɑ:d] 紧身舞衣
lingerie [ˌlænʒə'ri:] 贴身女内衣
swimsuit ['swɪmsu:t] 游泳衣
single-breasted ['siŋgl 'brestid] 单排扣
ruffle ['rʌfl] 荷叶边
neckband ['nekbænd] 立领
straight grain line [streit grein lain] 直线
cross grain line [krɔs grein lain] 横纹
bias grain line ['baiəs grein lain] 斜纹
dummy ['dʌmi] 假人

## *Exercises*

1. Translate the following terms into Chinese.

（1）blouse
（2）bell skirt
（3）fashion
（4）sleeveless
（5）culottes
（6）batwing sleeve
（7）trousers
（8）closed fitting
（9）frill
（10）modeling
（11）roll collar
（12）prototype
（13）scye
（14）bias grain
（15）selvedge
（16）weft
（17）warp
（18）bust
（19）dart
（20）ease
（21）S.A.
（22）calico
（23）cuff
（24）pleat

2. Explain the following fashion terms.

（1）Blouse
（2）Cape
（3）Batwing Sleeve
（4）Cowl Neckline
（5）Double-Breasted
（6）Armhole
（7）Basic Pattern
（8）Sleeve Head
（9）Seam Allowance

3. Draw the design sketch according to the following descriptions.

（1）Bolero: it is definite as a short jacket reaching the waist.
（2）Culottes: it is similar to a skirt but it is divided and usually flared from hips.
（3）Bishop sleeve: a long sleeve with fullness below the elbow, and must be gathered into a cuff.
（4）Jabot: a neck frill in the formal shirt.
（5）Raglan Sleeve: a sleeve with the armhole line extending to the neck so that the shoulder section is joined to the sleeve crown without a seam.

# 译文

# 第二章 设计与纸样

## 2.1 服装设计

对大多数人来说，“时尚”这个词意味着衣服。但是人们可能会问这个问题，“时尚是什么衣服?”他们用同样的方式来形容“时尚”这个形容词：“她穿着一件时髦的外套。”“他的衬衫真是一种时尚的颜色。”当然时尚包含很多东西，不只是指衣服。在假日、餐馆、电影和书籍中都有流行的时尚。而时装设计是指对一种风格进行独特的或个性化的处理。一种风格能通过多种设计进行传达，各种设计虽有不同但又有关联，因为它们有统一的风格。“时尚感”涵盖了明确哪一种服饰好看与否的能力。由于整个设计理念依赖于主观意识，时装设计师通过他们的想象与流行趋势来设计服装系列。时装草图和设计图展示了穿着的外观效果（图 2-1）。

**图 2-1 时装画**

### 2.1.1 著名时装周

（1）巴黎时装周：著名的巴黎时装周，每半年举办一次，包括春 / 夏季和秋 / 冬季活动。除了成衣秀之外，还有男士和高级定制时装秀。日期是由法国时装联盟决定（图 2-2）。

（2）米兰时装周：这是一场每年在意大利米兰举行的服装贸易展。每年 2 月至 3 月举行的春 / 夏季活动和每年 9 月至 10 月举行的秋 / 冬季活动。在活动期间，最重要的时装表演是致力于女性和男装的时尚（图 2-3）。

图 2-2　女装

图 2-3　时装表演

（3）伦敦时装周：这是一场服装贸易展，每年 2 月和 9 月在英国伦敦举行两次。它是由英国时装协会组织的。2010 年春季，伦敦时装周成为第一个全面拥抱数字媒体的重要时装周，它为所有在 T 台上展示自己时装的设计师提供了在互联网上直播节目的机会（图 2-4）。

图 2-4　T 台秀

（4）纽约时装周：它是世界四大时装周之一，每年 2 月和 9 月在纽约举行。有时，长裙款式是所有特征的集合，就像电影"性感与城市"中四种不同风格的女装一样（图 2-5）。

图 2-5　不同风格

### 2.1.2　时装图解

时装图解表达了设计师的设计理念，并支持着服装系列的创新和市场销售。在样板制作或成衣生产之前，时装画将通过CAD方法转变作业图或生产图。生产图能清楚准确地描绘出服装重要的裁剪细节，比如口袋、省道、缝骨、面缝线迹、大小比例，以及其他特殊的设计细节。

（1）女衬衫：女衬衫是一件柔软宽松的上衣，可以搭配短裙或裤子一起穿着。在今天的时装里，例如，领口与领子、长度、裁剪、细节与装饰部分与合适的原材料搭配，将设计出一种独特的款式。特别是束腰或束臀碎褶使衬衫宽松地穿在穿着者身体上（图2-6）。

（2）制服衬衫：此类衬衫属于工作制服，如不外穿夹克时，则要视穿着者职业而做出相应便捷实用的设计。例如，警察、紧急服务机构、保安等，其他一些行政人员、工作场所与学校也会因职责而穿制服。

（3）工作服：此类衬衫的设计首先要考虑穿着者从事的职业，特别是一些体力劳动的工作。工作服提供耐用性与安全性，其次才考虑时尚与款式方面，但不容忽视（图2-7）。

**图2-6　女装衬衫**　　**图2-7　职业款式**

（4）可选口袋：口袋分功能性与装饰性两种，并且有不同的款式，通常有三种：明贴袋、挖袋和缝骨袋。明贴袋是直接在服装表面缝制另外一块面料。明贴袋可以应用在男装、女装和童装上。小块面料三面缝合，顶部放开作为袋口。它们的形状可以是正方形、长方形、三角形、曲线的或者其他特殊形状，又或者使用面缝线迹、花边或饰带处理作为装饰之用。时装设计师经常在明贴袋上创造各种变化，例如，采用装饰面线、拉链、悬垂效果、装饰贴边、纽扣搭配和袋盖等设计元素（图2-8）。

（5）A型款式：在1955年，克里斯汀•迪奥设计了一款新风貌时装，在肩和胸部较窄，小腿较宽，给人留下大写字母“A”的外形印象。“A”字形的横切线一般在膝围或以下的位置，就好像荷叶边裙褶（图 2-9）。

图 2-8　各种口袋　　　　图 2-9　时尚款

（6）公主线款式：公主线或公主装描述了女装长裙或其他腰线没有任何缝骨的长衣片裁剪的服装。

（7）吊带领口款式：领口是带型或在前衣身直接裁出，一直连续延伸到衣服的后背。这款服装是无袖的，也许只有一带条环绕领圈或跨过肩线。

（8）马裤：用于骑马穿的裤子。马裤通常是贴身的，并沿着小腿以下不同长度甚至到膝围处作为开口，可采用纽扣或抽绳，或用一条或多条带条与皮带扣或别针进行扣紧。

（9）夹克：它是外套的基本款式，并且该款服装衣身长度是到腹部中位。有时，它特指男装中的订做款式，比如男西装。在广义上，它还包括牛仔外套和针织套衫，男性、女性和儿童都可以穿。面料可以在颜色、格子、条纹、大格等方面不断变化（图 2-10）。

（10）短裙：短裙是指自腰而下的服装，是女装的基本设计款式，与外套相配便可形成套装或礼服。许多短裙款式在流行时尚中不断变化，如牛仔裙。独特的款式可以从裙子的长度、宽度、外观轮廓、裁剪和细节等方面进行区分（图 2-11）。

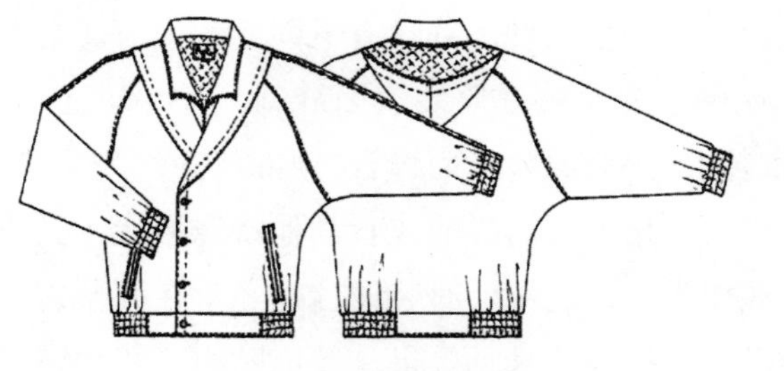

图 2-10　男装夹克

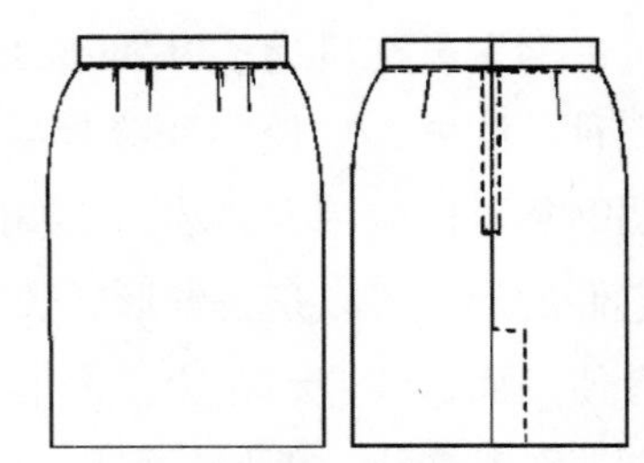

图 2-11　西装裙

（11）牛仔裤：牛仔裤常指各种紧身的裤子，如五袋款牛仔裤。最著名的牛仔裤就是李维斯牛仔裤，由世界上著名裤业制造公司制造。世界上第一条牛仔裤是在 1850 年由李维斯为美国的淘金者制造的。现在，牛仔裤已经成为男女休闲便装中的流行服装，而且也出现了各种合身款式，包括极瘦型、锥型、修腰型、直筒型、小喇叭型、窄脚型、喇叭型、低腰型和宽摆型等（图 2-12）。

（12）运动装：一种为不同运动而设计的服装款式，由上装和下装组成。从实用性考虑，从中可以发现，运动装的颜色和面料的设计尤为重要（图 2-13）。

（13）旗袍：是一款中式长裙。旗袍有一窄而直立的衣领，领尾在前领圈处相扣。最初的旗袍是宽松的款式，并且几乎覆盖了女性大部分身体，只露出了头、手以及脚尖。随着时间的推移，旗袍缝制变得更加贴身与暴露（图 2-14）。

（14）内衣：内衣是一件穿在外套下面的衣服，通常直接与皮肤接触，定型身体。内衣有不同的名称，如内衣（Unders）、内衣物（Underwear）、内衣（Underclothes）和内衣裤（Underclothing）等。女装内衣也可统称为内衣。如果由适合的面料或纺织品做成，一些内衣可用作睡衣或游泳衣（图 2-15）。

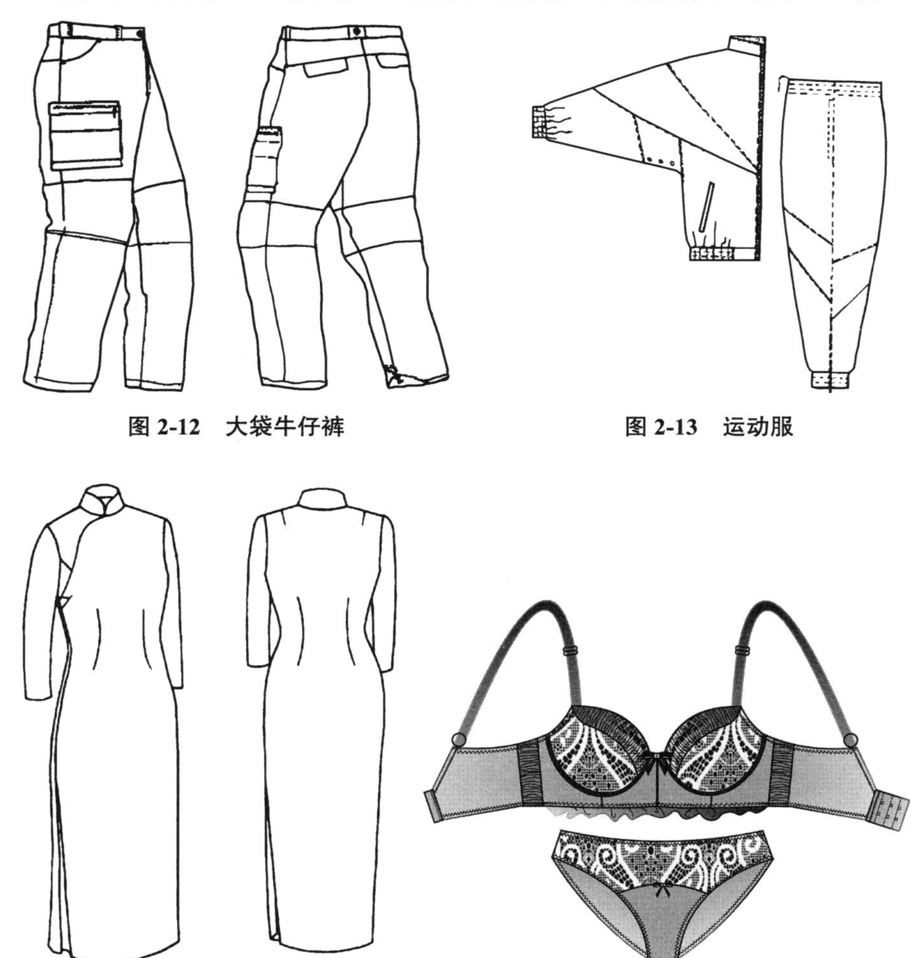

**图 2-12 大袋牛仔裤**

**图 2-13 运动服**

**图 2-14 旗袍**

**图 2-15 内衣**

## 2.2 流行趋势

### 2.2.1 20世纪80年代的时装风格

在20世纪80年代，时装千变万化。女装灵感来自活跃、自我意识强的女性。既有经典优雅的风格，也有休闲实用的风格。女装追求精制、富有魅力、奢华的风格。在休闲、文雅的西装和外套中又保留了男性的外表轮廓和细节。晚装很女性化、优雅，有的柔软飘逸，有的紧身贴体，配以紧身的或展开的或蓬起的裙子。裙子的长度根据款式，可以从膝盖变化到脚踝。迷你裙也出现了。腰线或被提高，或被降低。衣袖裁剪宽大，因此活动更加自由。肩部常常是被重视的，不同长度和宽度的组合形成了新的比例。

在男装中，传统套装风格优雅，裁剪经典，穿着舒适。腰部稍微修身，肩部稍加肩垫，翻领不是很宽。流行有绗缝效果和采用轻薄面料的外套。男士晚装是宴会装的生动组合。休闲装采用轻薄面料和休闲款式，但要求有复杂的功能性细节。如果设计师的预言准确，牛仔服的颜色将会丰富多彩。有人说，现在穿着棕色、绿色、红色、灰色和白色的牛仔服是普遍的。其他更时尚的明橙色和紫色将很快在连锁店出现。

### 2.2.2 设计师系列

淡粉色系和淡蓝色系正是符合卡尔·拉格斐系列的主要色调。外套采用平纹布、粗花呢、毛绉绸或软皮革。传统的外形轮廓。无论从哪方面看，香奈儿的箱型夹克，新款长裙和真丝女装衬衫都创造了优雅的整体感觉。带着一点幽默感的紧身夹克衫，穿在连有条带的露背裙或紧身高腰裙上。穿上粉色、黄色和蓝色的服装，甚至可以使最传统的工作服变得生动活泼。

至于裙子，卡尔·拉格斐采用了刚好高于膝盖线或脚踝线的下摆底边。它究竟会到哪个部位呢？香奈儿在长裙或迷你裙上展示了随处可见的大腿开衩，显示在腿部的任何长度似乎都有可能，并且晚礼服运用了裙褶皱边和荷叶边装饰。

### 2.2.3 夏季廓型

柔软的面料，灰色和平滑轻便的外形轮廓，预示着在告别粗犷厚实外观的冬装后，亮丽的服装重新到来。一款简单的毛纱风格的衬衣长裙或者是清爽的棉麻上衣，搭配一顶漂亮的帽子，手提包或者是一双平底鞋，在城市热浪中展现了一种随意舒适的感觉。

商务装的柔软外形轮廓，结合华丽的淡黄色与海军蓝，给人一种清新而引人注目的视觉效果。更加修长的夹克使用柔软的肩部线条，与紧身裤或裙子搭配穿着。开心的假期服装，采用明亮的颜色，华丽的印花布和清爽的面料，与潮流牛仔裤和裤袜搭配穿着。特大码的衬衫绑在腰间，更强调了修腰的效果。当鲜绿色、粉色或黄色的漂亮裙子加上一点迷人的黄昏色彩到各款夏季服装上时，会给性感的莎笼裙和实用的猎装增添更轻松的风格，既可穿着于商务场合，又可穿着于休闲场合。

### 2.2.4 2015 春夏牛仔系列

2015 年春夏牛仔系列聚焦在特殊裁剪的都市休闲风格上。天然面料如纯麻布和纯棉布，或者是混纺的麻布和棉布，担任着主角。轻松的风格、精湛的缝制和修身效果。加有风琴袋、肩部垫圈和随意的缝制效果在成衣夹克中真正受到关注。具有高科技成分的尼龙休闲衫来自运动灵感，在质地上具有水洗与漂白效果。经典款式的夹克是舒适的，保留了它们合体的纸样形式，加上使用质量好的服装染色，从而保证了一个潇洒外观。裤子再次以紧身款式为焦点，布料采用加棉混纺麻布，经过后整理同样可以创造出牛仔布的特殊洗水效果。采用泡泡纱的条纹面料也是非常重要的。这个系列介绍了牛仔裤的新变化，从松紧的经典款式和未经加工的牛仔布到漂白洗水效果的变化。

## 2.3 纸样绘制

### 2.3.1 基本方法

有两种主要的服装纸样制作方法：平面裁剪和立体裁剪。

（1）平面裁剪：是一种根据人体尺寸完成纸样绘制的方法，特指平面纸样。平面纸样设计过程中，先开发基于个体或标准服装号型的基本纸样，并且基本纸样已经初具廓型和放松量。相对于时髦服装而言，平面裁剪更多是用在固定款式的成衣制造中，采用平面纸样进行洋服裁剪是最成功的方法。

（2）立体裁剪：是指将面料披覆在人体模型或相符的人体上，制作出一套三维立体的布片纸样，所以也称为立体纸样。然后将这个布片纸样转变成纸张纸样，用来修改和裁制最终的确认纸样。立体裁剪能提供一个清晰的面料悬垂效果，可以在服装部件裁剪与缝制之前获得服装的整体设计效果。在立体裁剪过程中，最基本的放松量可以用别针固定在褶裥、碎褶、塔克裥等上面，松开之后，纸样就绘制出来了。立体裁剪技术最适合针织面料和各种柔软面料。它也常采用斜纹面料。

### 2.3.2 尺寸测量方法

在绘制纸样之前，必须获取正确的尺寸，无论是测量身体尺寸还是服装尺寸，且一般是采用软尺测量。所有的尺寸规格都是用厘米或英寸表示。服装应该放平并扫平表面测量，测量台必须足够大以便可容纳整件服装。测量针织衣服时，小心不要拉大尺寸，平整或轻拍服装直到没有褶皱。现以服装尺寸测量为例说明测量方法。

（1）上装尺寸测量。

（a）领围：敞开开口并放平领子，使服装里侧面向自己。从纽扣中位到最远扣眼的末端测量。

（b）胸围：袖窿底端下 2.5cm，测量两点之间的垂直宽度。以两倍量度的宽度作为胸围尺寸。有箱型褶的服装，必须展开褶裥进行测量。

（c）腰围：测量侧缝到另一边侧缝之间最细两点之间的宽度，以两倍量度作为腰部尺寸。有箱型褶的服装，必须展开褶裥进行测量。

（d）下摆：从一边侧缝末端测量到另一边侧缝末端的垂直宽度，以两倍的宽度作为下摆尺寸。有箱型褶的服装，必须完全展开褶裥进行测量。

（e）前身长：将衣领向上翻起，从肩线的最高点垂直向下测量到底边线的距离。

（f）肩宽：测量从右肩点到左肩点的直线距离。

（g）后衣长：从后片往下测量到底边线之间的垂直距离。

（h）后背宽：该宽度尺寸是测量左右袖窿下点之间的水平距离，也叫后背宽（“X-Back”）。

（i）前胸宽：该宽度尺寸是测量两个袖窿底点之间的水平距离。

（j）袖长（长袖）：从后中测量到袖口末端的垂直距离。

（k）袖长（短袖）：测量袖窿顶点到袖口的垂直距离。

（l）背长：从衣服领圈缝骨中位向下量度到腰缝之间的距离。

（m）袖窿深：从袖窿的顶点到同一个袖窿的底端，以对角的方式测量其长度。

（n）袖口：对折袖子，测量袖口底端的宽度。对于伸缩型袖口，测量时必须将松紧带或者针织布料完全展开进行测量。

（o）肘围：对折袖子，使袖口顶端与袖窿顶端在同一直线上，在手肘位测量折线的宽度，以两倍的宽度作为肘围尺寸。

（2）下装尺寸测量。

（a）腰围：从腰头中位测量其围度尺寸。但对有松紧带的腰头，测量时必须展开所有碎褶进行。

（b）臀围：根据尺码表的规定将褶裥展开，并抚平臀围位的皱褶。在裤裆点上 10.2cm（4 英寸）的位置，以直角方式从一边侧缝垂直横量到另一边侧缝。

（c）上臀围：测量位于腰线和臀围之间的中间位置的水平距离。

（d）腿围 / 股上围：将裤子侧边向上对折摆平，在内侧缝裤裆缝口下 5cm（2 英寸）处拉直测量。

（e）膝围：摆平裤子（侧面），从裤腿内侧裆缝向下 30.5cm（12 英寸）垂直横量。

（f）脚阔：摆平裤子（侧面）。在裤脚位从裤腿内侧横量到另一边的宽度。

（g）前裆：轻轻推平裤子，使裆位在同一直线上，沿着前裆从腰头顶端测量到裆缝位的距离（包括腰头）。

（h）后裆：与前裆测量法一样，但必须是沿着后裆缝测量。

（i）外长：沿着外侧缝从腰头的顶端测量裤脚末端的距离，也被称为“外侧缝长”。

（j）内长：沿着内侧缝从裆缝顶点测量到裤脚末端的距离，也被称为“内侧缝长”。

（k）裙长：测量腰围线到裙子底边线之间的长度。

### 2.3.3 上身基本纸样

（1）所需尺寸：在制作基本纸样时，身体尺寸是必备的，包括身高、胸围、肩宽、背长，以及后背宽等，如女装上身基本纸样。

（2）纸样说明：对于上身基本纸样，一般采用平面纸样方法以胸围尺寸的半围原理进行绘制（半围 = 胸围 /2）。以下显示详细绘制说明（图 2-16）。

1—0：身高 /8−4cm。

2—0：后腰长尺寸。

3—0：后腰长 /2+3.5cm。

4：该点是 0—1 的中点。

5—0：所有尺码均为 3cm，并垂直向下延伸到所有制作点。

6—0：半围 /8+1.6cm。

7—6：半围 /16−1cm。

8—4：背宽 /12+1.3cm，并垂直延长得到点 9 和 10。

9：此相交点在肩线。

10：此相交点在袖窿线。

11—10：所有尺码均为 2.2cm，连接点 7 和点 9 相交于点 12。

12：此相交点在后背宽线。

13—12：所有尺码均为 3cm，以点 13 为圆心，以 7—13 的距离为半径向右画弧线。

14—11：在这一步骤中，将肩宽 +0.6cm 确定点 14。

15—9：3/8 半围 −4.7cm，并垂直延长得到点 16。

16：此相交点在肩线的延长线上。

17—3：半围 +5cm，垂直延长得到点 18 和 19，并确定点 18。

18：此相交点为腰线下延长 1.3cm。

19：此相交点在前颈线处。

20—19：3/16 半围 −6cm。

21—20：半围 /8+0.6cm。

22—20：半围 /8+0.6cm。

23—17：3/16 半围 +1.5cm，以点 23 为圆心，以 22—23 的距离为半径画弧线。

24—16：所有尺码均为 2.5cm。

25—24：所有尺码均为 4cm。

26—25：等于肩长 11—14 加 0.6cm 的尺寸，并以点 23 上 1.3cm 作为褶点，以点 22 至点 26 的距离画弧形成胸省。

27—1：半围 /2+1.5cm，并垂直延长至基本腰线下 0.6cm 得到点 28。

28：此相交点在腰侧点。

29—28：与 30—28 相等，所有尺码均为 2.7cm，连接得侧缝位。

30：当缝合侧缝时，点 30 必须重叠。

31—2：半围 /8+5.5cm，确定后腰省的形状。

32—31：所有尺码均为 4cm，两个点必须在基本腰线下 0.3cm。

33—1：半围 /8+5.5cm，确定后腰省的形状。

34—18：3/16 半围 −5.5cm 并低于腰线下 1.5cm，然后以 23 为圆心，以 23—34 的距离画弧线。

35—34：所有尺码均为 5cm，在弧线上连接点 34 点 35，在点 23 下 1.3cm 得到前腰省尖点。

当连接腰围线时，必须先连接腰省和侧缝。然后检查肩端的袖窿形状和后领圈，以及肩缝等。

（3）基本纸样应用：纸样是采用身体尺寸绘制一套基本或原板纸样，并且通过移褶省处理，将所有尺寸、曲线和省道等细节都集合进这套纸样里。也就是说，基本纸样已经有省道，这些省道被分为腰省和胸省进行应用。在省道有效应用前，纸样师傅已经确认省道的位置和尺寸，所以纸样师傅不必描述该省道究竟是胸省还是腰省。

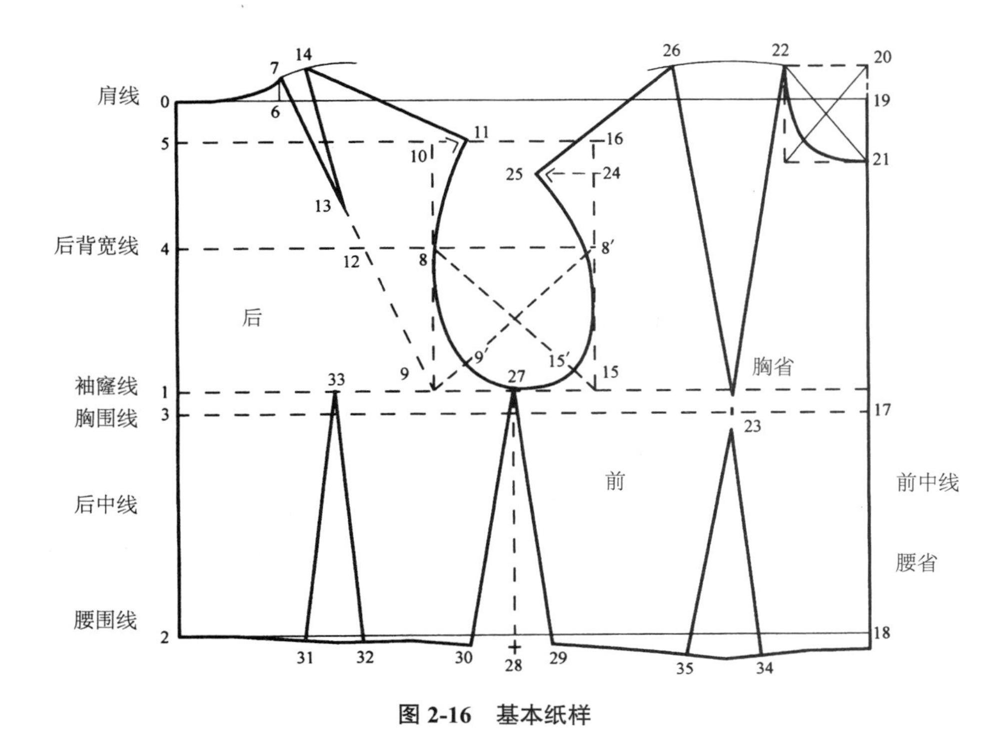


图 2-16 基本纸样

## 2.4 相关术语

### 2.4.1 设计词汇

（1）长披风：最初，披风是一种给神父穿的服装。披风是披在穿着者的背、臂、胸之上的无袖外套或短装斗篷，并在领圈位扣紧。

（2）围裙：一种系在背部的罩裙。围裙是一件覆盖前面身体的保护性外套。围裙是为了卫生甚至保护衣服不被磨损，或者为了其他象征性意义而穿着。

（3）钟型裙：裙子形状如钟型，腰部臀位处较窄，下摆较宽大。目前，裙子的底边线可以从短装到地面各种不同长度，也可以根据文化观念与穿着者的个人爱好而不同，一般受时装潮流与社会背景等因素影响。

（4）短上衣：长度到腰部的短夹克。波莱罗短上衣最初适用于慢速拉丁音乐与相关舞会穿着。

（5）圆台裙：一块圆形面料在中间裁成一个合适腰围的开口。

（6）流行：法语字眼，意为“流行的”。

（7）裙裤：一种分衩的裙子，一般从臀部开始形成喇叭外形。最初，裙裤一般是在大腿到膝盖处用纽扣、绳带与带扣或拉绳等方式紧扣封闭的，现在裙裤已经是一种女士流行服装。

（8）紧身腰裥服装：在腰间收碎褶或打褶裥圆形裁剪的裙子，有腰头，长度下降到膝围。紧身腰裥裙是一种在德国、俄罗斯等地穿着的传统服装。松身的紧身腰裥裙设计灵感来自乡村时装。

（9）褶襞短裙：在腰位的臀部短皱边，一般系于贴身夹克、女式紧身衣、罩衣和长裙上。

（10）主教袖：在袖口位打碎褶，以便在肘部以下形成较大宽松位的长袖。主教袖是一款宽大罩衣型的衬衫款式，通常在前幅或袖口有装饰性大荷叶边。

（11）蝙蝠袖：在袖窿处有许多松位的长袖子。这款袖子的裁剪在袖窿是很大的，几乎一直延伸到腰线，袖口细小。它通常不是“绱袖型”袖子，但属于衣服的一部分。

（12）开襟领：一种笔直的三尖领，常用于男式衬衫和女式衬衫，可以像燕子领一样张开穿着，也可以像立领闭合上穿着，与夏威夷领相似。

（13）垂褶领圈：一种柔软下垂的领圈，上衣的前幅面料采用斜裁，并通过柔软的折叠效果就形成。

（14）露胸式衣服：它是一款非常低胸裁剪的紧身上衣，女装上部分，由颈、肩、背、胸组成，衣服的领口暴露。袒胸装在舞会礼服、晚礼服、体操衣、贴身女内衣和游泳衣中的低胸领口线应用最为常见。

（15）斗篷袖：袖窿很低的一片袖，事实上，袖窿可以延伸到腰线，有些情形在罩衣上是没有侧缝骨的，一些袖窿是圆弧或椭圆形的，也有的底部是方形。最初，多尔曼袖子是指袖子窄小，前面开口，长而宽松服装。

（16）双排扣：前门襟有两排纽扣或系结物的服装。最现代的双排扣外套，一排纽扣起作用，另一排纽扣或系结物只是装饰之用。相比之下，单排扣只是采用窄小重叠位和一排纽扣。

（17）帝国式腰线：它是指女装长裙的高腰线，低于胸线的紧身上衣，从而给出高腰外观，并且配打碎褶松身长裙。

（18）伊顿式阔翻领：一款可翻折的白色硬领，曾由一位叫伊顿的男孩穿着。

（19）试身：试穿衣服以便调节服装合体时尚。

（20）男装高领：一种高大的可向下翻的领子，有时高度可达到耳朵处。

（21）胸前装饰花边：一款在衬衫前片的领圈装饰花边或荷叶边。它已演变为装饰性服装配件，包括花边或其他从喉咙处下垂面料，从领边或领子处悬吊下或连接，或者简单用针别在喉咙处。

（22）和服袖：袖子与衣身一片裁裁剪，通常在后幅处有缝骨，且与传统的中国长袍相似。

（23）轮廓线：指服装的外形轮廓，使服装看起来具有时尚感或不时尚，轮廓是指服装的外形。

（24）立体裁剪：在人体模型上披盖布料，以便实现服装的形状。

（25）马扎尔袖：袖子与衣身一片裁剪，在腋下缝骨处缝有三角插布的袖子。

（26）牛角袖：袖子连同袖窿线延伸至领圈位，导致肩部和袖头连接在一起而没有缝骨。

（27）翻领（青果领）：领子柔软卷起可自然翻转，且不被烫平，与需要烫折痕的领子决然相反。

（28）袖窿袖：袖头与袖窿能合体缝合的袖子。

（29）样板：设计样板，比如原型板。

（30）V 形领圈：在前领圈处裁剪成“V”字的形状，“V”字的深度有各种不同。

## 2.4.2 纸样词汇

（1）袖窿：也叫作夹圈或袖孔。

（2）基本纸样：一套反映人体体型和形态，但不包含款式特点的纸样。这个主码纸样可以变化形成不同款式。

（3）斜裁：与面料布边成一定角度的裁剪。布料的经向与纬向呈 45° 角。

（4）胸围线：作为基本样的结构线，也是胸围尺寸测量的线，胸点在这条线上可以找到，且所有前幅省道必须从胸点发射出来。

（5）折线：在领子和把襟贴折反而形成的折线。

（6）裤裆：把腿部分开而形成的一个角度。且裤裆深是从前腰头的中心点至被大腿分开而形成的角度的垂直距离。

（7）袖头：袖子的顶端部分。

（8）移省：将省道从一个位置移到另一个位置，并且省尖点用来移褶。

（9）省位：位于上衣的原始省位，如肩省、胸省、腰省等。这些省位可以根据服装的潮流与款式进行选择。

（10）纸样绘制：一套纸样的外形绘制，可通过复制方式得到一套完整的纸样。

（11）放松量（容位）：比如袖头的尺寸比袖窿尺寸要大，这个“差异”就叫作放松量。容位对机织服装在直纹、横纹的裁剪非常重要，允许有一些或没有拉伸。

（12）放码：从一套主码纸样产生一套不同尺码的纸样。

（13）布纹：用来说明布料的纱线方向，机织面料的布纹有直纹、横纹和斜纹。

（14）排料图（唛架）：在裁剪之前，将整套服装纸样平铺在裁床上的一张纸。

（15）布纹不直：用来描述机织布的经编和纬编不成直角时的变形扭曲。

（16）打板师：把服装概念或款式转化成用于服装制作的纸样形式的师傅。

（17）尺码表：基于基本纸样绘制的尺寸表。

（18）缝份：缝份是指在两块或多块面料缝合时，缝合线与边位之间的份量，有时也叫 Inlay 或 Turnings。例如，领圈线、袖窿、贴边缝骨等车缝子口为 1cm。

（19）样衣：服装样式的胚布纸样。“Toil” 原意也叫布料，一块测试服装缝合的同一种材料。

（20）折反位：用作裤子或裙子翻边。

（21）宽松位：比身体尺寸多出的服装尺寸。

（22）立体人模：用于立体裁剪的泡沫人模或假人。

## 本章小结

- 以不同案例形式简单介绍了服装设计效果图的分类与款式描述，同时引出大量的服装款式、服装部件等专业名称。
- 解读多篇具有代表性时装信息文章，明白有关时装信息词汇的理解与描述。
- 对纸样制作方法进行简单划分，并对身体尺寸测量做了详细描述。同时简单介绍纸样制作的方法与应用。
- 针对时装设计与纸样构造方面的专业知识，对有关专业名词做了详细解释。
- 本章节重点提高学生对服装款式描述与表达能力，以及对时装信息文章阅读理解能力。

CHAPTER 3

# GARMENT MANUFACTURING 服装生产

**课题名称：** GARMENT MANUFACTURING 服装生产

**课题内容：** Sample Manufacturing 样衣制作

Production Sheet 生产制作通知单

Making Up Men's Shirt 男式衬衫制作

Pants Construction 裤子结构

Flow Chart of Garment Manufacturing 服装制作流程图

Garment Part Explaining 部件解释

**课题时间：** 8 课时

**教学目的：** 让学生了解制板通知单、生产指示书等服装生产文件的格式与应用，并了解服装制作与工艺流程。掌握服装制作工艺的基本描述与表达以及对各种服装部件的解释与描述。重点掌握并熟记服装生产与流程、工序与部件名称以及成衣尺寸部位等相关术语。

**教学方式：** 结合 PPT 与音频多媒体课件，以教师课堂讲述为主，学生结合案例分析讨论与学习相关工艺流程。

**教学要求：** 1. 了解制板通知单、生产指示书等各种生产文件的格式与应用，并熟悉相关内容的表达。

2. 熟悉服装生产与工艺结构的描述与表达。

3. 明确各类服装生产工艺流程图。

4. 熟悉各相关服装部件的解释与描述。

**课前（后）准备：** 结合专业知识，课前预习课文内容。课后熟读课文主要部分，熟悉各服装生产工艺与流程的表达，并熟记服装生产与流程、工序与部件名称，以及成衣尺寸部位等相关术语。

# CHAPTER 3

# GARMENT MANUFACTURING
# 服装生产

①在服装厂，生产部门通过生产流程对生产的日程安排和进程负责。

In a garment factory, the manufacturing department is usually responsible for scheduling and expediting work through the manufacturing process. ① The manufacturing process can be divided into few areas: pattern drawing, markers making, cutting, sewing, pressing and packing, delivery. Production activities involve the process of converting materials into garments, which meet the needs of the customers. The production scheduling provides for production to maximize manufacturing efficiency and to produce finished products that meet delivery schedules. In the sewing area, sewing and pressing equipments using various attachments to reduce operator skill requirements have significantly increased productivity and quality.

## 3.1 Sample Manufacturing 样衣制作

Sample room makes out a sample order and finishes the sample, and then the sample is sent to the customer for approval. The size specification of the sample and raw materials is listed on the sample order.

### 3.1.1 Sample Order 制板通知单

A sample order should carry the dates of preparation, issuance, schedule of start and completion. 4 copies are prepared for the following distribution: one to sales, one to the sample room, and two to the production department (Table 3-1~Table 3-3).

### 3.1.2 Sample Card 样板卡

Comments to the sample can be recorded. Incorrect sewing methods and garment parts may also be mentioned. Much more correct information must be marked in the sample card so as to let everyone have a clear picture of this sample. Furthermore, people can clearly know what kind of sample (such as sales sample, approval sample, size sample, shipment sample, and so on.) it belongs to (Table 3-4).

**Table 3–1 Sample Order(Ⅰ)**

**Sample Order Form**

Style: Culottes with Yoke Order No.:W-16 Client: Tommy Qty: 2 Pcs. Order Date: 2020-2-1

| Points \ Size | M | | L | | Production Sketch |
|---|---|---|---|---|---|
| | Spec. | Act. | Spec. | Act. | |
| Waist | 68 | −1 | 71 | 0 | |
| Hips/ Seat | 102 | 0 | 106 | +0.5 | |
| Front Rise | 30 | +0.4 | 31 | 0 | |
| Back Rise | 43 | 0 | 44 | 0 | |
| In-Seam | 64 | −1 | 65 | -1 | |
| Out-Leg | 94 | 0 | 96 | -0.5 | |
| Thigh | 26 | −1 | 27 | 0 | Front View Back View |
| Bottom | 56 | 0 | 58 | +1 | |

| Garment Manufacturing | Fabric Swatch |
|---|---|
| 1. Front Dart: 7cm length | |
| 2. Back Dart: 9cm length | |
| 3. Waistband: 5cm width with edge-stitching | |
| 4. Beltloop:4 belt-loops with 0.8cm × 4cm | |
| 5. Yoke: double needle top stitching with 0.8 cm | |
| 6. Front fly: 3.4cm J shape stitching | Color: Printed Fabric |
| 7. Pocket: straight pocket in 16cm | Construction: Plain 60 × 52/28S × 28S |
| 8. Bottom: 8cm | Content: 100% Cotton |

| Accessory | |
|---|---|
| Button: 24L, 4 holes plastic | Lining: 100% viscose rayon, grey color |
| Thread: PP 603, match color | Zipper: Nylon 17cm |
| Interlining: fusible interlining for waistband | Other: |

**Finished: Garment Wash** **Pressing & Packing: Flat**

Department: Sample Room Approved By: Andy Liu Prepared By: Sam Chen

**Table 3–2 Sample Order（Ⅱ）**

**SAMPLE ORDER FOR KNITWEAR**

DATE: 2020-3-8

INITIAL SAMPLE: √　BRAND LABEL: JEANSWEST　STYLE: ROUND NECK

APPROVAL SAMPLE: —　SEASON: WINTER/ FALL　MONTH OF STORE: 2016-5-20

DESCRIPTION: 100% WOOL MEN'S LONG SLEEVE SWEATER WITH ROUND NECK

BODY REF. NO.: JW-008　SPEC. NO.: AU-090　SUPPLIER: DA NAN FACTORY

CATEGORY: MENS: √　WOMENS: —　SIZE: M　CLR.: BLACK & GREY

FAB./ YARN: 100% WOOL　WEIGHT/ YARN COUNT: 60S　TRIMMINGS: —

**ACCESSORIES**

THREAD TYPE: PP 604　CLR.: BLACK & GREY

ZIPPER TYPE & SIZE: —　LENGTH: —　CLR.: —

BUTTON: PLASTIC　SIZE: 30L　QTY.: 2PCS　CLR.: BLACK & GREY

SPARE BTN QTY.: 1 PC　POSITION: LEFT HAND INSIDE SIDE SEAM

WOVEN LABEL: JEANSWEST　POSITION: BACK NECK CENTER

PO.NO. LBL POS.: 4″ABOVE HEM, LHD INSIDE SIDE SEAM.

CARE LABEL POSITION（W/SIZE INDICATION）: 1）LHD SIDE OF CB INSIDE NK.
2）CAUGHT AT BTM OF MAIN LBL.

PRICE TICKET TAG POS.: PIN THRU. CARE LABEL WITH PLASTIC PIN

HANGTAG POS.: PIN THRU. MAIN/ CARE LBL.

OTHER: —

**CONSTRUCTION**

| | | | | |
|---|---|---|---|---|
| NECK/ COLLAR: | SNL.: — | DBL.: √ | COLLAR/ NECK BAND: √ | |
| C.F.: | SNL.: — | DBL.: — | WIDTH: — | |
| ARMHOLE: | SNL.: √ | DBL.: — | OVRLK.: √ | OTHER: — |
| SLEEVE: | SNL. — | DBL. √ | OVRLK.: √ | OTHER: — |
| CUFF: | SNL.: — | DBL.: √ | WIDTH: 5 CM | |
| SHOULDER: | SNL.: √ | DBL.: — | OVRLK.: √ | |

SEAM FORWARD: 2 CM

**POCKET**

PLACEMENT: FM. SIDE SEAM: —　FM. HIGH SHOULDER PNT.: 17 CM

PKT. SIZE: 11CM　SNL.: √　DBL.: —　OVRLK.: —

BOTTOM: STRAIGHT

FINISHING: GARMENT WASH　PRESS: FLAT √　STAND: —

SPECIAL INSTRUCTION: ______

MERCHANDISER: ALICE KANG 2020-3-8

**Table 3–3 Sample Order（Ⅲ）**

**MEN'S WEAR KNIT**
**TOPS MEASUREMENT CHART**

DATE: 2020-3-8

STYLE DESCRIPTION: 100% WOOL MEN'S LONG SLEEVE SWEATER WITH ROUND NECK
STYLE NO.: JWS-108 SEASON: WINTER/ FALL SPECIAL INSTRUCTIONS: —

| SIZE SPECIFICATION（CM） | GRADING | S | M | L | XL |
|---|---|---|---|---|---|
| A. SHOULDER WIDTH—SEAM TO SEAM | 2.5 | | 46 | | |
| B. CHEST—MEASURE 2.5 CM UNDER ARMHOLE | 5 | | 102 | | |
| C. BOTTOM—CIRCUMFERENCE GIRTH | 5 | | 96 | | |
| D. SLEEVE WIDTH 2.5 CM BELOW ARMHOLE—FLAT & PERPENDICULAR MEASURE | 1.5 | | 23 | | |
| E. ARMHOLE CURVE MEASURE—CIRCUMFERENCE | 1.5 | | 50 | | |
| F. SLV. LENGTH—FM. CB. THRU. SHOULDER PT. TO CUFF | 2.5 | | 78 | | |
| G. C/B LENGTH—NOT INCL. NECK BAND | 1.5 | | 62 | | |
| H. COLLAR LENGTH—CLOSED 1ST BTN MEASURE | 1.5 | | — | | |
| I. NECK OPENING—SEAM TO SEAM | 0.6 | | 20 | | |
| J. FRONT NECK DROP—TOP TO TOP | 0.2 | | 5 | | |
| K. BACK NECK DROP—HIGH PT. TO C/B NK. | 0 | | 3 | | |
| L. COLLAR WIDTH—AT C/B | 0 | | — | | |
| M. NECK BAND WIDTH—AT C/B | 0 | | 4 | | |
| N. CUFF—CIRCUMFERENCE | 1.5 | | 18 | | |
| O. CHEST PKT. PLACEMENT 1—FM. C/F EDGE | 0.2 | | 8 | | |
| P. CHEST PKT. PLACEMENT 2—FM. HIGH PT. TO PKT. OPENING | 0.6 | | 17 | | |

MERCHANDISER: ALICE KANG 2020-3-8

**Table 3–4 Sample Card**

| RYKIEL<br>SAMPLE: Approval Sample |
|---|
| Date: 2020-2-1 |
| Our Ref.: TQ-010 |
| Style: Five-pockets Jeans |
| Customer: Tommy |
| Contract No.: TR-086 |
| Size: M |
| Description: Ladies' 5-Pkt Jeans with 100% Cotton |
| Fabric: Denim, 100% Cotton |
| Remark: Attached Sample |
| Counter Sample/ Initial Sample/ Size Sample/ PP Sample Approval Sample/ Shipment Sample/ IC Sample Sales Sample |

### 3.1.3 Approval Sample Card 核准样卡

The method of construction is determined by the kind of garment and its price. The two main methods of construction are as follows:

（1）Section work：each operator on the assembly line sews only one part of the garment.

（2）Complete garment construction：one operator sews the whole garment. Other workers do special work, such as over-locking seams, hemming, buttonholes, etc. This method is often used for better garments, especially tailored jackets.

Workers are paid based on the number of finished garments in a clothing factory. The price paid for a worker is determined by the difficulty of the task and the time needed to complete the task. This is called piecework. Workers detach the segment of the work-ticket that corresponds to the task of completing these garments. At the end of the pay period, each worker's salary is determined by the number of tickets he turns in.

There are also other workers in charge of trimming threads, checking for sewing errors, hanging garments on hangers, attaching hang tags, and covering each garment with a plastic bag. When garments have been incorrectly sewn, the quality controller returns them to the factory or contractor to be repaired. The manufacturing department is responsible for mass-production, establishing the line based on various sizes and colors. ②

②生产部门负责批量生产，根据不同的颜色和尺寸来安排生产线。

The sample is approved for bulk production, the instructions and comments should be noticed on the approval sample card. ③ ( Table 3-5, Table3-6). It is a very important process for confirming that all details are correct. The information is as follows:

③样衣经确认可进行大货生产以后，应在核准样板卡上说明各部分结构和生产指示。

（1）Client & Factory.

（2）Style number & contract number.

（3）Fabric & trimming.

（4）Size specification, color and measurement.

（5）Garment construction and measurement.

**Table 3–5 Approval Sample Card**

<table>
<tr><td colspan="5">AUDITING SAMPLE CARD<br>Men's Category     Date: March 3, 2020</td></tr>
<tr><td colspan="5">Client: Kmart    Factory: DaYang Manufacturer    Style No.:TQ—008<br>Shell Fabric: Denim    Contract No.: 8502UK    Size Spec.: 28 ~ 36 Range<br>Description: Slim Straight Fit    Commenting on: 2nd PP SAMPLE    Status: REJECT</td></tr>
<tr><td colspan="5">NOTE: Approved for bulk production subject to the instructions and comments.</td></tr>
<tr><td rowspan="2">Checked Points</td><td rowspan="2">Size (cm)</td><td>Requested Spec</td><td>2nd PPSMPL</td><td>Corrections</td></tr>
<tr><td>32</td><td>32</td><td>32</td></tr>
<tr><td colspan="2">1.Waist girth: at top of waistband</td><td>88</td><td>88</td><td>OK</td></tr>
<tr><td colspan="2">2. Waistband width</td><td>4.4</td><td>4.5</td><td>—</td></tr>
<tr><td colspan="2">3. Belt loop length</td><td>5.5</td><td>5.5</td><td>—</td></tr>
<tr><td colspan="2">4. Belt loop width</td><td>1.2</td><td>1.2</td><td>—</td></tr>
<tr><td colspan="2">5. Belt loop from front pocket edge</td><td>1.5</td><td>1.5</td><td>—</td></tr>
<tr><td colspan="2">6. Belt length excluding Dring</td><td>115</td><td>Missing</td><td>—</td></tr>
<tr><td colspan="2">7. Belt width</td><td>4</td><td>Missing</td><td>—</td></tr>
<tr><td colspan="2">8. Low hips girth 6.5cm above crotch</td><td>108</td><td>109.5</td><td>Reduce 1.5cm</td></tr>
<tr><td colspan="2">9. Thigh girth: 2.5 cm below crotch</td><td>64</td><td>64.6</td><td>Reduce 0.6cm</td></tr>
<tr><td colspan="2">10. Knee 5cm above mid inseam</td><td>45</td><td>45.6</td><td>Reduce 0.6cm</td></tr>
<tr><td colspan="2">11. Leg opening circumference</td><td>45</td><td>43.5</td><td>Add 1.5cm</td></tr>
<tr><td colspan="2">12. Front rise including waistband</td><td>26</td><td>25.6</td><td>OK</td></tr>
<tr><td colspan="2">13. Back rise including waistband</td><td>29</td><td>29</td><td>—</td></tr>
<tr><td colspan="2">14. Crotch width 2.5cm over crotch inseam</td><td>14.5</td><td>15</td><td>—</td></tr>
<tr><td colspan="2">15. Inseam length</td><td>84</td><td>84.5</td><td>OK</td></tr>
<tr><td colspan="2">16. Out-seam incl. W.B.</td><td>107</td><td>106.5</td><td>OK</td></tr>
<tr><td colspan="2">17. Fly length from waistband top edge to bottom stitch</td><td>19</td><td>19</td><td>—</td></tr>
<tr><td colspan="2">18. Fly stitch width</td><td>4</td><td>4</td><td>—</td></tr>
<tr><td colspan="2">19. Front pocket length along side seam from top of WB</td><td>11</td><td>11</td><td>OK</td></tr>
<tr><td colspan="2">20. Front pocket width along waistband</td><td>12</td><td>12</td><td>—</td></tr>
<tr><td colspan="2">21. Front pocket bag width</td><td>18</td><td>18</td><td>—</td></tr>
<tr><td colspan="2">22. Front pocket bag length</td><td>28</td><td>26</td><td>Add 1cm</td></tr>
<tr><td colspan="2">23. Coin pocket distance from WB bottom edge at centre front</td><td>7</td><td>7</td><td>—</td></tr>
</table>

**Continued**

| | | | |
|---|---|---|---|
| 24. Coin pocket distance from WB top edge at side-seam | 7.5 | 7.5 | — |
| 25. Coin pocket width at top edge | 7.5 | 7.5 | OK |
| 26.Coin pocket distance from side-seam | 1.5 | 2.1 | Reduce 0.6cm |
| 27.Back yoke length at centre back | 8 | 8 | OK |
| 28. Back yoke length at side seam | 3 | 3 | — |
| 29.Back pocket distance from yoke at centre back | 3 | 3 | — |
| 30.Back pocket distance from yoke at side seam | 2.5 | 2.5 | — |
| 31.Back pocket width at top | 18 | 18 | OK |
| 32. Back pocket width at bottom | 15 | 15 | — |
| 33. Back pocket length at center | 16 | 16 | — |
| 34. Back pocket length at side | 14.5 | 14.5 | — |
| DATE | 2020-3-8 | 2020-3-8 | 2020-3-8 |

**COMMENTS**

1. Reduce low hips 1.5 cm to be 106cm total.
2. Reduce thigh with 0.6 cm to be 64cm total.
3. Reduce knee with 0.6 cm.
4. Add leg opening with 0.6 cm.
5. Add 1cm front pocket bag length.
6. Reduce 0.6 cm coin pkt. distance from side-seam.
7. Pls make all corrections & resubmit for the 3rd PP SAMPLE approval.

**PRODUCTION SKETCH**

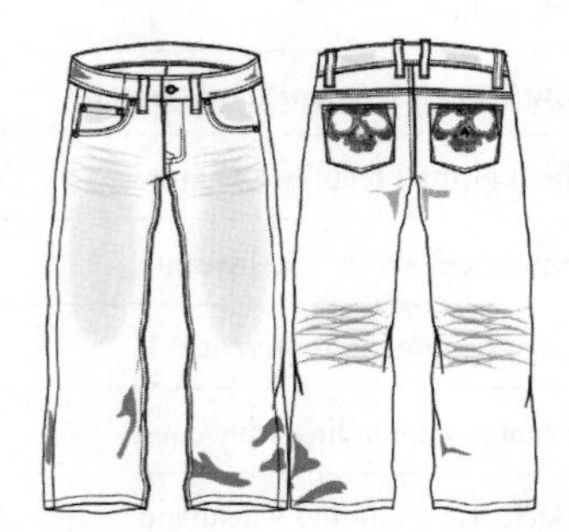

**ATTENTION IN MASS PRODUCTION:**

1. Either the correction on the rises have not made, or done incorrectly, front and back rises still puckered due to too much access fabric, very unsightly. These corrections on the front and back rises must be done before we can approve for bulk.
2. In addition to the correction on the rises, pls correct six points on other areas of the garment listed above. We would like to modify the low hip, thigh width and pocket bags. Pls remove 1.5cm at low hip from the back rise by 1.5cm, reduce 0.6cm total at thighs width as sketch below. As well, we would like to change the pocket bag to be at 27cm in length.
3. We cannot approve for bulk until we the good fit sample. Pls re-submit the 3rd pp sample ASAP.
4. Pls be advised, to save time and to ensure that the next garment will fit correctly, we will send you our pattern block today.

Merchandiser: May Liu 2020-3-8　　Department Manager: CK Zhang 2020-3-8

**Table 3–6 Approval Sample Form**

## APPROVAL SAMPLE

**1. GARMENT CONSTRUCTION**

| | SIZE | DIMENSIONS |
|---|---|---|
| (1) FRONT POCKET | (M) | 5 CM × 12 CM × — |
| (2) BACK POCKET | (M) | 16 CM × 18 CM × 14 CM |

(3) BACK YOKE SIZE 2.5 CM AT SEATSEAM.
7 CM AT SIDESEAM.

(4) BACK POCKET PLACEMENT

A. 2 CM BELOW + PARALLEL TO BACK YOKE EDGE.

B. 9 CM BELOW + PARALLEL TO BOTTOM OF WAISTBAND.

C. BELOW BOTTOM OF WAISTBAND 8 CM AT SEATSEAM.
7 CM AT SIDESEAM.

**2. COLOUR CO–ORDINATION**

| BODY | TOP-STITCHING | EMB. | LABEL | METALWARE |
|---|---|---|---|---|
| FEONT | YELLOW | RED & WHITE | MATCHED | BRASS |
| BACK | YELLOW | — | — | — |

**3. FABRIC CONSTRUCTION**

(1) NAME / CODE: 1/3 TWILL "Z" DIRECTION FABRIC CONTENT: 100% COTTON

(2) WEIGHT 18 OZ. PER SQ.FT. — LBS. PER DOZ.

(3) CONSTRUCTION: $\frac{60 \times 50}{16S \times 16S}$

**4. TRIMMINGS**

| | ACCEPTED | UNACCEPTED |
|---|---|---|
| (1) TOP-ST. THREAD SIZE | √ | |
| (2) BUTTON / SNAP | √ | |
| (3) RIVET | √ | |
| (4) INTERLINING | √ | |
| (5) LINING | √ | |
| (6) PIPING / INSERT | | √ |
| (7) RIBBING | √ | |
| (8) OTHER | √ | |

## Words and Expressions

manufacturing [ˌmænju'fæktʃəriŋ] 制造
production schedule [prə'dʌkʃən 'ʃedju:l] 生产进度表
pattern grading ['pætən greid] 纸样放码
expediting ['ekspidaitiŋ] 加快，促进
pattern drawing ['pætən 'drɔ:iŋ] 纸样绘制
marker making ['mɑ:kə 'meikiŋ ] 排料（广东话：唛架制作）
FAB./fabric ['fæbrik] 面料
LBL./ label ['leibl] 商标
specialize ['speʃəlaiz] 专业化
special effect ['speʃəl i'fekt] 特殊效果
packing ['pækiŋ] 包装
converting [kən'və:tiŋ] 改变，转换
efficiency [i'fiʃənsi] 工作效率
attachment [ə'tætʃmənt ] 附件
significantly [sig'nifikəntli] 重要地
productivity [prədʌk'tiviti] 生产效率
quality ['kwɔliti] 质量
showroom ['ʃəurum ] 样板陈列室
transmitted [trænz'mitid] 传输
approval [ə'pru:vəl] 核准，核可
specification/ Spec. [ˌspesifi'keiʃən ] 规格
raw material [rɔ: mə'tiəriəl] 原材料
distribution [distri'bju:ʃən] 分配
collar ['kɔlə] 领子
shoulder ['ʃəuldə] 肩宽
bust/ chest [bʌst / tʃest] 胸围
armhole ['ɑ:mhəul] 袖窿（广东话：夹圈）
X-back [eks-bæk] 后背宽
C/B length [si:/bi: leŋθ] 后中长，后中心线
Slv. length [sli:v leŋθ] 袖长
cuff ['kʌf] 袖级、袖口克夫
waist [weist] 腰围
hips/ seat [hips / si:t] 臀围（广东话：坐围）
front rise/ F.R. [frʌnt raiz] 前裆（广东话：前浪）
back rise/ B.R. [bæk raiz] 后裆（广东话：后浪）
in-seam [in-si:m] 内接缝（广东话：内长）
out-leg [ɑut-leg] 外长
thigh [θai] 大腿围
bottom ['bɔtəm] 下摆
front part [frʌnt pɑ:t] 前片，前幅
back part [bæk pɑ:t] 后片，后幅
sleeve opening [sli:v 'əupəniŋ] 袖口
pocket ['pɔkit] 口袋
waistband ['weɪstbænd] 裤腰
fastening ['fɑ:sniŋ] 系结物
button ['bʌtn] 纽扣
thread [θred ] 缝纫线
interlining ['intə'lainiŋ] 里衬（广东话：衬朴）
lining ['lɑiniŋ] 里料
shoulder pad ['ʃəuldə pæd] 垫肩
metal-ware ['metl-weə] 金属辅料
ribbing ['ribiŋ] 罗纹
knitwear ['nitˌwɛə] 针织服装
yarn [jɑ:n] 纱线
count [kɑunt] 支数
CLR./ color ['kʌlə] 颜色
POS. / position [pə'ziʃən] 位置
initial sample [i'niʃəl 'sæmpl] 初板
brand label [brænd 'leibl] 主商标
approval sample [ə'pru:vəl 'sæmpl] 核准板，核可样
month of store [mʌnθ ɔvˌ stɔ: (r)] 存货期
LHD side/ left hand side [left hænd said] 左手边
NK./ neck [nek] 领圈
CB/ center back ['sentə bæk] 后中
prices ticket tag [prɑisiz 'tikit tæg] 电脑价钱牌
THRU./ through [θru:] 通过，穿过
plastic pin ['plæstik pin] 塑料大头针
CF/ center front ['sentə frʌnt] 前中
SNL/ single thread ['siŋgl θred] 单线 / 单针
DBL/ double threads ['dʌbl θredz] 双线 /

双针
OVRLK/ over-lock ['əuvə-lɔk] 包缝（广东话：锁边）
FM./ from [frɔm,] 由，从
flat package [flæt 'pækidʒ] 平包，平板包装
high shoulder point [hɑi 'ʃəuldə pɔint] 高肩点
spare Qty. [spɛə 'kwɔntiti ] 损耗数
measurement chart ['meʒəmənt tʃɑ:t] 尺寸表
circumference [sə'kʌmfərəns] 全围
perpendicular [pə:pən'dikjulə] 直角
shoulder width ['ʃəuldə widθ] 肩宽
collar / neck band ['kɔlə nek bænd] 领围
collar stand width / neck band width ['kɔlə stænd widθ /nek bænd widθ] 领底宽
woven label ['wəuvən 'leibl] 机织商标
salesman sample ['seilzmən 'sæmpl] 销售样
size set sample [saiz set 'sæmpl] 尺码确认样
shipping sample ['ʃipiŋ 'sæmpl] 船样（广东话：船头板）
counter sample ['kɑuntə 'sæmpl] 回样
garment construction ['gɑ:mənt kən'strʌkʃən] 服装结构
category ['kætigəri] 种类
hemming ['hemiŋ] 缝边脚，卷边
tailored jackets ['teiləd 'dʒækit] 洋服
piecework ['pi:swək] 计件工作
hangtag ['hæŋtæg] 吊牌
mass-production [mæs-prə'dʌkʃən] 大量生产
retailer [ri:'teilə] 零售商
bulk production [bʌlk prə'dʌkʃən] 批量生产
client ['klɑiənt] 客人，买方
size specification [saiz ˌspesifi'keiʃən ] 尺码规格
measurement['meʒəmənt] 尺寸
contract ['kɔntrækt,] 合同
bottoms ['bɔtəmz] 下装
tops [tɔps] 上装
out-seam [ɑut-si:m] 侧缝，外长
zipper ['zipə] 拉链
neck drop [nek drɔp] 领深
neck opening [nek 'əupniŋ] 领开口
Act./ Actual ['æktjuəl] 实际的
yoke [jəuk] 育克，过肩（广东话：担干，机头）
front pocket [frʌnt 'pɔkit] 前口袋
back pocket [bæk 'pɔkit] 后口袋
seat-seam [si:t-si:m] 坐围缝迹
top- stitching [tɔp-'stitʃiŋ] 面缝线迹（广东话：缉面线）
EMB./ Embroidery [im'brɔidəri] 绣花
Dimension [di'menʃən] 尺寸
piping/ insert ['paipiŋ in'sə:t] 嵌边
work-in-process/ WIP [wə:k in-'prəuses] 半成品
pre-production sample/ PP Sample [pri:-prə'dʌkʃən 'sæmpl] 产前样，PP 板
fit sample [fit 'sæmpl] 合身样
inspection certificate [in'spekʃən sə'tifikit] 检查确认证
inspection certificate sample / IC Sample [ɪɔ 'sæmpl] 客人确认样，IC 板
body rise ['bɔdi raiz] 裤裆（广东话：直浪）
plastic bag ['plæstik bæg] 塑胶袋
run-stitching [rʌn -'stitʃiŋ ] 初缝线迹（广东话：运线）
edge-stitching [edʒ-'stitʃiŋ] 缝边线迹（广东话：缉边线）
hip pocket [hip 'pɔkit] 后口袋
Denim ['denim] 牛仔布
belt loop [belt lu:p] 裤耳
crotch [krɔtʃ] 裤裆，小浪，浪顶
leg opening [leg 'əupniŋ] 裤脚围
waistband [ 'weɪstbænd] 腰头
fly [flai] 暗门襟
low hips [ləu hips] 下臀围
thigh [θai] 髀围，大腿围

yoke [jəuk] 育克

pucker [ˈpʌkə] 皱纹，起皱

ASAP /As Soon As Possible 尽快

## *Exercises*

1. List main areas of the manufacturing function.

   Example: sewing

2. List the information on an approved sample card.
3. Translate the following terms into Chinese.

（1）pattern grading
（2）packing
（3）quality
（4）approval sample
（5）size specification
（6）collar
（7）shoulder
（8）bust
（9）armhole
（10）C/B length
（11）cuff
（12）waist
（13）hips
（14）front rise
（15）in-seam
（16）out-leg
（17）bottom
（18）back part
（19）sleeve opening
（20）waistband
（21）fastening
（22）button
（23）thread
（24）interlining
（25）lining
（26）shoulder pad
（27）metal-ware
（28）ribbing
（29）knitwear
（30）yarn
（31）count
（32）initial sample
（33）brand label
（34）center back
（35）over-locking
（36）flat package
（37）measurement chart
（38）scye
（39）neck band width
（40）woven label
（41）sales sample
（42）size set sample
（43）shipment sample
（44）counter sample
（45）hemming
（46）tailored jackets
（47）hang tag
（48）retailer
（49）bottoms
（50）tops
（51）out-seam
（52）zipper
（53）neck drop
（54）yoke
（55）front pocket
（56）top-stitching
（57）rivet
（58）piping
（59）WIP
（60）PP Sample
（61）fit sample
（62）IC sample
（63）body rise
（64）run-stitching
（65）edge-stitching
（66）hip pocket
（67）care label
（68）first sample
（69）initial sample
（70）advertised sample
（71）promotion sample
（72）brand label

4. List the major measurement points of the following styles（Men's Jacket, Five-pocket Jeans）.

   Example: chest, collar, etc.

   waist, front rise, etc.

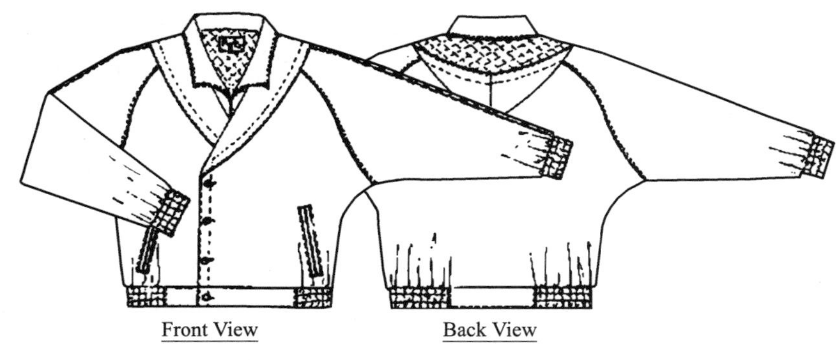


**Men's Jacket**

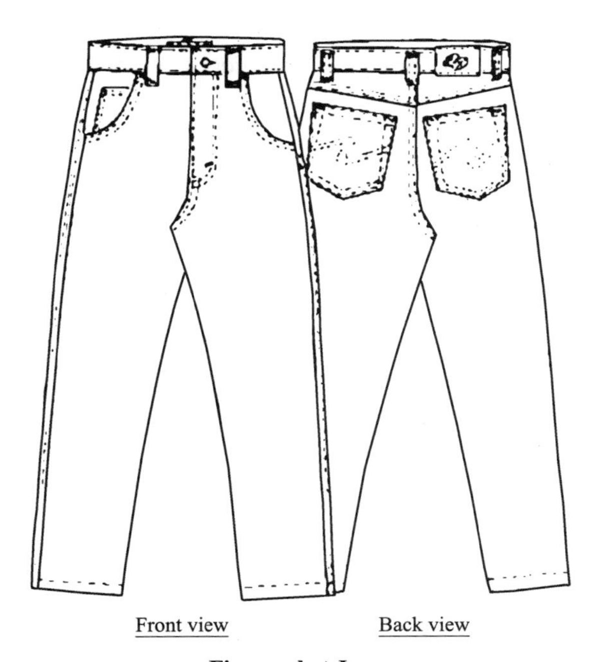


**Five-pocket Jeans**

## 3.2 Production Sheet 生产制作通知单

### 3.2.1 Introduction 简介

In the apparel company, cutting and manufacturing of products is planned on orders or on a forecast of these orders. These activities are done within the purchasing, manufacturing and production control departments. ①

①服装公司的产品裁剪和制造是依据订单或订单预测来安排的，并在采购、生产与产品控制部门范围内进行。

In order to produce a style, the purchasing department must have ordered and ensured delivery of necessary materials. Production control must have been sched-

②生产控制必须安排在工厂的计划工作内，并保证与生产这类产品相关的职能部门了解产品要求。

uled into the factory plan and ensure that other functional departments involved in the process of producing this product are aware of its requirements. ② The manufacturing department must have ensured its capacity. And get enough labor and machinery. In order to accomplish these tasks, the personnel and engineering departments have become involved in staffing, and ensuring all equipment is available.

After goods have been finished and sent to the warehouse, the order may be released for shipment to the customer. From this point, the distribution center begins to process the order. In order to meet the customer's requirements, the distribution center must be aware of the destination and shipping date of the order. Then it must prepare staff for picking and packing goods. Once goods are completed, they are transferred to the warehouse. Through the in-process monitoring system, which has tracked work through the production cycle, the production control department and the warehouse are notified.

### 3.2.2 Case Analysis 案例分析

③通过回顾前面的生产职能和活动，我们就能了解各种职能怎样相互联系，以及怎样采用计算机系统进行监督和控制，并简化涉及不同工作范围的沟通。

From the preceding overview of the production functions and activities, we can see how the various functions interrelate and how the uses of a computer system to monitor and control this work, and simplify the communications between the various areas are involved. ③ The production order indicates the working procedure and other details. The following information should be shown out:

（1）Fabrics and accessories needed.

（2）Style & quantity.

（3）Color assortment.

（4）Size distribution & size specification.

（5）Shipping marks & packing method.

（6）Delivery date and other production detail.

**Table 3–7 Production Order**

**PRODUCTION ORDER**

CASUAL WEAR ( MEN'S ) DATE:2-MAR-2020

| ORDER NO | STYLE NO | DESCRIPTION | QUANTITY | SHIPMENT DATE |
|---|---|---|---|---|
| RKY-001 | RYKIEL | SHORTS/ PLEAT | 125DOZ | 9-MAY-2020 |

**SIZE SPEC:** SZ-001 ( REGULAR W.B. ) COLOR CODE:

| | | | | | | |
|---|---|---|---|---|---|---|
| GARMENT WASH ( ADD SOFT ) | 30 | 32 | 34 | 36 | 38 | |
| WAIST ( INSIDE MEAS. ) | 30 | 32 | 34 | 36 | 38 | INCH |
| SEAT ( 8cm ABOVE CROTCH OPENDED PLEAT ACROSS MEAS. ) | 43 | 45 | 47 | 49 | 51 | |
| THIGH ( AT CROTCH ) | 27 | 28 | 29 | 30 | 31 | |
| BOTTOM | 23 | 24 | 25 | 26 | 27 | |
| FRONT RISE ( INCLUDE WAISTBAND ) | 11 | 12 | 13 | 14 | 15 | |
| BACK RISE ( INCLUDE WAISTBAND ) | 16 | 17 | 18 | 19 | 20 | |
| INSEAM | 9 | 9 | 9 | 9 | 9 | |
| ZIPPER | 7 | 7 | 8 | 8 | 8 | |
| **SIZE ASSORTMENT** | | | | | | |
| NAVY | 60 | 240 | 600 | 360 | 240 | =1500PCS |

**SHIPPING MARK**

CASUAL WEAR
MODEL NAME: RYKIEL
COLOUR:
LOT MODEL NO.: 4115-4130-90
SALES ORDER NO.: 20398
QUANTITY: 36PCS.
SIZE:
CARTON NO.: 1-UP
MADE IN CHINA

**SIDE MARK**

MODEL NAME: RYKIEL
COLOUR:
SIZE:
GR. WT.:
NET WT.:
MEAS.:

● THE WEIGHT MARKED IN SHIPPING MARK MUST BE THE SAME AS THE REAL WEIGHT

| FABRIC | FABRIC CONSTRUCTION | | FABRIC COLOR |
|---|---|---|---|
| COTTON/PLAID | 68 × 54/16S × 16S 100%COTTON | | BLUE |
| THREAD | ZIPPER | LABEL | BUTTON |

**Continued**

<table>
<tr><td>PP604</td><td>BC-360 #3 BRASS</td><td rowspan="2">RYKIEL</td><td>25L RYKIEL</td></tr>
<tr><td>WASH COLOR</td><td>AUTO #560</td><td>4-HOLES NAVY PLASTIC</td></tr>
<tr><td colspan="4">（GARMENT CONSTRUCTION）</td></tr>
<tr><td colspan="4">LABEL PLACEMENT:<br>● LA0100 RYKIEL（WINE RED GROUND KHAKI LETTERING）MAIN LABEL: MATCHING LABEL COLOR THREAD. SEW ALL SIDES OF LABEL. AT RIGHT CORNER ABOVE RIGHT BACK POCKET（AS SKETCH）<br>● CARE LABEL ON TOP P.O.+STYLE NO. LABEL ON BOTTOM（ALL OF BEIGE GROUND BLUE LETTERING）ALL TOGETHER INSERT AT INSIDE OF CENTER BACK WAISTBAND</td></tr>
<tr><td colspan="4">FRONT PANEL: TWO PLEATS（AS SKETCH）</td></tr>
<tr><td colspan="4">FRONT POCKET:<br>● OUTSEAM PKT. WITH 1/4" SINGLE NEEDLE, 1 ½" WIDTH FACING<br>● EDGE OF FACING AND BEARER CLEAN FINISH WITH EDGESTITCH, WHICH STITCH DOWN WITH PKT BAG<br>● WHITE T/C PKT. BAG, WITH CLEAN FINISH WITH 1/8" SINGLE NEEDLE AND SHOULD BE EXTENDED TO FRONT FLY</td></tr>
<tr><td colspan="4">FRONT RISE: OVERLOCK AND EDGESTITCH</td></tr>
<tr><td colspan="4">BACK PANEL: TWO EDGESTITCH DARTS（AS SKETCH）</td></tr>
<tr><td colspan="4">BACK POCKET: SINGLE WELT PKT WITH EDGESTITCH. CLEAN FINISH WITH 1/8"<br>SINGLE NEEDLE WHITE T/C PKT BAG</td></tr>
<tr><td colspan="4">BACK RISE: OVERLOCK WITH 5 THREADS. RIGHT COVER LEFT AND EDGESTITCH</td></tr>
<tr><td colspan="4">OUTSEAM: OVERLOCK WITH 5 THREADS. BACK COVER FRONT AND EDGESTITCH</td></tr>
<tr><td colspan="4">INSEAM: OVERLOCK WITH 5 THREADS</td></tr>
<tr><td colspan="4">BOTTOM: CLEAN FINISH WITH DOUBLE NEEDLE WHICH 1 ½"ABOVE BOTTOM.</td></tr>
<tr><td colspan="4">WAISTBAND: SEW ONE PIECE 1 ½" DOUBLE NEEDLE WAISTBAND NOT SEAM ALLOWANCE</td></tr>
<tr><td colspan="4">BELTLOOP:<br>● 6 PCS.（2"×1/2"）BELTLOOP WITH DOUBLE NEEDLE. ALL OF BELTLOOPS INSERTED AT BOTTOM OF WAIST-BAND<br>● 2 PCS. EACH TOUCH TOGETHER NEAR OF THE CENTER FRONT PLEAT<br>● 2 PCS. AT CENTER BACK IN WHICH APARTS 2 ½"<br>● 2 PCS. AT BACK PANEL, WITH CENTER BETWEEN FRONT BELTLOOP AND CENTER BACK BELTLOOP（AS SKETCH）</td></tr>
</table>

**Continued**

| **BARTACK:**<br>● 12 PCS AT TOP AND BOTTOM OF BELTLOOPS ( BOTTOM OF BELTLOOP WITH CONCEALED BARTACK )<br>● 4 PCS AT FRONT PKT. OPENING<br>● 4 PCS AT BACK PKT. OPENING WITH 1/2" BARTACK<br>● 2 PCS AT FRONT FLY ( AS SKETCH )<br>● 1 PC AT CROTCH<br>● TOTAL: 23 PCS | |
|---|---|
| **BUTTON HOLE:**<br>1 PC. AT WAISTBAND<br>2 PCS. AT CENTER OF BACK PKT<br>TOTAL: 3 PCS. 11/16" CUT OPENING MEAS | |
| **BUTTON:** TOTAL 3 PCS. CO-ORDINATION WITH BUTTONHOLE | |
| **PKT BAG:** WHITE T/C PKT BAG ( USE MATCHING PKT BAG ) | |
| **INTERLINING:** #53915 INTERLINING USED AT INSIDE WAISTBAND<br>GP-3 INTERLINING USE AT BACK WELT PKT | |
| **REMARK:** THE POCKET BAG AND INTERLINING COLOUR STANDARD CANNOT SHOW ON THE OUTSIDE | |
| **PRODUCTION SKETCH & DETAILS** | |
| **FRONT POCKET BAGS DEPT:**<br>● MEAS. FROM BOTTOM OF PKT. OPENING TO BOTTOM OF PKT. BAG. ALL SIZES ARE 5 1/2"<br>**BACK POCKET BAG DEPT:**<br>● MEAS. FROM BOTTOM OF WELT TO BOTTOM OF PKT. BAG. ALL SIZES ARE 5 1/2"<br>**OTHER DETAILS:**<br>● LEFT AND RIGHT AT CENTER FRONT OF WAISTBAND SHOULD MATCH WITH HORIZONTAL PLAID<br>● FRONT RISE AND BACK RISE SHOULD BE MATCHED AT HORIZONTAL PLAID<br>● LEFT & RIGHT SIDE OF BACK PKT. WELT'S PLAID MUST BE EQUAL<br>● BACK DART FACING OUT-SEAM | 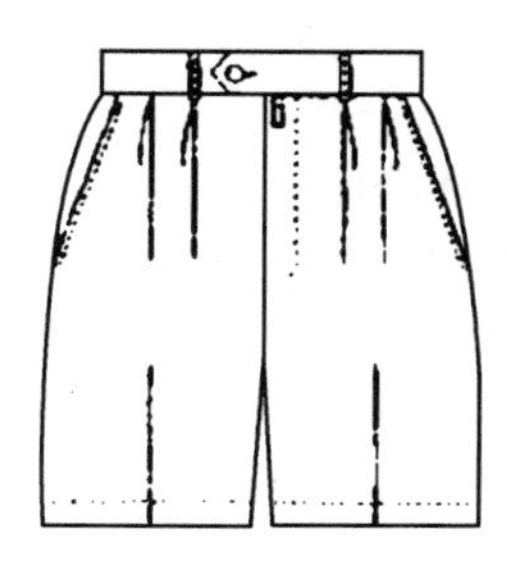<br>FRONT PANEL<br>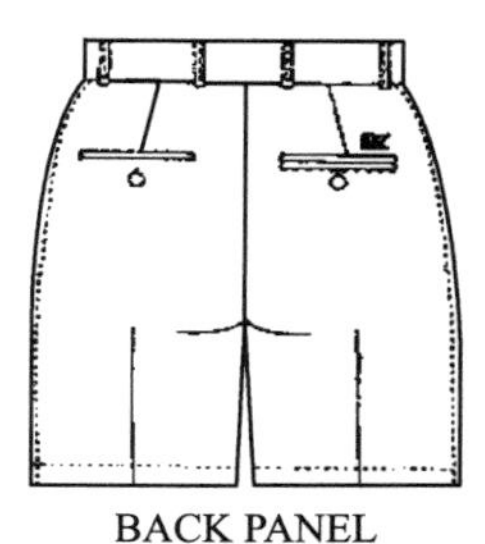<br>BACK PANEL |

**Continued**

- **LS00500 WAIST TAG:** USE WHITE COLOUR THREAD TACKS 1 PC. CLIP AT LEFT BACK OUTSIDE WAISTBAND. CENTER BETWEEN ABOVE POCKET OPENING（AS SKETCH）
- **CT–36 PKT FLASHER:** USE WHITE COLOUR THREAD TACKS 2 PCS. CLIP AT LEFT BACK PKT（AS SKETCH）
- **LS08 PRICE TICKET:** USE WHITE COLOUR THREAD TACKS 4 PCS

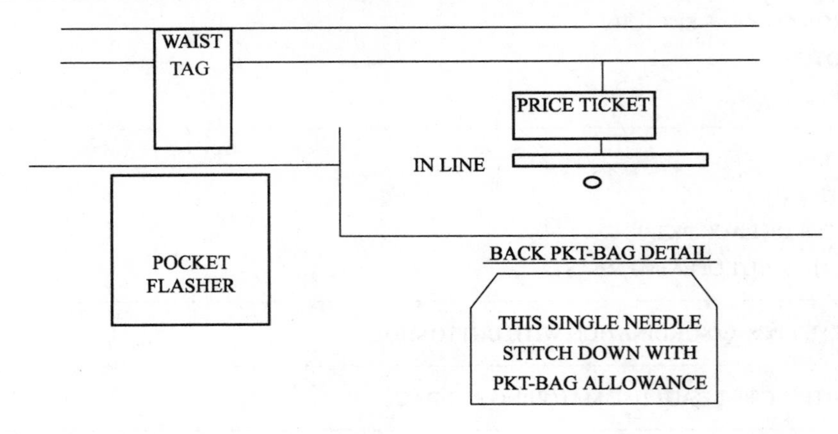


*****RYKIEL–CASUALS PACKING METHOD FOR SHORTS *****

- BEFORE PACKING, THE GARMENT SHOULD BE BUTTONED AND THE WAISTBAND BUTTONED
- PRESSING: THE CENTER CREASE PLEAT UP WITH CENTER CREASE LINE
- PACKING: EACH PIECE CENTER CREASE FOLD（THE RYKIEL PRICE TICKET TOWARD UP）. PUT INTO A NON-SLIDE POLYBAG WITH WARNING PRINTED AT BACK. 12 PCS SOLID COLOUR SOLID SIZE PUT INTO A POLYWARP. CARDBOARD PAPER ON TOP AND BOTTOM INSCRITION PAPER, 36 PCS.（3 POLYWARP）SOLID COLOUR SOLID SIZE PUT INTO A 3-PLY EXPORT CARTON

**PACKING REMARKS:**

- BEFORE DIVIDING THE SIZE OF BULK, WE MUST MEASURE WAISTBAND AND CHECK THE SIZE AND INSEAM. IF THE SIZE AND INSEAM ARE WRONG, WE MUST CORRECT THEM
- BEFORE PACKING INTO A POLYBAG, WE MUST CHECK THE TICKET, WAIST TAG, CARE LABEL, THE SIZE AND INSEAM MUST MATCH
- WE MUST CHECK THE DIRT FOR THE BULK
- BEFORE PUTTING THE GARMENT INTO A POLYBAG, MUST FINISH ALL ABOVE PROCEDURES

DEPARTMENT: PRODUCTION　　　　PREPARED BY: ALICE LIU

DATE: 2-MAR-2020

## Words and Expressions

be aware of [bi: ə'weə ɔv.] 知道的，注意到的
communication [kə'mju:nikeiʃən] 沟通，联络
style [stail] 款式
production order [prə'dʌkʃən 'ɔ:də] 生产通知单
pocket opening ['pɔkit 'əupəniŋ] 袋口
casual wear ['kæʒjuəl weə] 休闲装
pleat [pli:t] 褶裥（广东话：活褶）
dart [dɑ:t] 省（广东话：死褶）
garment wash ['gɑ:mənt wɔʃ] 成衣水洗
crotch /crutch [krɔtʃ / krʌtʃ] 裤裆
waistband ['weɪstbænd] 腰头
inseam ['insi:m] 内接缝（广东话：内长）
size assortment [saizə'sɔ:tmənt] 尺码分配
carton box ['kɑ:tən bɔks] 纸箱
shipping mark ['ʃipiŋ mɑ:k] 正箱唛
side mark [said mɑ:k] 侧箱唛
plaids [plædz] 格子布
khaki ['kɑ:ki] 卡其色
matching color ['mætʃiŋ 'kʌlə] 配色
production sketch [prə'dʌkʃən sketʃ] 生产图样
measurement / meas. ['meʒəmənt] 尺寸
facing ['feisiŋ] 贴边
bearer ['bɛərə] 袋衬，背带
clean finish [kli:n 'finiʃ] 净加工，还口
over-lock stitch ['əuvə-lɔk stitʃ] 包缝线迹
edge-stitched [edʒ- stitʃ] 缝边线迹（广东话：缉边线）
pocket-bag ['pɔkit-bæg] 袋布
front fly [frʌnt flai] 前门襟（广东话：前纽牌）
seam allowance/ S.A. [si:m ə'lauəns] 缝份，缝头（广东话：子口）
belt-loop [belt-lu:p] 襻带（广东话：裤耳）
bartack [bɑ:tæk] 打套结（广东话：打枣）
horizontal [hɔri'zɔntl] 水平的
thread tacking [θred 'tækiŋ] 打线结
waist tag [weist tæg] 吊牌，腰卡
pocket flasher ['pɔkit 'flæʃə] 袋卡
price ticket [prɑis 'tikit] 价格卡
packing method ['pækiŋ 'meθəd] 包装方法
center crease line ['sentə kri:s lɑin] 中心折缝线，中骨线
cardboard ['kɑ:dbɔ:d ] 纸板
button-holing ['bʌtn -'həuliŋ] 打扣眼（广东话：打纽门）
poly-warp [ 'pɔli-wɔ:p] 塑胶袋
side-seam pocket [said -si:m 'pɔkit] 侧口袋
pocket-bag depth ['pɔkit-bæg depθ] 袋布深
single needle ['siŋgl 'ni:dl] 单针
double needle ['dʌbl 'ni:dl] 双针
procedure [prə'si:dʒə] 程序
sticker ['stikə] 标签
packing list ['pækiŋ list] 包装表
solid color ['sɔlid 'kʌlə] 单色
solid size ['sɔlid saiz] 单码
inscription paper [in'skripʃən 'peipə] 横贴纸
case pack label [keis pæk 'leibl] 外箱贴纸
engineering department [endʒi'niəriŋ di'pɑ:tmənt] 工程部
leisure wear ['leʒə weə] 休闲装 *
slant pocket [slɑ:nt 'pɔkit] 侧缝斜袋 *
cross crotch [krɔs krɔtʃ] 十字裆 *
lined crotch [lɑind krɔtʃ] 一字裆 *
fish dart [fiʃ dɑ:t] 鱼形褶，腰间橄榄省 *
box-pleat [bɔks-pli:t] 箱形褶裥（广东话：工字褶）*
inverted-pleat [in'vətid-pli:t] 暗褶（广东话：内工字褶）*
tuck [tʌk] 裥 *
pin tuck [pin tʌk] 针裥 *
accordion pleat [ə'kɔ:djən - pli:t] 风琴褶 *

sunray pleat ['sʌnrei pli:t] 太阳褶 *
gathers ['gæðəz] 折裥，皱褶（广东话：碎褶）*
frill [fril] 小绉边褶 *
ruffle ['rʌfl] 大皱边，皱褶 *
seam pocket [si:m 'pɔkit] 摆缝袋，缝骨袋 *
straight pocket [streit 'pɔkit] 直袋 *
tape [teip] 卷尺，胶带，带条
curved pocket [kə:vd 'pɔkit] 弯袋 *
coined/ cashed pocket [kɔind /kæʃt 'pɔkit] 裱袋 *
jetted/ welt pocket ['dʒetid / velt 'pɔkit] 滚边袋 *
hexagonal pocket [hek'sægənəl 'pɔkit] 六角袋 *
“J” shaped pocket [j ʃeipt 'pɔkit] “J” 型袋 *
patch pocket [pætʃ 'pɔkit] 贴袋 *
peach shaped pocket [pi:tʃ ʃeipt 'pɔkit] 杏型袋 *
ruler shaped pocket ['ru:lə ʃeipt 'pɔkit] 曲尺袋 *
round cornered pocket [raund 'kɔ:nəd 'pɔkit] 圆角袋 *
bellows pocket ['beləuz 'pɔkit] 风琴袋 *
square shaped pocket [skwεə ʃeipt 'pɔkit] 方角袋 *
three pointed pocket [θri: 'pɔintid 'pɔkit] 三尖袋 *
zipper pocket ['zipə 'pɔkit] 拉链袋 *
concealed zipper [kən'si:ld 'zipə] 隐形拉链
covering stitch ['kʌvəriŋ stitʃ] 绷缝线迹 *
chain stitch [tʃein stitʃ] 链式线迹 *
blind stitch [blɑind stitʃ] 暗缝线迹 *
wrap-seam [ræp - si:m ] 包缝缝型
zig-zag stitch ['zigzæg stitʃ] Z 形线迹
joining crotch ['dʒɔiniŋ krɔtʃ] 接缝裤裆 *
pins/ clips [pinz / klips] 针，夹 *
folding size [fəuldiŋ saiz ] 折衣尺寸 *
pocket flap ['pɔkit flæp] 袋盖 *
flat-seam [flæt -si:m] 平缝 *
fell seam [fel si:m] 对折缝（广东话：埋夹缝）*
knife pleat [nɑif pli:t] 剑褶 *

## Exercises

1. Translate the following terms into Chinese.

（1）center crease line
（2）bearer
（3）clean finish
（4）over-lock
（5）engineering
（6）pocket-bag
（7）front fly
（8）side panel
（9）seam allowance
（10）belt-loop
（11）bartack
（12）catch facing
（13）cross crotch
（14）pocket opening
（15）straight pocket
（16）production order
（17）slant pocket
（18）casual wear
（19）leisure wear
（20）coined pocket
（21）pleat
（22）dart
（23）box-pleat
（24）inverted-pleat
（25）patch pocket
（26）tuck
（27）pin tuck
（28）accordion pleat
（29）sunray pleat
（30）bellows pocket
（31）packing method
（32）seam pocket
（33）three pointed pkt.
（34）garment wash
（35）concealed zipper
（36）waistband
（37）thread tacking
（38）inseam
（39）waist tag
（40）size assortment
（41）pocket flasher
（42）carton

(43) price ticket
(44) frill
(45) plaids
(46) matching color
(47) tape
(48) cardboard
(49) design sketch
(50) button-holing
(51) poly-warp
(52) solid color
(53) solid size
(54) double needles
(55) case pack label
(56) wrap-seam
(57) zig-zag stitch
(58) packing list
(59) covering stitch
(60) chain stitch
(61) blind stitch
(62) pocket flap
(63) facing
(64) flat-seam
(65) fell seam
(66) accessory
(67) buckle loop
(68) stitch type
(69) shipping mark

2. List the major information shown in the production order.
3. Try to prepare a sample order based on the following production sketches (Closed Fitting Jeans).

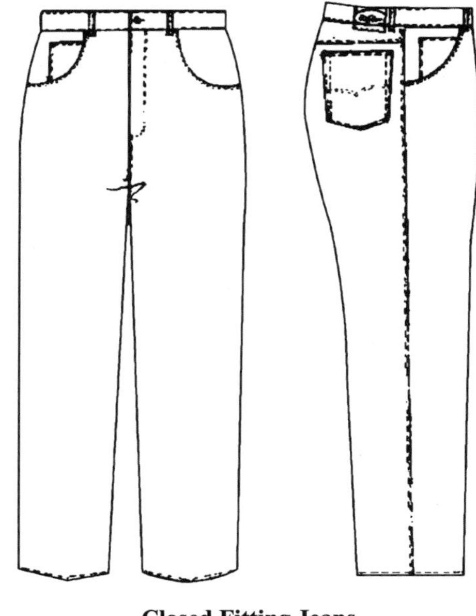

**Closed Fitting Jeans**

## 3.3 Making up Men's Shirt 男式衬衫制作

### 3.3.1 Description of Dress Shirt 款式描述

A man's shirt is a garment for the upper body, it is a short or long sleeved shirt with a dress collar, and long sleeves with cuffs, a full vertical opening with buttons or snaps (Figure 3-1). North Americans would call that a "dress shirt". It is generally made in a solid color, but it may have a striped or checked pattern. ① It is usually worn with a neck-tie for business, street or semiformal purpose.

①男装衬衫是一件上身服装，它是一件短袖或长袖且有传统领子的衬衫，且长袖有袖级，有纽扣或按钮的全长直开口。在北美叫"礼服式衬衫"，一般是素色，但也可能是条纹或格纹。

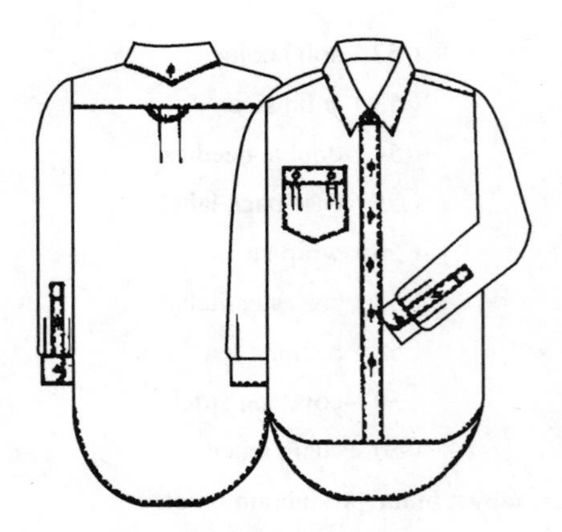

Figure 3-1 Men's Shirt

### 3.3.2 Application of Sewing Machines 缝纫机的应用

The sewing machine has greatly improved the efficiency and productivity of the clothing industry. Industrial sewing machines, by contrast to domestic machines, are larger, faster, and more varied in their size, cost, appearance and task. In shirt making up, there are major semi-automatic equipments.

（1）Single needle lockstitch machine with thread trimmer.②

②带剪线刀的单针平缝纫机（车）。

（2）Double needle lockstitch machine.

（3）3 or 5 threads overlock machine.

（4）Bar tacker.

（5）Button-holing machine.

（6）Single chain-stitch button-sewer.

（7）Chain stitch machine.

（8）Cuff turning & pressing machine.

（9）Pocket creasing machine.

（10）Collar pressing & turning machine.

（11）Sewing attachment: guide, hemmer, piper & hemmer foot, binder and folder, etc.

### 3.3.3 Sewing Sequence 缝纫工序

（1）Semi-assembling section.

● Pocket

Op.1 serge pocket mouth

Op.2 crease pocket shape

● Collar

Op.3 sew top & under collar with interlining

* sew leaf edge and collar end

* topstitching collar edge

Op.4 sew inside and outside stand with interlining[3]

Op.5 join collar and neckband

● Cuff

Op.6 sew cuff with interlining

* topstitching cuff edge

Op.7 set button holes

Op.8 sew buttons

● Front

（Right side）

Op.9 fold front edge

Op.10 sew buttons

（Left side）

Op.11 sew placket with interlining

Op.12 sew button-holes

Op.13 set pocket

● Back

Op.14 set pleat & loop tab

Op.15 sew yoke

● Sleeve

Op.16 sew placket

Op.17 fix pleat

（2）Assembling section.

Op.18 join shoulder with front and back part

*topstitching

Op.19 attach collar with label

Op.20 set in sleeves

Op.21 close side and sleeve seam

Op.22 attach cuff

Op.23 sew neckband buttons

Op.24 sew neckband button-holes

Op.25 sew hem

③将领面、领底与里衬车缝在一起。

## 3.4 Pants Construction 裤子结构

### 3.4.1 Specification Descriptions 规格描述

The garment model is the type of pants indicate following details:

（1）Jeans constructed waistband.

（2）Two open reverse pleats.

（3）On-seam front pockets.

（4）Two single jetted back pockets, left as worn with button through flap.④

（5）Double back darts with edge-stitching.

（6）“Rykiel” label set above right as keep back pocket.

（7）3.2cm (1 1/4" ) cuffed bottom hem.

（8）Pattern fits to the style #168.

④两个单嵌线（单唇）后袋，效果如穿起时，左边袋用纽扣扣起袋盖。

### 3.4.2 Construction Details 部件细节

（1）Waistband: a piece of jeans constructed waistband.

（a）Self-fabric waistband, folded with fusible non- woven interlining.

（b）Finished waistband is 3.8cm（1 1/2"）wide, 1.6mm（1/16"）stitch margin at top and bottom of waistband. ⑤

⑤完成后的裤腰宽 是 3.8cm（1 1/2 英寸），且在裤腰顶端和底部缉 1.6mm (1/16 英寸) 边线。

（c）12 SPI. Button-holes, located at left center front.

（2）Belt-loops: six belt-loops:

（a）2 pcs. at front, 1cm（3/8"） × 5.1cm（2"）finished dimension with 3/16" margin top-stitching. Located adjacent to first pleat.

（b）4 pcs. at back, 3/8" × 2" finished dimension with edge-stitching, and 2 pcs. are placed on the back, 1.3cm（1/2"）from side seam. 2 pcs. are placed on back 7.6cm（3"）from center back rise seam.

（c）Belt-loops set into waistband seam, dropped 1.3cm（1/2"）with hidden bartack to pants.

（d）Horizontal bartack to top of belt-loop.

（3）Left fly: self fabric facing.

（a）Fly facing fuse with fusible non- woven interlining, edge overlocked.

（b）Zipper tape stitched to fly with double needle, and fly joined to body and single needle edge stitched through all ply, 1.6mm（1/16"）margin. ⑥

⑥将拉链带布用双针缝于门襟上，并用单针穿过所有裁片，将门襟缝合于裤身上，缉 1.6mm（1/16 英寸）边线。

（c）Single needle & “J” shape stitched with 3.8cm（1 1/2"）margin to fly.

（d）YKK metal zipper with antique brass coating.

（4）Right facing: one piece of self fabric facing.

（a）Catch facing folded for 2 ply, no interlining, and edge overlocked.

（b）Join to body with zipper tape. Single needle edge -stitching 1.6mm（1/16"）from tape through all ply.

（c）One horizontal bartack located at base of "J" shape stitched fly and one vertical bartack located at 2.5cm（1"）above base. ⑦

⑦一个水平套结位于"J"形门襟缉线的底部，并有一个垂直套结位于门襟底部2.5cm (1 英寸)以上的位置。

（5）Pleat: two reverse pleats open to waistband.

（a）The pleat faces the side seam. First pleat depth is 3.2cm（1 1/4"）, Second pleat depth is 1.9cm（3/4"）, with 3.2cm（1 1/4"）between pleats.

（b）First pleat placement graded by size, see size specs.

（6）Front pockets: on-seam front pockets.

（a）Seam pockets with 1cm（3/8"）single needle top-stitching. Horizontal bartack at top and bottom of pocket opening.

（b）Pocket bags with self-fabric top and bottom facing. Facing edge overlocked and single needle stitched to pocketing.

（c）Pocket placed 2.5cm（1"）down from bottom of waistband seam16.5cm（6 1/2"）finished opening. 2 pocket bags, and pocketing seam joined with overlocking, turned and single needle 0.6cm（1/4"）margin for clean finish interior. Pocket bag dept is 30.5cm（12"）（from bottom of waistband to bottom of pocketing）.

（7）Front rise: seam joined with 5-thread safety overlock seam, and edge stitched with single needle.

（8）Back rise: seam joined with 5-thread safety overlock seam. Back rise with no exterior top-stitching.

（9）Back pocket: two single jetted pockets, left pocket as worn with a button through flap.

（a）Pocket located at 6.4cm（2 1/2"）from waistband joined at seam to start pocket. Self-fabric facing at top and bottom with edge overlock and single needle edge-stitching to pocketing. ⑧ Pocket bags closed with overlock seam. Pocket bag length is 22.9cm（9"）from finished waistband seam.

⑧衣袋位于裤腰缝合线下6.4cm（2 1/2英寸）起的位置，作为口袋起点位。原身布袋贴的顶部和底部锁边，并用单针边线缝合在袋布上。

（b）Single needle edge stitched all around the pocket, 1.6mm（1/16"）margin. Vertical bartacks located at each corner of the pocket.

（c）2 plies pocket flap with light-weight interlining. Single needle edge stitched all around. Flap set into top pocket seam and single needle edge- stitching. Vertical button- hole located at center of flap and beginning 1.3cm（1/2"）from edge of pocket flap. Vertically stitched button sew.

（d）"Rykiel" label placed above right as worn back pocket centered between double darts. All edges folded under and single needle edge stitched four sides 1.6mm（1/16"）margin. Thread color match label ground. Finished dimension is 2.8cm（1 1/8"）（Width）×2.5cm（1"）(High）.

（10）Dart: two back dart on each side.

（a）Darts spaced 1.9cm（1 3/4"）at waistband seam and 5.7cm（2 1/4"）at top of pocket mouth seam. ⑨

⑨省道在裤腰缝合线的距离为1.9cm (1 3/4英寸），在袋口上端缝和线的距离为5.7cm (2 1/4英寸)。

（b）Each dart with single needle and edge stitched on outside edges.

（11）Out-seam: seam joined with 5-thread safety overlock and turned towards back of pants. Edge-stitching with single needle.

（12）Inseam: seam joined with 5-thread safety overlock stitch, no top-stitching.

（13）Hem: finished cuff height is 3.2cm（1 1/4"）. Open bottom with overlock edge, then single turned and single needle top stitching for 5.7cm（2 1/4"）pre-hem height. Pre-hem single turned to the outside to form 3.2cm（1 1/4"）finished cuff. Single needle vertical tacks at inseam and outseam to secure cuff.

（14）Finished: press entire garment with pleats. First pleat is in line with center crease line（Figure 3-2）.

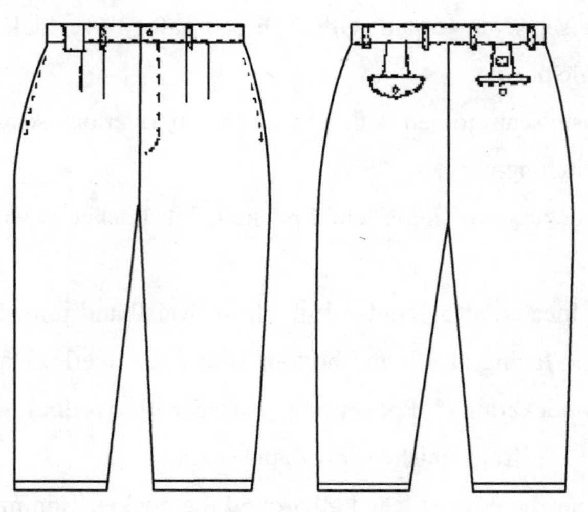

**Figure 3-2 Pants**

## 3.4.3 Accessory Details 辅料细节

（1）Pocket: 65% polyester / 35% cotton. Off- white pocketing for all colors except using white on white. Fabric construction: 22.2tex（45S）, density is

110 × 76.

（2）Interlining: light-weight, fusible non-woven interlining. For waistband, pocket welts, pocket flap, and left fly.

（3）Thread: dyed to match body color, cotton wrapped with polyester.

（4）Main label: woven label with edge turned and pressed, single needle sewn to interior ply, located at center back waistband. ⑩ Thread color to match label.

⑩折烫“机织商标”的边位，用单针缝合于后中裤腰内层位。

（5）Care label/ Size label/ Content label: the label folded and inserted into center back, bottom of waistband seam.

### 3.4.4 Packing Method 包装方法

（1）Fold pants: folded in half along front crease then in half at knee. 27.9cm × 50.8cm（11" × 20"）cardboard folding board with foam strip placed inside fold. Right as worn side should be on exterior of fold to show hang-tag.

（2）Hang-tag: “Rykiel” printed and placed on right as worn waistband, centered over back pocket. Two bartacks placed in top corners at 0.6cm（1/4"）from edge.

（3）Bar coded sticker: placed at the back right corner of polybag in lower side.

（4）Polybag: plain polybag, 47cm × 69.2cm（18 1/2" × 27 1/4"）with open end securely sealed.

（5）Shipping carton: 44.5cm（*W*）× 68.6cm（*L*）× 30.5cm（*H*）（17 1/2"*W* × 27"*L* × 12"*H*）heavy weight cardboard. Quantity 24 pcs./ carton. The weight is not to exceed 18.1kg（40lbs）.

## *Words and Expressions*

making up ['meikiŋ ʌp] 制作

dress shirt [dres ʃə:t] 传统男衬衫

pattern ['pætən] 图样，纸样

solid color ['sɔlid kʌlə] 净色，单色

single needle lockstitch machine ['siŋgl 'ni:dl 'lɔkstitʃ mə'ʃi:n] 单针平缝机

flat M/C [flæt em/si:] 平缝机

thread trimmer [θred 'trimə] 剪线刀

double needle lockstitch machine ['dʌbl 'ni:dl 'lɔkstitʃ mə'ʃi:n] 双针平缝机

overlock machine ['əuvə'lɔk mə'ʃi:n] 包缝机（广东话：锁边机）

bar tacker [bɑ: 'tækə] 打结机（广东话：打枣车）

button holing machine ['bʌtn 'həuliŋ mə'ʃi:n] 扣眼机

button sewer ['bʌtn 'sjuə] 钉扣机

chain stitch machine [tʃein stit' mə'ʃi:n] 锁链机

cuff turning & pressing machine ['kʌf 'tə:niŋ ənd 'presiŋ mə'ʃi:n] 翻袖口机

pocket creasing machine

['pɔkit kri:s mə'ʃi:n] 烫袋机
collar pressing & turning machine ['kɔlə 'presiŋ ənd 'tə:niŋ mə'ʃi:n] 翻烫领机
attachment [ə'tætʃmənt] 附件
guide [gɑid] 比尺
hemmer ['hemə] 卷边器
binder ['baində] 滚边器
folder [fəuldə] 折边器
hemmer foot ['hemə fut] 卷边压脚
piper [pɑipə] 镶边器
sewing sequence ['səuiŋ 'si:kwəns] 缝纫次序
assembling section [ə'sembl 'sekʃən] 合成部分
serge [sə:dʒ] 哔叽，锁毛边
crease pkt. shape [kri:s 'pɔkit ʃeip] 烫袋形
Op./ Operation [ɔpə'reiʃən] 工序
top & under collar [tɔp ənd 'kɔlə] 底面领
top stitching [tɔp 'stitʃiŋ] 面缝线迹（广东话：缉面线）
run stitching [rʌn 'stitʃiŋ ] 初缝线迹（广东话：运线）
edge stitching [edʒ 'stitʃiŋ] 缝边线迹（广东话：缉边线）
collar band/ neckband/ collar stand ['kɔlə bænd / 'nekbænd / 'kɔlə stænd] 立领
collar fall ['kɔlə fɔ:l] 翻领
set pocket [set 'pɔkit] 装袋
fix pleats [fiks pli:ts] 打褶，打裥
yoke [jəuk] 育克（广东话：担干，机头）
placket ['plækit] 开襟，袖衩（广东话：明筒、三尖袖衩）
join shoulder seam [dʒɔin 'ʃəuldə si:m] 缝肩线
set in sleeve [set in sli:v] 绱袖
attach collar [ə'tætʃ 'kɔlə] 绱领
attach cuff [ə'tætʃ kʌf] 绱袖口
sew hem/ hemming [sju: hem /'hemiŋ] 缝合下摆
constructed [kən'strʌktid] 制作，装配
cuffed bottom hem [kʌft 'bɔtəm hem] 翻贴边裤脚
waistband ['weistbænd] 腰头
non-woven ['nɔ-'wəuvən ] 非机织物
fusible ['fju:zəbl] 黏合性的
interlining ['intə'lainiŋ] 里衬
margin ['mɑ:dʒin] 边缘
SPI/ stitch per inch ['stitʃ pə: intʃ] 每英寸线迹数
bartack [bɑ:tæk] 打套结（广东话：打枣）
dimension [di'menʃən] 尺寸
horizontal [hɔri'zɔntl] 水平的
vertical ['və:tikəl] 垂直的
fly facing [flɑi 'feisiŋ] 门襟，遮扣贴边（广东话：纽牌）
antique [æn'ti:k] 旧式的，古代的
brass coating [brɑ:s 'kəutiŋ] 镀黄铜的
catch facing [kætʃ 'feisiŋ] 拉链贴（广东话：纽子）
zipper tape ['zipə teip] 拉链带
overlock ['əuvə'lɔk] 包缝（广东话：锁边）
pocketing/ pocket bag ['pɔkitiŋ / 'pɔkit bæg] 袋布
exterior [eks'tiəriə] 外面，外部
interior [in'tiəriə] 内部，里面
single jetted pocket ['siŋgl 'dʒetid 'pɔkit] 单嵌线袋
clean finish [kli:n 'finiʃ] 还口
5 thread safety overlock stitch [faiv θred 'seifti 'əuvə'lɔk stitʃ] 五线保险包缝线迹
W./ width [widθ] 宽度
H./ height [hɑit] 高度
L./ length [leŋθ] 长度
single needle stitch ['siŋgl 'ni:dl stitʃ] 单针线迹
out-seam/ side seam [ɑut-si:m / said si:m] 侧骨，外缝
hem [hem] 卷边，脚口折边
center crease line ['sentə kri:s lɑin] 中缝线
off-white [ɔ:f-wait] 非纯白色，黄白色

white on white [wait ɔn wait] 白底上有白花样

fabric construction ['fæbrik kən'strʌkʃən] 面料组织结构

polyester/ cotton [' pɔliestə / 'kɔtn] 涤棉混纺物

main label [mein 'leibl] 主标，主唛

care label [keə 'leibl] 洗水标，洗唛

size label [saiz 'leibl] 尺码标，尺码唛

content label ['kɔntent, 'leibl] 成分标，成分唛

fold pants [fəuld pænts] 折叠裤子

hang-tag [hæŋ- tæg] 吊牌

bar coded sticker [bɑː(r) kəudid 'stikə] 条形码标签

sealed [siːld] 封口的

plain [plein] 平的，平纹布

lbs/ pounds [pɑundz] 磅（重量）

shipping carton ['ʃipiŋ 'kɑːtən] 装运箱，出口箱

cuff-less bottom ['kʌf-les 'bɔtəm] 不翻边裤脚 *

short-ship [ʃɔːt-ʃip] 短数，少出货 *

water streak ['wɔːtə striːk] 洗水痕 *

collar turner ['kɔlə 'təːnə] 反领机 *

zig-zag lockstitch machine ['zig-zæg 'lɔkstitʃ mə'ʃiːn mə'ʃiːn] 人字平缝机 *

blinding ['blaindiŋ] 挑缝（广东话：挑脚）*

## Exercises

1. Translate the following terms into Chinese.

（1）piper
（2）dress shirt
（3）sewing sequence
（4）pattern
（5）semi-assembling
（6）solid color
（7）single needle
（8）lockstitch
（9）serge
（10）thread trimmer
（11）crease pkt. shape
（12）under collar
（13）overlocking
（14）top stitching
（15）bar-tacker
（16）run stitching
（17）button holes
（18）edge stitching
（19）neckband
（20）button sewer
（21）collar stand
（22）chain stitch
（23）collar fall
（24）cuff pressing
（25）pocket
（26）fix pleats
（27）yoke
（28）placket
（29）join shoulder seam
（30）set in sleeve
（31）attachment
（32）attach cuff
（33）hemmer
（34）hemming
（35）binder
（36）zig-zag lockstitch
（37）folder
（38）hemmer foot
（39）blinding stitch
（40）handling
（41）stitch formation
（42）thread
（43）stitch type
（44）mechanism
（45）fabric width
（46）shrink-resistant
（47）cuff bottom hem
（48）cotton plaids
（49）clean finished
（50）care label
（51）outseam
（52）bartack
（53）interlining
（54）jetted pkt.
（55）waistband
（56）jeans
（57）fell seam
（58）hang-tag
（59）waist-tag
（60）zipper
（61）knitted fabric
（62）suiting
（63）bust
（64）bottoms
（65）shorts
（66）cuff-less bottom

(67) polyester
(68) ironing
(69) pocket bag
(70) specification
(71) shipping mark
(72) shipping carton
(73) short-ship
(74) water streak
(75) raw material
(76) polybag
(77) apparel
(78) grain

2. List 4 types of sewing attachments.
   Example: Folder.
3. List 10 pieces of garment parts of 5-pkt jeans.
   Example: Waistband, Belt-loop, etc.
4. Illustrate a men's shirt with production sketch and then construct the manufacturing sequences.

## 3.5 Flow Chart of Garment Manufacturing 服装制作流程图

### 3.5.1 Flow Chart of the Men's Trousers 男西裤工艺流程图 (Figure 3–3)

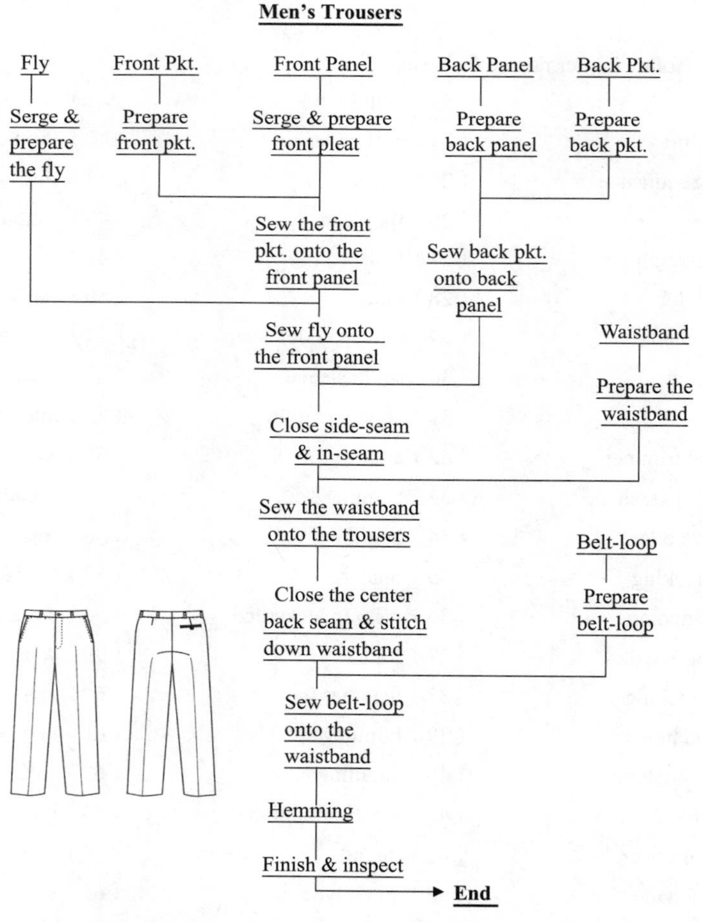


Figure 3-3 Flow Chart of the Men's Trousers

### 3.5.2 Flow Chart of the T-Shirt T 恤衫工艺流程图( Figure 3-4 )

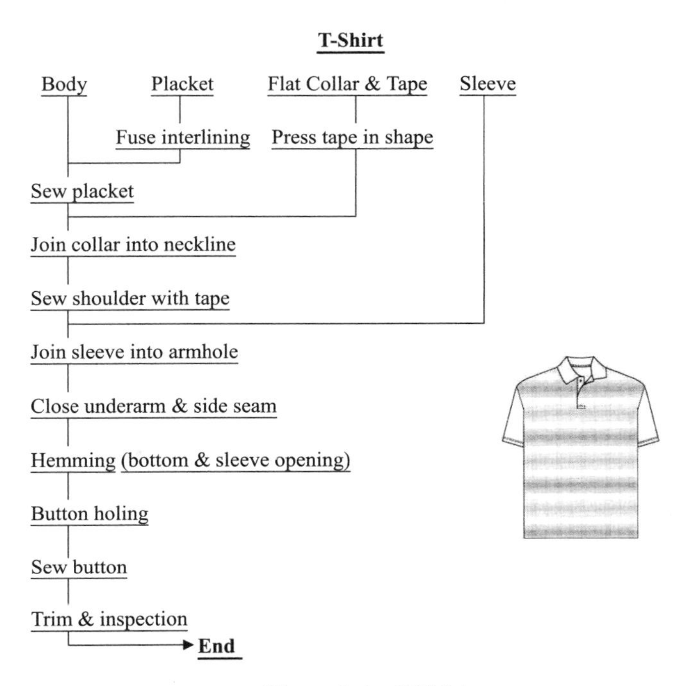


**Figure 3-4 T-Shirt**

### 3.5.3 Flow Chart of Fully Fashioned Sweater 羊毛衫工艺流程图( Figure 3-5 )

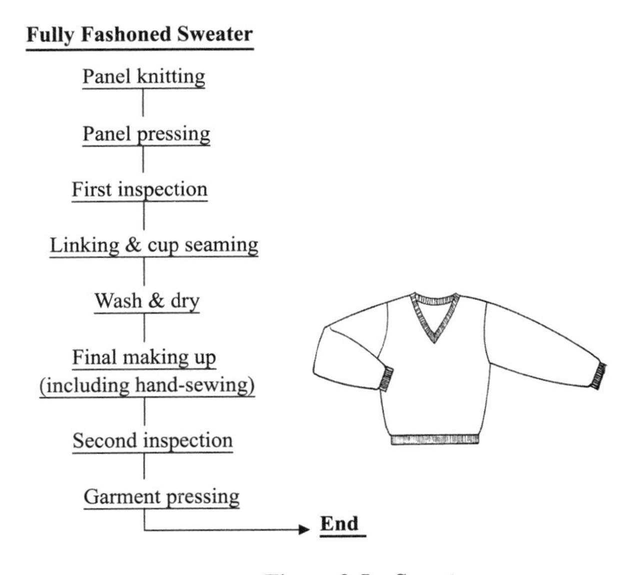


**Figure 3-5 Sweater**

### 3.5.4 Flow Chart of the Women's Blouse 女装衬衫工艺流程图（Figure 3–6）

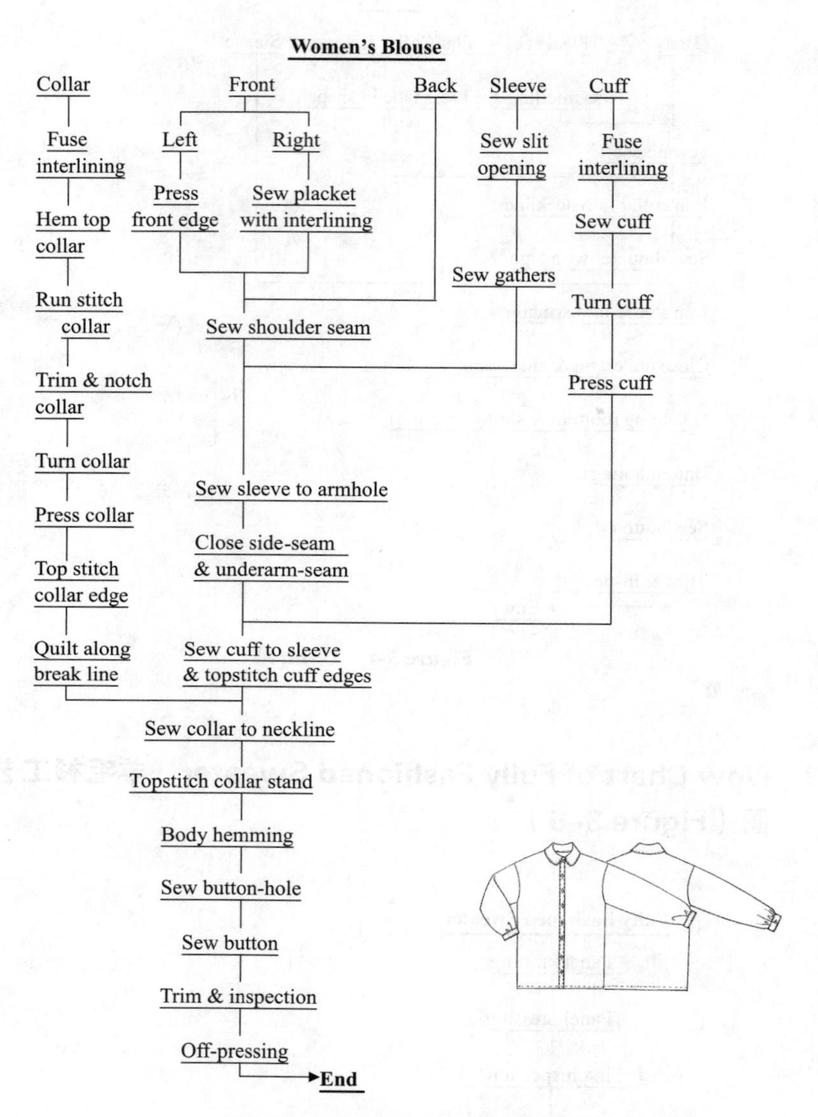


**Figure 3-6 Women's Blouse**

### 3.5.5 Flow Chart of the Jeans 牛仔裤工艺流程图（Figure 3–7）

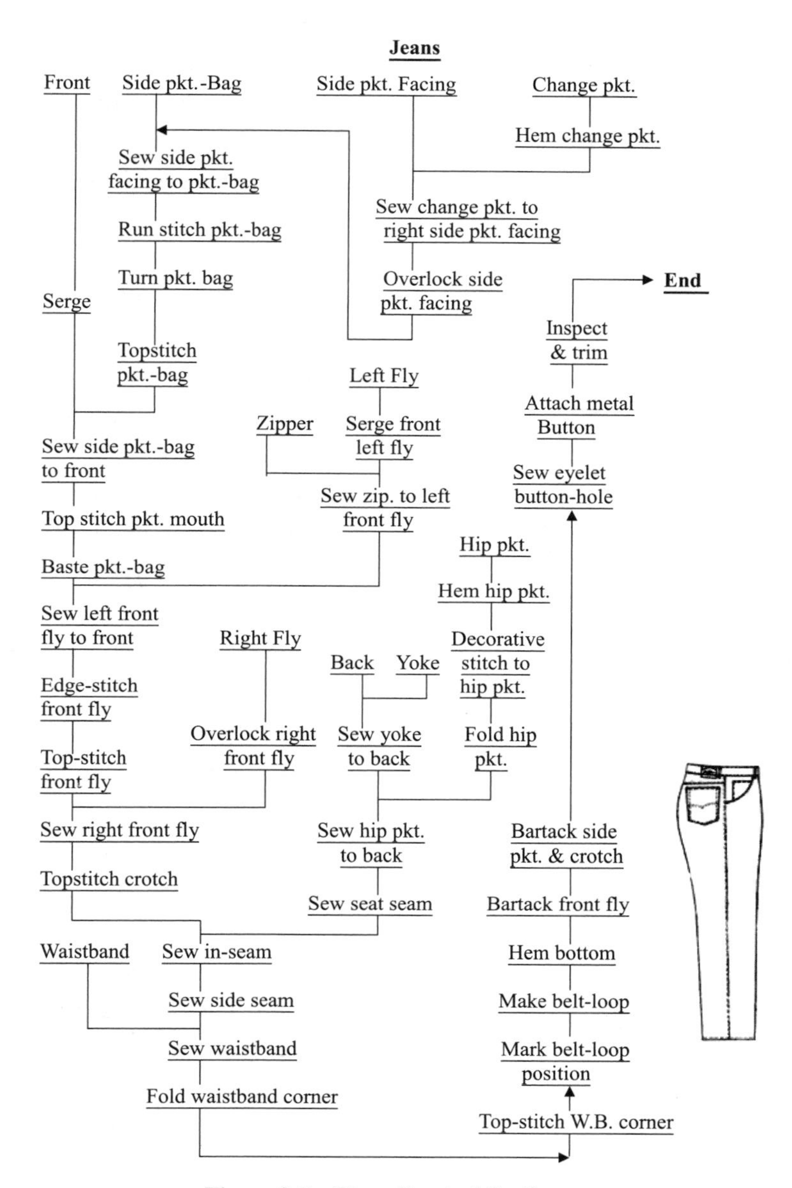


**Figure 3-7 Flow Chart of the Jeans**

## Words and Expressions

flow chart [fləu tʃɑ:t] 流程图
fly [flɑi] 门襟（广东话：纽牌）
front panel [frʌnt 'pænl] 前幅，前片
back panel [bæk 'pænl] 后幅，后片
waistband/ W.B. [weistbænd] 裤腰，腰头
close side-seam [ kləuz, said-si:m] 缝合侧缝（广东话：埋侧骨）
in-seam [in-si:m] 内接缝（广东话：内浪、内长）
belt-loop [belt-lu:p] 襻带，裤襻（广东话：裤耳）
hemming ['hemiŋ] 卷边（广东话：车脚）
pkt./ pocket ['pɔkit] 口袋
T-shirt/ Tee shirt ['ti:-ˌʃə:t] T 恤衫
Tape [teip] 带条
rib tape [rib teip] 罗纹带条
inspect [in'spekt] 检查
trim [trim] 剪线
fully fashioned sweater ['fuli 'fæʃənd 'swetə] 全成型羊毛衫
panel knitting ['pænl 'nitiŋ] 织片
panel pressing ['pænl 'presiŋ] 烫片
linking & cup seaming ['liŋkiŋ ənd kʌp si:miŋ] 缝盘缝制
notch [nɔtʃ] 刀口，刻口（广东话：扼位）
slit opening [slit 'əupəniŋ] 狭长开口（广东话：袖侧）
gathers ['gæðəz] 碎裥（广东话：碎褶）
quilt [kwilt] 衍缝
break line [breik lɑin] 衍缝线，驳口线（广东话：反襟线）
underarm-seam ['ʌndərɑ:m-si:m ] 袖底缝迹
under-pressing ['ʌndə-presiŋ ] 中烫
neckline ['neklain] 领圈线
off-pressing/ final pressing [ɔ:f-presiŋ /'fɑinl presiŋ] 终烫
bartack [bɑ:tæk] 套结（广东话：打枣）
seat seam [si:t si:m] 后裤裆缝（广东话：后浪骨）
jeans [dʒeinz] 牛仔裤
baste [beist] 假缝，粗缝
corner ['kɔ:nə] 边角位
final inspection ['fɑinl in'spekʃən] 终检 *
in-process inspection [in-'prəuses in 'spekʃən] 中检 *
covering stitch ['kʌvəriŋ stitʃ] 拉覆线步 *
join crotch [dʒɔin krɔtʃ] 接缝裤裆 *
cross crotch [krɔs krɔtʃ] 十字裤裆 *
fold back facing [fəuld bæk 'feisiŋ] 原身出贴 *
elastic waistband [i'læstik weistbænd] 松紧带裤头 *
back stitch [bæk stitʃ] 回针 *
press open seam [pres 'əupən si:m] 烫开缝骨 *
right side/ R.S. [rait said] 正面 *
wrong side/ W.S. [rɔŋ said] 反面 *
hemming foot ['hemiŋ fut] 包边压脚 *
embroidery [im'brɔidəri] 绣花 *
6 panel skirt [siks 'pænl skə:t] 6 幅裙 *
side panel [said 'pænl] 侧面嵌片（广东话：小身，侧片）
clean finish/ turn finish [kli:n 'finiʃ / tə:n 'finiʃ] 还口

## Exercises

1. Translate the following terms into Chinese.

(1) flow chart
(2) fold line
(3) fly
(4) front panel
(5) underarm-seam
(6) back panel
(7) under-pressing
(8) side panel
(9) neckline
(10) panels skirt
(11) off-pressing
(12) waistband
(13) close side-seam
(14) bartack
(15) belt-loop
(16) hemming
(17) final inspection
(18) pocket
(19) Tee shirt
(20) in-line inspection
(21) stitch
(22) adhesive tape
(23) rib tape
(24) join crotch
(25) cross crotch
(26) facing
(27) sweater
(28) panel knitting
(29) basting
(30) back stitch
(31) press open seam
(32) clean finish
(33) jeans
(34) slit opening
(35) gathers
(36) quilting
(37) front fly
(38) catch facing
(39) box pleat
(40) stud
(41) metal-ware
(42) pocket opening
(43) template
(44) cutting piece
(45) slacks
(46) close fitting
(47) pants
(48) fastening
(49) design sketch
(50) crotch point
(51) bias cut
(52) remnant
(53) stripe matched
(54) buckle

2. Write out the operation flow chart for men's shorts (Men's Shorts).

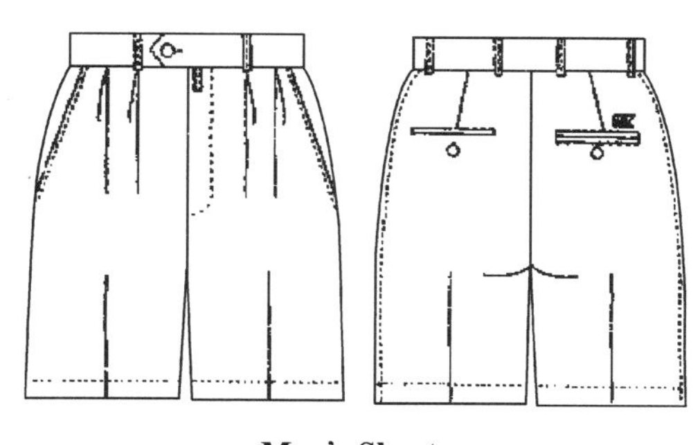

**Men's Shorts**

# 3.6 Garment Part Explaining 部件解释

## 3.6.1 Garment Components 服装部件

(1) Collar: neckband upright or turned over of coat, dress, shirt etc. It is a completing upper part of garment. The collar should be classified into several types, such as shirt collar, tailor collar, stand collar, roll collar and flat collar.

（2）Cuff: a band applied to the end of the sleeve to give a finished appearance, for example, an extra layer of fabric at the lower edge of the sleeve in shirt, coat, jacket.

（3）Fall: the outside or face side of a collar, but collar stand is the inside part of the collar that rises from the garment. Usually found collar fall and collar stand in the shirt collar.

（4）Flap: covering for pockets, so it is named pocket cover.

（5）Fly: front opening of trousers may be buttoned/ zippered.

（6）Fly Fastening: concealed fastening as on coat or jacket, where an underlay of material buttons to the other side of the garment.

（7）Gusset: a triangular piece of fabric inserted to adjust fit to give ease and shaping, usually found in a Kimono or knickers.

（8）Godet: an inset triangular piece of material.

（9）Gore: panel set in a skirt, e.g. gored skirt/ panel skirt.

（10）Hem: an edge finish formed by turning under fabric and stitching, named as bottom.

（11）Lapel/ Rever: the part of a jacket or coat that folds from the neck.

（12）Leaf Edge: outer edge of a collar.

（13）Pockets: can be either functional or decorative. There are three basic kinds, namely patch pocket, seam pocket and set in pocket.

（14）Patch Pocket: an extra piece lay on top of garment.

（15）Placket: the opening in a shirt or a jacket made to give enough room for the garment to be put on, plackets can be finished with a zipper, or can be press- stud fastened/ buttoned.

（16）Set-in Pocket: made with a special slit in the garment.

（17）Seam Pocket: make the pocket mouth in seam.

（18）Sleeve: cover arm from the shoulder, it should be divided into set in sleeve and grown on sleeve.

（19）Vent: the lapped finished opening at sleeve, jacket or skirt.

（20）Waistband: material around the top of trousers or skirt.

（21）Yoke: separately made shoulder piece of shirt, coat or blouse, or waist-piece of jeans, or skirt, from which the rest is suspended. ①

①将衬衫、外套或罩衫的肩部分离，或将牛仔裤或裙子的腰部形成悬挂部位。

### 3.6.2 Types of Pleat & Dart 褶的类型

（1）Dart: triangular shaped figure used to suppress fabric in order to give shape to projection on the body. Such as bust, hips, shoulder and elbow, etc.

（2）Frill: a narrow ruffle or edging gathers onto another edge.

（3）Flounce: a gathered strip applied to a garment.

（4）Flare: to widen at the bottom or hemline.

（5）Gather: to draw up a fabric by one or more threads.

（6）Pin-Tuck: a very narrow tuck stitched close to the edge.

（7）Pleat: a fold of fabric used to control fullness.

（8）Accordion Pleat: very narrow straight pleats.

（9）Box Pleat: two flat pleats with edges meeting inside, and the opposite of box pleats is the inverted pleat, the edges meeting outside the garment.

（10）Cartridge Pleat: un-pressed round pleats, usually fairly narrow and used as decoration.

（11）Kick Pleat: a small pleat in a skirt to give ease of walking.

（12）Dior Pleat: to lay an extra piece of fabric under the turned back edges of the seam. The extra fabric is stitched to the garment at the top only, the sides being left free.

（13）Knife Pleat: similar to an accordion pleat but must all face in the same direction.

（14）Sunray Pleat: pleats from a central point radiating to the edge of the garment.

（15）Tuck: a straight fold of material, evenly spaced and stitched..

## *Words and Expressions*

Pkt. flap/ pkt. cover ['pɔkit flæp / 'pɔkit 'kʌvə] 袋盖

fly fastening/ concealed fastening [flɑi 'fɑ:sniŋ /kən'si:ld 'fɑ:sniŋ] 暗门襟（广东话：暗纽牌）

underlay ['ʌndə'lei] 下层

front opening [frʌnt 'əupəniŋ] 前幅开口

gusset ['gʌsit] 衣袖插片

knickers ['nikəz] 灯笼裤

godet/ gore [gəu'det / gɔ:] 三角布

panel ['pænl] 嵌镶片

inset ['in'set ] 插布

lapel/ rever [lə'pel / rev] 襟贴，挂面

neck [nek] 领圈

leaf edge [li:f edʒ] 领边

decorative ['dekərətiv] 装饰的

patch pocket [pætʃ 'pɔkit] 贴袋

seam pocket [si:m 'pɔkit] 缝骨袋

set in pocket [set in 'pɔkit] 嵌入袋

press- stud [pres-stæd] 揿扣（广东话：工

字扣）
fastening ['fɑ:sniŋ] 系结物
slit [slit] 剪开口，衩位
collar stand/ neckband ['kɔlə stænd/ 'nekbænd] 下级领
vent [vent] 衩位
waistband ['weistbænd] 裤腰
blouse [blɑuz] 罩衣
suspender [sə'spendə(r)] 吊带
suppression [sə'preʃən] 收褶
frill [fril] 装饰边
ruffle ['rʌfl] 荷叶型装饰边
flounce [flauns] 裙褶
flare [fleə] 宽摆的
hemline ['hemlain] 底边线
pleat [pli:t] 褶裥（广东话：活褶）
fullness ['fulnis] 松位
accordion pleat [ə'kɔ:djən pli:t] 风琴褶
box pleat [bɔks pli:t] 箱形褶裥（广东话：工字褶）
cartridge pleat ['kɑ:tridʒ pli:t] 子弹褶
inverted pleat [in'vətid pli:t] 暗褶（广东话：内工字褶）
kick pleat [kik pli:t] 开衩（广东话：踢褶）
Dior pleat [diəu pli:t] 迪奥褶
knife pleat [nɑif pli:t] 剑褶，刀褶
sunray pleat ['sʌnrei pli:t] 太阳褶
radiate ['reidieit] 发射
kimono sleeve [ki'məunəu sli:v ] 和服袖

## Exercises

1. Translate the following terms into Chinese.

（1）neckband
（2）dress
（3）garment
（4）sleeve
（5）collar
（6）flap
（7）trousers
（8）concealed fastening
（9）jacket
（10）buttons
（11）fabric
（12）Kimono
（13）material
（14）panel skirt
（15）stitching
（16）lapel
（17）neck
（18）pockets
（19）decoration
（20）patch pocket
（21）seam pocket
（22）placket
（23）zipper
（24）slit
（25）seam
（26）sleeve
（27）shoulder
（28）collar stand
（29）vent
（30）skirt
（31）waistband
（32）yoke
（33）coat
（34）blouse
（35）suspender
（36）dart
（37）bust
（38）hips
（39）hemline
（40）gathering
（41）pin-tuck
（42）fullness
（43）pleats
（44）box pleat
（45）inverted pleat
（46）kick pleat
（47）ruffle
（48）collar fall

2. List the major garment components for following production sketch ( Five-pocket Jeans ).

Example: Waistband

**Five-pocket Jeans**

3. Explain the following garment terminology.

( 1 ) collar
( 2 ) pockets
( 3 ) sleeve
( 4 ) waistband
( 5 ) gather
( 6 ) pin-tuck

# 译文

# 第三章 服装生产

在服装厂，生产部门是通过生产流程来对生产的日程安排和进程负责。生产功能可以分为几个环节：纸样绘制、排料、裁剪、缝纫、包装等。生产活动包括将面料转变为成衣的程序，以达到顾客的需求。生产进度表是为了使生产效率最大化以及确保产量能达到交货排期的要求。在缝纫部，缝纫和整烫设备通过使用各种不同的附件降低操作人员的技术要求，对于产量和质量的提高有重大意义。

## 3.1 样衣制作

样板房编制样板制造单并完成样板的制作，然后将样板送给客户确认。样板的尺寸规格和原材料被列在板单上。

### 3.1.1 制板通知单

样板制造单的内容包括准备日期、注意事项、开始和完成的日期。复印四份做如下分配：一份给销售部，一份给样板房，其他两份给生产部门（表 3–1 ～表 3–3）。

**表 3–1 样板制造单（Ⅰ）**

| 样板制造单 | | | | | |
|---|---|---|---|---|---|
| 款式：育克裙裤 | 订单编号：W-16 | 客户：Tommy | 数量：2 件 | 订单日期：2020-2-1 | |
| 尺码<br>部位 | M | | L | | 生产图 |
| | 标准尺寸 | 实际尺寸与标准尺寸之间的差值 | 标准尺寸 | 实际尺寸与标准尺寸之间的差值 | |
| 腰围 | 68 | −1 | 71 | 0 | |
| 臀围 | 102 | 0 | 106 | +0.5 | |
| 前裆 | 30 | +0.4 | 31 | 0 | |
| 后裆 | 43 | 0 | 44 | 0 | |
| 内长 | 64 | −1 | 65 | -1 | |
| 外长 | 94 | 0 | 96 | -0.5 | |
| 股上围 | 26 | −1 | 27 | 0 | 前视图 后视图 |
| 下摆围 | 56 | 0 | 58 | +1 | |

续表

| 工艺结构 | 布板 |
|---|---|
| 1. 前片省：7cm 长 | |
| 2. 后片省：9cm 长 | |
| 3. 腰头：5cm 宽，缉边线 | |
| 4. 裤襻：4 个裤襻，0.8cm × 4cm | |
| 5. 育克：0.8cm 双针面线 | 颜色：印花布 |
| 6. 前门襟：3.4cm J 型面线 | 结构：平纹 |
| 7. 口袋：16cm 直插袋 | 60 × 52/28S × 28S |
| 8. 下摆接驳：8cm 接驳 | 成分：100% 棉 |
| **辅料** | |
| 纽扣：24L、4 孔塑料扣 | 里料：100% 人造丝，灰色 |
| 缝纫线：PP 603、配色线 | 拉链：尼龙 17cm |
| 里衬：腰头用黏合衬 | 其他：— |
| **后整：成衣洗水** | **熨烫包装：平烫** |
| 部门：板房　　核准：Andy Liu | 制表：Sam Chen |

**表 3–2　样板制造单（Ⅱ）**

| 针织服装样板制造单 | | | |
|---|---|---|---|
| | | | 日期：2020-3-8 |
| 初板：√ | 品牌商标：JEANSWEST | | 款式：圆领 |
| 核准板：— | 季节：秋冬 | | 交货日期 2020-5-20 |
| 款式描述：全羊毛男装长袖圆领针织外套 | | | |
| 衣身编号：JW—008 | 规格编号：AU-090 | | 供应商：DA NAN 工厂 |
| 种类：男装：√ | 女装：— | 尺码：M | 颜色：黑 / 灰 |
| 面料 / 纱线：100% 羊毛 | 重量 / 纱线支数：60 英支 | | 配件：— |
| **辅料** | | | |
| 缝纫线种类：PP 604 | 颜色：黑 / 灰 | | |
| 拉链种类与尺码：— | 长度：— | 颜色：— | |
| 纽扣：塑料 | 号数：30L | 数量：2 粒 | 颜色：黑 / 灰 |
| 损耗扣数量：1 粒 | 位置：左手内侧缝 | | |
| 机织商标：JEANSWEST | 位置：后领圈中位 | | |
| 合同款号商标位置：左手内侧缝，下摆上 4 英寸 | | | |
| 洗水商标（腰围 / 尺码指示）：1）后中领圈内侧左手边<br>2）挂于主商标下端 | | | |
| 价钱牌的位置：用胶针穿挂在洗水商标上 | | | |
| 吊牌位置：用胶针穿挂在主商标与洗水商标上 | | | |
| 其他：— | | | |

续表

| 结构 | | | | |
|---|---|---|---|---|
| 领口 / 领子 | 单针：— | 双针：√ | 领子 / 领脚：√ | |
| 前中： | 单针：— | 双针：— | 宽度：— | |
| 袖窿： | 单针：√ | 双针：— | 锁边：√ | 其他：— |
| 袖子： | 单针：— | 双针：— | 锁边：√ | 其他：— |
| 袖级： | 单针：— | 双针：√ | 宽度：5 cm | |
| 肩缝： | 单针：√ | 双针：— | 锁边：√ | 缝骨前移：2cm |
| **口袋** | | | | |
| 位置：从侧缝计算：— | | 从高肩点计算：17 cm | | |
| 口袋尺寸：:11cm | | 单针：√ | 双针：— | 锁边：— |
| 下摆：直脚 | | | | |
| 后整理：成衣洗水 | | 熨烫：装烫√ | 企装烫 — | |
| 特殊指示：√ | | | | |
| | | | | 跟单员：ALICE KANG 2020-3-8 |

**表 3–3　样板制造单（Ⅲ）**

男装针织
上装尺寸表

日期：2020-3-8

款式描述：100% 全羊毛男装长袖圆领针织外套
款号：JWS-108　　季节：秋冬　　特殊指示：—

| 尺码规格（cm） | GRADING | S | M | L | XL |
|---|---|---|---|---|---|
| A. 肩宽——肩点至肩点 | 2.5 | | 46 | | |
| B. 胸围——袖窿下 2.5cm 测量 | 5 | | 102 | | |
| C. 下摆——圆周测量一圈 | 5 | | 96 | | |
| D. 袖窿 2.5cm 袖宽——扫平后垂直测量 | 1.5 | | 23 | | |
| E. 袖窿弯线测量——圆周一圈 | 1.5 | | 50 | | |
| F. 袖长——从后中量起经过肩点到袖口位 | 2.5 | | 78 | | |
| G. 后中长——不包括领脚 | 1.5 | | 62 | | |
| H. 领长——扣上第 1 个扣测量 | 1.5 | | — | | |
| I. 领口——缝骨之间的距离 | 0.6 | | 20 | | |
| J. 前领深——前领深尺寸 | 0.2 | | 5 | | |
| K. 后领深——后领到高肩点的尺寸 | 0 | | 3 | | |

续表

| L. 领宽——后中测量 | 0 | | — | | |
|---|---|---|---|---|---|
| M. 领脚宽——后中测量 | 0 | | 4 | | |
| N. 袖级——半围或圆周测量 | 1.5 | | 18 | | |
| O. 胸袋位置 1——测量从前中边位到袋边的距离 | 0.2 | | 8 | | |
| P. 胸袋位置 2——测量从高肩点袋袋口的距离 | 0.6 | | 17 | | |
| 跟单员：ALICE KANG 2020-3-8 | | | | | |

## 3.1.2 样板卡

样板卡记录了生产指示。错误的缝纫方法和服装部位也可能提及。更多的正确信息应该被记录在样板卡上，以使每个人都对样板有清晰的认识。并且，员工应该清楚知道样板的所属类型（如销售样、核准板、样本尺寸、船样，等）。

表 3–4 样板卡

| **RYKIEL**<br>**样板：核准板** |
|---|
| 日期：2020-2-1 |
| 参考编号：TQ—010 |
| 款式：五袋款牛仔裤 |
| 客户：Tommy |
| 合同编号：TR-086 |
| 尺码：M |
| 描述：100% 棉女装五袋款牛仔裤 |
| 布料：牛仔布，100% 棉 |
| 备注：附板 |
| 回样 / 初样 / 尺码样 /PP 样 / 核准样 / 船头样 / IC 样 / 营销样 |

## 3.1.3 核准样卡

制制作的方法是由服装类型和价格决定，一般有以下两种制作方式：

（1）分工生产：流水线上的每个工人只负责服装某个部件的制作。

（2）单独整件生产方式：一名工人完成整件衣服的缝纫过程。其他工人负责完成特殊工序，如锁边缝、卷边、扣眼缝等。这种方法通常用来制作高档服装，特别是西装。

在服装厂，工人的薪酬是根据完成服装的件数进行计算。工价的高低是由工序的难易程度和所需时间决定，这叫作计件工资。工人完成某服装零部件的制作后就会剪取相应的工票。在计算工资时，每个工人的工资是由他提交的工票数量决定。

有一些工人负责剪线、检查缝纫疵点、挂衣服、贴吊牌以及将每件衣服装进塑料袋。如果衣服缝纫错误，质检部人员将退回给工厂或承包商进行翻工。生产部门负责批量生产，根据不同的颜色和尺寸来安排生产线。

样衣经确认可进行大货生产以后，应在核准样板上说明各部分结构和生产指示（表3–5、表3–6）。确认样板卡所有资料准确无误是非常重要的。具体信息如下：

（1）客户和工厂。

（2）款号和合同编号。

（3）面料和辅料。

（4）规格尺寸、颜色和测量部位。

（5）服装款式结构和尺寸。

**表3–5　核准样卡**

<table>
<tr><td colspan="5" align="center"><u>核板卡</u></td></tr>
<tr><td colspan="3">男装</td><td colspan="2" align="right">日期：2020-3-3</td></tr>
<tr><td colspan="2">客户：Kmart<br>面料：牛仔布<br>款式描述：修长直筒裤</td><td colspan="2">工厂：DaYang生产厂<br>合同编号：8502UK<br>注解：二次PP板</td><td>款号：TQ—008<br>尺码规格：28~36<br>状态：拒接</td></tr>
<tr><td colspan="5">注意：主要针对生产指示与样板的评语进行记录，审核后供大货生产指导之用。</td></tr>
<tr><td rowspan="2">测量部位</td><td rowspan="2">尺码（cm）</td><td>规格要求</td><td>二次PP板</td><td>修改</td></tr>
<tr><td>32</td><td>32</td><td>32</td></tr>
<tr><td colspan="2">1. 腰围：腰头顶一圈</td><td>88</td><td>88</td><td>OK</td></tr>
<tr><td colspan="2">2. 腰头宽</td><td>4.4</td><td>4.5</td><td>—</td></tr>
<tr><td colspan="2">3. 裤耳长</td><td>5.5</td><td>5.5</td><td>—</td></tr>
<tr><td colspan="2">4. 裤耳宽</td><td>1.2</td><td>1.2</td><td>—</td></tr>
<tr><td colspan="2">5. 前袋边到裤耳的距离</td><td>1.5</td><td>1.5</td><td>—</td></tr>
<tr><td colspan="2">6. 皮带长连D扣</td><td>115</td><td>遗漏</td><td>—</td></tr>
<tr><td colspan="2">7. 皮带宽</td><td>4</td><td>遗漏</td><td>—</td></tr>
</table>

续表

| 8. 裆上 6.5cm 下臀围 | 108 | 109.5 | 减少 1.5cm |
|---|---|---|---|
| 9. 大腿围：裆下 2.5 cm | 64 | 64.6 | 减少 0.6cm |
| 10. 膝围：内长中点位上 5 cm | 45 | 45.6 | 减少 0.6cm |
| 11. 裤脚围 | 45 | 43.5 | 加大 1.5cm |
| 12. 前裆连腰头 | 26 | 25.6 | OK |
| 13. 后裆连腰头 | 29 | 29 | — |
| 14. 裆宽：裆下 2.5cm 从内长位测量 | 14.5 | 15 | — |
| 15. 内长 | 84 | 84.5 | OK |
| 16. 外长连腰头 | 107 | 106.5 | OK |
| 17. 门襟长：从腰头顶位到末端线缉 | 19 | 19 | — |
| 18. 门襟线缉宽 | 4 | 4 | — |
| 19. 前口袋长：沿着侧缝从腰头底部量 | 11 | 11 | OK |
| 20. 前袋口宽：沿腰头测量 | 12 | 12 | — |
| 21. 前袋布宽 | 18 | 18 | — |
| 22. 前袋布长 | 28 | 26 | 加大 1cm |
| 23. 前中位从腰头顶端到表袋口的距离 | 7 | 7 | — |
| 24. 侧缝位从腰头顶端到表袋口的距离 | 7.5 | 7.5 | — |
| 25. 表袋口宽 | 7.5 | 7.5 | OK |
| 26. 从侧缝到袋边位的距离 | 1.5 | 2.1 | 减少 0.6cm |
| 27. 后中位后育克长 | 8 | 8 | OK |
| 28. 侧缝位后育克长 | 3 | 3 | — |
| 29. 后中位育克到后袋口的距离 | 3 | 3 | — |
| 30. 侧缝位育克到后袋口的距离 | 2.5 | 2.5 | — |
| 31. 后袋口宽 | 18 | 18 | OK |
| 32. 后袋底宽 | 15 | 15 | — |
| 33. 后袋中长 | 16 | 16 | — |
| 34. 后袋侧长 | 14.5 | 14.5 | — |
| 日期 | 2020-3-8 | 2020-3-8 | 2020-3-8 |

续表

<table>
<tr><td>

**评语**

1. 下臀围减少 1.5cm，完成后为 106cm。
2. 大腿围减少 0.6cm，完成后为 64cm。
3. 膝围减少 0.6cm。
4. 下摆围加大 0.6cm。
5. 前袋布长加 1cm。
6. 从侧缝到袋边位的距离减少 0.6cm。
7. 请完成所有修改并重新进行 3 次 PP 板批复。

</td><td>

**生产图**

</td></tr>
<tr><td colspan="2">

**大量生产注意事项：**

1. 裆位无论是没有修正，还是做的不正确，由于过多面料，前裆和后裆仍然会起皱，非常难看。在我们批准大货生产之前，前裆与后裆问题必须修正完成。
2. 除了裆位修正外，请改正上表列出的 6 点服装部位。我们想修改下臀部、大腿宽和口袋布。请从后裆位移除 1.5cm 下臀部尺寸，大腿围宽度减少 0.6cm，方法如图所示。同样，我们也想将袋布长变为 27cm。
3. 我们不能批准大货生产，直到我们有好的合身板。请尽快重新提交第三次 PP 板以便进行批复。
4. 为了节省时间，并确保下一件服装正确合体，我们今天将纸样寄给你们。

跟单员：May Liu 2020-3-8　　　　部门经理：CK Zhang 2020-3-8

</td></tr>
</table>

**表 3–6　核准样表**

<table>
<tr><td colspan="5">

**核准样**

**1. 服装结构**

| | 尺码 | 尺寸 |
|---|---|---|
| （1）前袋 | （M） | 5cm × 12cm × — |
| （2）后袋 | （M） | 16cm × 18cm × 14cm |

（3）后育克尺寸　2.5cm 在后裆骨。
　　7cm 在侧缝骨。

（4）后袋的位置
A. 2cm 裤腰下并平行育克边。
B. 9cm 裤腰下并平行腰头底部。
C. 腰头底部下　8cm 在后裆骨。
　　7cm 在侧缝骨。

</td></tr>
<tr><td colspan="5">

**2. 颜色搭配**

</td></tr>
<tr><td>衣身</td><td>面缝线迹</td><td>刺绣</td><td>商标</td><td>金属附件</td></tr>
<tr><td>前幅</td><td>黄色</td><td>红和白</td><td>配色线</td><td>黄铜色</td></tr>
<tr><td>后幅</td><td>黄色</td><td>—</td><td>—</td><td>—</td></tr>
</table>

续表

**3. 面料结构**

（1）名称 / 编码：$\frac{1}{3}$ 斜纹、Z 方向　　布料成分：100% 棉

（2）重量：18 盎司 / 平方英尺　　— 磅 / 打

（3）结构 $\frac{\text{（经纱密度 60）×（纬纱密度 50）}}{\text{（经纱支数 16 英支）×（纬纱支数 16 英支）}}$

**4. 辅料**

| | 接受 | 不接受 |
|---|---|---|
| （1）面缝线迹的缝线号数 | √ | |
| （2）纽扣 / 按扣 | √ | |
| （3）铆钉 | √ | |
| （4）里衬 | √ | |
| （5）里子 | √ | |
| （6）嵌边 | | √ |
| （7）罗纹 | √ | |
| （8）其他 | √ | |

## 3.2 生产制作通知单

### 3.2.1 简介

服装公司的产品裁剪和制造是依据订单或订单预测来安排的，并在采购、生产与产品控制部门范围内进行。

为了生产一款服装，采购部必须负责订购所需材料并确保其货期。生产控制必须安排在工厂的工作计划内，并保证与生产这类产品相关的职能部门了解产品要求。生产部门要确保其生产能力，并且有足够的劳动力和设备。为了完成生产任务，人事部与工程部应该积极参与工作，并且确保所有设备能够运作。

当所有的产品完成并送到仓库后，订单应该通过船运发送给顾客。从此刻起，配送中心开始处理订单。为了满足顾客的需要，配送中心必须了解订单目的地与装船日期。然后准备人员进行分码包装。产品一旦完成，将被转移到仓库。通过制程监控系统跟踪整个生产过程，并将情况告知生产控制部门和仓库。

### 3.2.2 案例分析

通过回顾前面的生产职能和活动，我们就能了解各种职能怎样相互联系，

以及怎样采用计算机系统进行监督和控制，并简化涉及不同工作范围的沟通。生产订单显示了工作程序和其他细节。以下信息将显示：

（1）所需面料和辅料。

（2）款式和数量。

（3）颜色分配。

（4）尺码分配和规格尺寸。

（5）正箱唛和包装方法。

（6）交货期和其他生产细节。

**表3–7 生产制作通知单**

生产制作通知单

便装（男装） 日期：2020-3-2

| 订单编号 | 款式编号 | 款式描述 | 数量 | 交货期 |
|---|---|---|---|---|
| RKY—001 | RYKIEL | 褶裥短裤 | 125打 | 2020-5-9 |

| 尺码规格：SZ—001（常规裤腰） | 颜色编号： | | | | |
|---|---|---|---|---|---|
| 成衣洗水（加软剂石磨） | 30 | 32 | 34 | 36 | 38 |
| 腰围（内量） | 30 | 32 | 34 | 36 | 38 |
| 臀围（裆顶上8cm张开褶裥横量） | 43 | 45 | 47 | 49 | 51 |
| 脾围（裆顶位量度） | 27 | 28 | 29 | 30 | 31 |
| 脚围 | 23 | 24 | 25 | 26 | 27 |
| 前浪（连腰头量度） | 11 | 12 | 13 | 14 | 15 |
| 后浪（连腰头量度） | 16 | 17 | 18 | 19 | 20 |
| 内长 | 9 | 9 | 9 | 9 | 9 |
| 拉链 | 7 | 7 | 8 | 8 | 8 |
| 尺码/颜色分配 | | | | | |
| 海军蓝 | 60 | 240 | 600 | 360 | 240 =1500条 |

正箱唛

便装

模型名称：RYKIEL

颜色：

批量编号：4115—4130—90

销售订单编号：20398

数量：36条

尺码：

纸箱编号：从1号起编码

中国制造

侧箱唛

模型名称：RYKIEL

颜色：

尺码：

毛重：

净重：

纸箱尺寸：

● 注：装箱资料标明的重量必须与纸箱的实际重量一致

**续表**

<table>
<tr><td>布料</td><td colspan="2">布料结构</td><td>布料颜色</td></tr>
<tr><td>格子棉布</td><td colspan="2">68 × 54/16S × 16S　100% 棉</td><td>蓝色</td></tr>
<tr><td>缝纫线</td><td>拉链</td><td>主商标</td><td>纽扣</td></tr>
<tr><td>PP 604</td><td>BC—360 #3 黄铜牙</td><td rowspan="2">RYKIEL</td><td>25 号 RYKIEL</td></tr>
<tr><td>配洗水后颜色</td><td>自动锁 #560</td><td>4 孔蓝色塑胶</td></tr>
<tr><td colspan="4">缝制结构</td></tr>
<tr><td colspan="4">商标位置：<br>● 主商标 LA0100 RYKIEL（酒红色底卡其色字体）：采用商标底色的缝纫线，车缝 4 边，位置在右后袋的右上角（参照图示）<br>● 洗水商标在上、合同 P.O. 和款号商标在下一起摄入后中裤腰内侧（全部商标都是杏色底蓝色字体）</td></tr>
<tr><td colspan="4">前幅：2 个褶裥（参照图示）</td></tr>
<tr><td colspan="4">前袋：<br>● 1/4" 单针侧缝袋，1 1/2" 宽袋贴<br>● 袋衬、袋贴用边线还口车于袋布处<br>● 漂白 T/C 袋布采用全袋布的方式伸到前纽牌位，并用 1/8" 单针还口封袋布底</td></tr>
<tr><td colspan="4">前浪：锁边并缉边线</td></tr>
<tr><td colspan="4">后幅：2 个缉边线的省道（参照图示）</td></tr>
<tr><td colspan="4">后袋：缉边线的单嵌线袋，漂白 T/C 袋布采用 1/8" 单针还口封袋布底</td></tr>
<tr><td colspan="4">后裤裆：5 线锁边，右盖左缉边线</td></tr>
<tr><td colspan="4">侧缝：5 线锁边，后幅盖前幅缉边线</td></tr>
<tr><td colspan="4">内裤裆：5 线锁边</td></tr>
<tr><td colspan="4">裤脚：脚上 1 1/2" 双针面线</td></tr>
<tr><td colspan="4">裤腰：双针车拉一片 1 1/2" 裤腰</td></tr>
<tr><td colspan="4">襻带：<br>● 6 个（2" × 1/2"）双针襻带，全部襻带摄入裤腰缝骨底端，其中：<br>● 2 个在左、右前中裥的附近<br>● 后中处 2 个襻带距离 2 1/2"<br>● 2 个在后幅位置，前幅襻带与后幅襻带的中间（参照图示）</td></tr>
<tr><td colspan="4">套结：<br>● 襻带上下共 12 个（襻带下端用隐形套结）<br>● 前袋口位 4 个<br>● 后袋口位 4 个，宽度为 1/2"<br>● 前门襟位 2 个（参照图示）<br>● 1 个在底裆内侧<br>● 总共：23 个</td></tr>
</table>

续表

<table>
<tr><td colspan="2">扣眼：<br>● 1 个在腰头位<br>● 2 个在后袋中位<br>● 总共：3 个，11/16" 开口</td></tr>
<tr><td colspan="2">纽扣：总共 3 个，必须配合扣眼大小</td></tr>
<tr><td colspan="2">袋布：漂白 T/C 袋布（用配色线缝制）</td></tr>
<tr><td colspan="2">里衬：<br>● #53915 里衬用于腰头内侧<br>● GP—3 里衬用于后袋嵌线位</td></tr>
<tr><td colspan="2">备注：袋布、里衬的颜色不能外露作为标准</td></tr>
<tr><td colspan="2">生产图示与细节</td></tr>
<tr><td>前袋布深：<br>● 测量从袋口底部到袋布底端的深度，所有尺码的尺寸均为 5 1/2"<br>后袋布深：<br>● 测量从袋口底部到袋布底端的深度，所有尺码的尺寸均为 5 1/2"<br>其他细节：<br>● 裤腰前中左右必须对水平格子<br>● 前后裤裆位必须对水平格子<br>● 左右后袋嵌线的格子必须一致<br>● 后省道的方向必须倒向侧缝</td><td>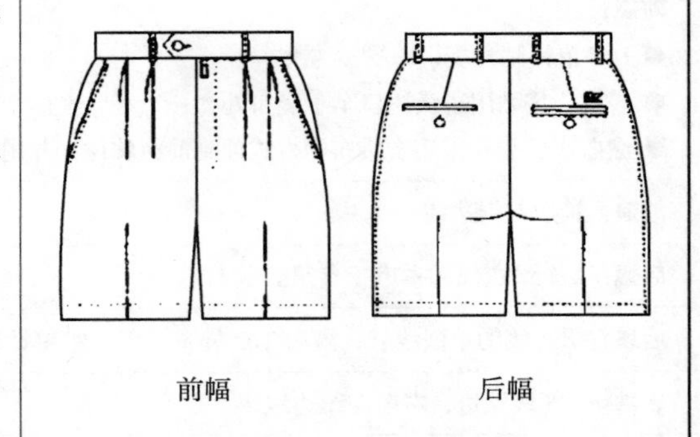<br></td></tr>
<tr><td colspan="2">● LS00500 腰卡：夹于左后后袋口中，并用白色线打 2 个线结<br>● CT-36 袋卡：夹于左后袋中，并用白色线打 2 个线结<br>● LS08 价钱牌：置于左后袋口上，用白色线打 4 个线结</td></tr>
<tr><td colspan="2">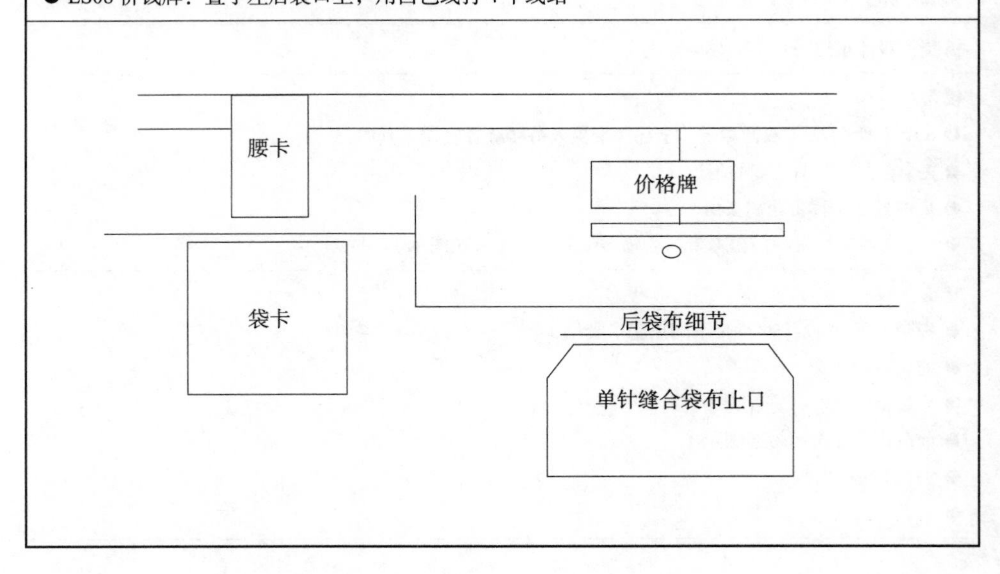<br></td></tr>
</table>

**续表**

| |
|---|
| *****RYKIEL 便装短裤的包装方法 *****<br>● 包装前，裤身上包括腰头在内的所有纽扣都必须扣上<br>● 熨烫：前中裥反向裤中骨线<br>● 包装：每件中骨对折（RYKIEL 价格牌向上），放入在背面印有警告语的防滑透明包装袋内，然后 12 件单色单码放入一个胶纸包，胶纸包上下必须垫用硬纸皮，并用客户要求的贴纸封包，最后 36 件（3 个胶纸包）单色单码放入一个规格为 3 坑的出口纸箱 |
| **包装备注：**<br>● 成品大货分码前，必须测量腰围尺寸，并检查尺码与内长，如果尺码与内长错误，必须更正<br>● 成品入胶袋前，必须检查价格牌、腰卡、洗水商标等辅料上的尺码、内长等资料是否一致<br>● 我们必须检查成品大货是否有脏污现象<br>● 成品入胶袋前，必须完成以上各项工作<br><br>部门：生产部　　　　制表：ALICE LIU<br>日期：2020-3-2 |

# 3.3 男式衬衫制作

## 3.3.1 款式描述

男装衬衫是一件上身服装，它是一件短袖或长袖且有传统领子的衬衫，且长袖有袖克夫，有纽扣或按纽的全长直开口（图 3–1）。男式衬衫在北美叫作“礼服式衬衫”，一般是纯色，但也可能是条纹或格子布的。衬衫通常是在商务场合、逛街或在半正式场合佩戴一条领带穿着。

## 3.3.2 缝纫机的应用

纫机已大幅提高服装行业的效率和生产率。相比之下，国内的工业缝纫机在规模、成本、外观和任务方面显得更大、更快、更多样。在衬衫制作中，主要使用半自动设备。

（1）带剪线刀的单针平缝纫机（车）。

（2）双线平缝机。

（3）三线或五线锁边机。

（4）打枣机。

（5）扣眼机。

（6）单线锁链钉扣机。

（7）锁链机。

（8）翻袖口机。

（9）烫袋机。

（10）翻烫领机。

（11）缝纫附件：比例尺、卷边器、镶边器和卷边压脚，滚边器和折边器等。

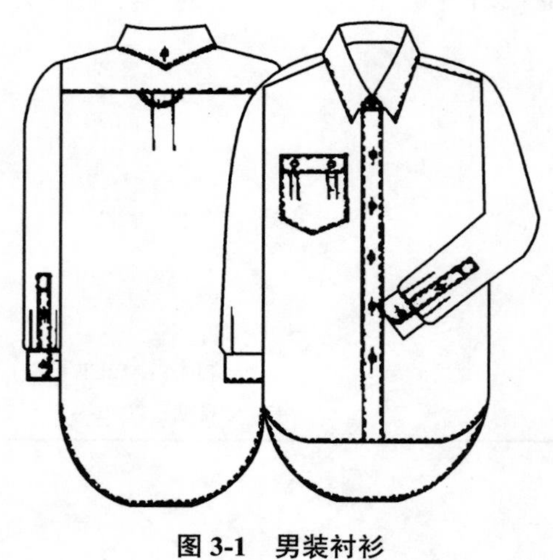

图 3-1 男装衬衫

### 3.3.3 缝纫工序

（1）半合并部件

●口袋

工序 1 辑袋口

工序 2 烫袋形

●领子

工序 3 缝合底面领连里衬

* 缝合领边与领嘴位

* 缉领面边线

工序 4 将领面、领底与里衬车缝在一起

工序 5 缝合领子与领座

●袖级

工序 6 缝袖级连里衬

* 缉袖级面线

工序 7 开扣眼

工序 8 钉纽扣

●前片（右前片）

工序 9 折门襟边位

工序 10 钉纽扣

●前片（左前片）

工序 11 缝合门襟连里衬

工序 12 开扣眼

工序 13 装袋

●后片

工序 14 缉褶裥与环圈贴

工序 15 缝育克

●袖子

工序 16 缝袖衩

工序 17 缉袖口褶裥

（2）合并缝制

工序 18 合肩缝

* 缉面线

工序 19 绱领连商标

工序 20 绱袖

工序 21 合袖底缝和摆缝

工序 22 绱袖级

工序 23 钉领口纽扣

工序 24 开领口扣眼

工序 25 缉下摆

## 3.4 裤子结构

### 3.4.1 规格描述

这款裤子显示以下细节为服装模型：

（1）牛仔裤腰结构。

（2）2个反向褶裥。

（3）缝骨前袋。

（4）两个单嵌线（单唇）后袋，效果如穿起时，左边袋用纽扣扣起袋盖。

（5）缉边线八字后省。

（6）“Rykiel”商标缝在后袋的正上方位置。

（7）翻贴边裤脚为3.2cm（11/4英寸）。

（8）纸样符合款式#168。

### 3.4.2 部件细节

（1）裤腰：一片牛仔裤腰结构。

（a）原身布裤腰，与非机织黏合衬一起折合黏实。

（b）完成后的裤腰宽是3.8cm，且在裤腰顶端和底部缉1.6mm边线。

（c）以每英寸12针的线迹密度，在左前中门襟位开扣眼。

（2）襻带：6个襻带。

（a）2个襻带在前幅，完成后的尺寸为1cm × 5.1cm，缉0.5cm的边线，位于第一个褶裥的旁边。

（b）4个襻带在后幅，完成后的尺寸为1cm × 5.1cm，缉边线。其中两个襻带在后幅离侧缝1.3cm位置；另外2个襻带位于离后裤裆中缝7.6cm的位置。

（c）将襻带绱于裤腰缝骨上，在腰头下1.3cm处打隐形套结。

（d）在襻带顶端打水平套结。

（3）左门襟：原身布门襟贴。

（a）门襟覆非机织黏合衬，锁边。

（b）将拉链带布用双针缝于门襟上，并用单针穿过所有裁片，将门襟缝合于裤身上，缉1.6mm边线。

（c）以单针缉3.8cm“J”形面线于门襟上。

（d）镀古铜YKK金属拉链。

（4）右门襟贴：一片原身布贴。

（a）纽子链贴双折，不加里衬，锁边。

（b）将拉链带缝于裤身。并用单针以1.6mm边线缝合所有布片。

（c）一个水平套结位于“J”形门襟缉线的底部，并有一个垂直套结位于门襟底部2.5cm以上的位置。

（5）褶裥：2个反向褶裥。

（a）褶裥向侧缝，第一个褶深 3.2cm，第二个褶深 1.9cm，两褶相距 3.2cm。

（b）第一个褶的位置由尺寸规格放码决定，参照尺寸表。

（6）前袋：缝骨前袋。

（a）缝骨前袋单针 1cm 袋口面线，在袋口的顶部和底部缝合水平套结。

（b）袋布用原身布落袋贴与垫袋布；将袋贴与垫袋布锁边后以单针缝合于袋布上。

（c）口袋位于裤腰缝底端下 2.5cm 的位置。袋口宽 16.5cm；两片袋布，袋布缝骨采用锁边缝合，并以单针还口缉 0.6cm 的边线；袋布深 30.5cm（从裤腰底端到袋布低端）。

（7）前裆：用五线保险包缝线迹锁缝，并且用单针缉边线。

（8）后裆：用五线保险包缝线迹锁缝，后裆无面缝线迹。

（9）后袋：两个单嵌线后袋，效果如穿起时，左边袋用纽扣扣起袋盖。

（a）衣袋：位于裤腰缝合线下 6.4cm 起的位置，作为口袋起点位；原身布袋贴和垫袋布锁边，并用单针边线缝合在袋布上；锁边缝合袋布，袋布长从裤腰缝口下 22.9cm。

（b）袋口单针缉 1.6mm 边线，袋角打垂直套结。

（c）两片袋盖覆黏合轻身衬；单针缉边线；袋盖插入袋口缝骨上端，并用单针缉边线；垂直扣眼位于带盖中位，离袋盖边 1.3cm；垂直方向缝合纽扣。

（d）“Rykiel”商标位于裤子穿起时右后袋上面的两个省道中间；商标四边向里折止口，并用单针沿四边缉 1.6cm 边线；缝线颜色与商标颜色搭配，完成的尺寸是宽 2.8cm，高 2.5cm。

（10）省道：后幅左右片各有两个后省道。

（a）省道在裤腰缝合线的距离为 1.9cm，在袋口上端缝合线的距离为 5.7cm。

（b）各省道用单针在外沿缉边线。

（11）侧缝：用五线保险包缝线迹缉外侧缝并且将止口反向后衣片，然后用单针缉边线。

（12）内裆：用五线保险包缝线迹缉内裆，无须缉面线。

（13）反脚折边：完成后的折边高度为 3.2cm。首先将底部脚口锁边，然后单折并用单针缉 5.7cm 面线，作为第一折下摆的高度。将第一折向外折 3.2cm 的反脚折边。在侧缝与内裆位用单针垂直打线结固定反脚折边。

（14）后整理：连褶裥整烫整件服装。第一个褶在裤子的中缝线上（图 3–2）。

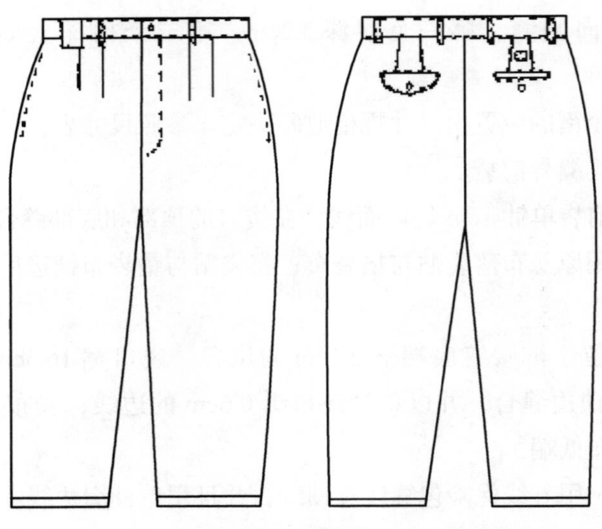

图 3-2 裤子

### 3.4.3 辅料细节

（1）袋布：65% 涤纶，35% 棉。除了白底上有白花样的面料采用纯白袋布外，其他颜色的面料均采用黄白色袋布。布料结构：纱线号数为 22.2tex，密度为 110 × 76。

（2）里衬：轻薄衬布，非机织黏合衬。用于裤腰头、口袋嵌边、袋盖、左门襟。

（3）缝纫线：颜色与衣身颜色一致，涤纶包棉的包芯线。

（4）主商标：折烫“机织商标”的边位，用单针缝合于后中裤腰内层位。缝纫线颜色要与商标颜色一致。

（5）洗涤商标 / 尺码商标 / 成分商标：将商标折叠并缝于后中裤腰缝骨的底端。

### 3.4.4 包装方法

（1）折叠裤子：沿着中缝线对折后，再沿膝围线对折。在折叠过程中塞进尺寸为 27.9cm × 50.8cm 的硬纸板和泡沫条。裤子穿起后的右边应该在外面，以便显示吊牌。

（2）吊牌：印有“Rykiel”吊牌应在裤子穿起后的右边裤腰处，刚好在后袋上方正中处，两个套结位于离边线 0.6cm 的顶角处。

（3）条形码标签：位于包装袋背面右下角较低处。

（4）包装袋：采用可封口的平板包装袋，47cm × 69.2cm。

（5）出口纸箱：采用宽 44.5cm × 长 68.6cm × 高 30.5cm 厚纸板箱，每箱数量为 24 件。重量不能超过 18.1 千克（约 40 磅）。

# 3.5 服装制作流程图

## 3.5.1 男西裤工艺流程图（图 3-3）

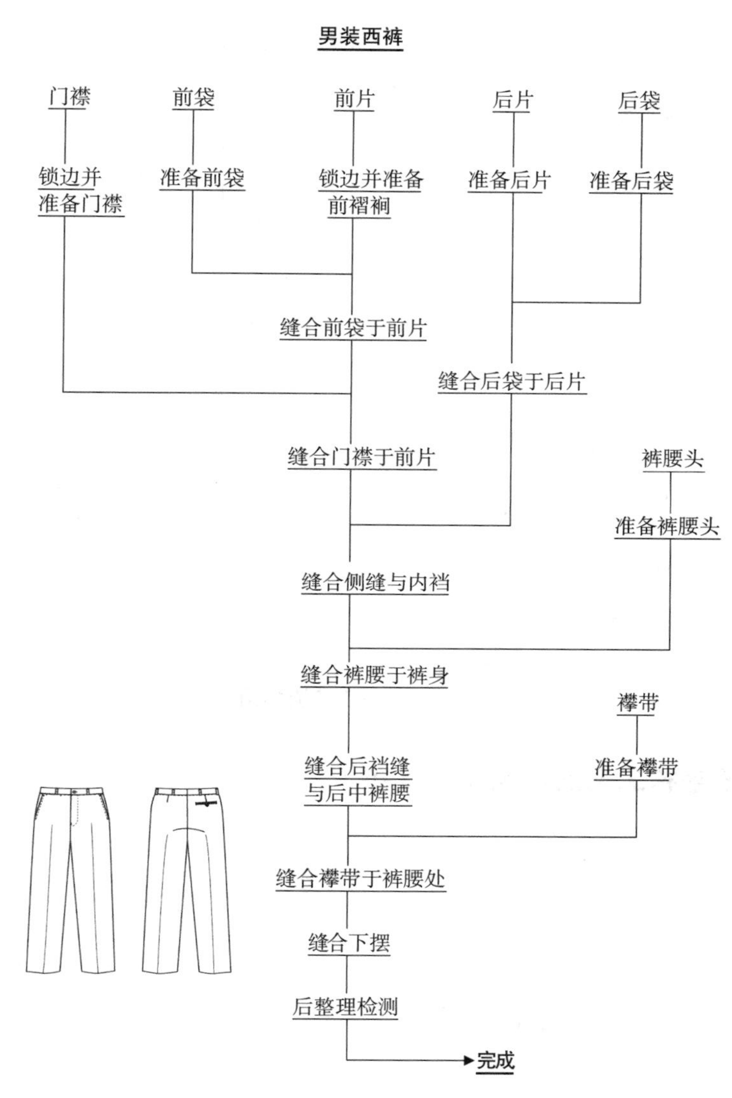


图 3-3 男装西裤工艺流程图

### 3.5.2 T 恤衫工艺流程图（图 3-4）

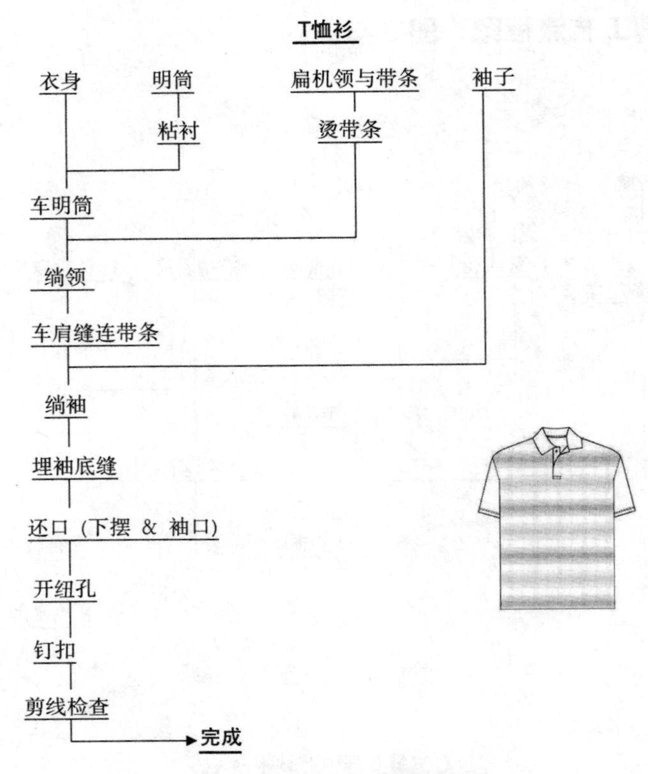


图 3-4　T 恤衫工艺流程图

### 3.5.3 羊毛衫工艺流程图（图 3-5）

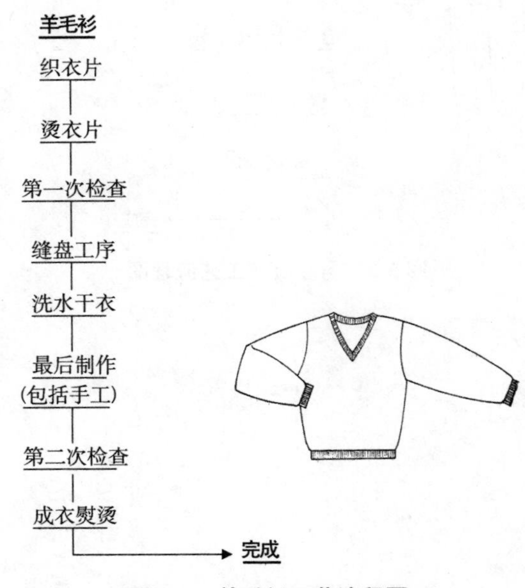


图 3-5　羊毛衫工艺流程图

### 3.5.4 女装衬衫工艺流程图（图 3-6）

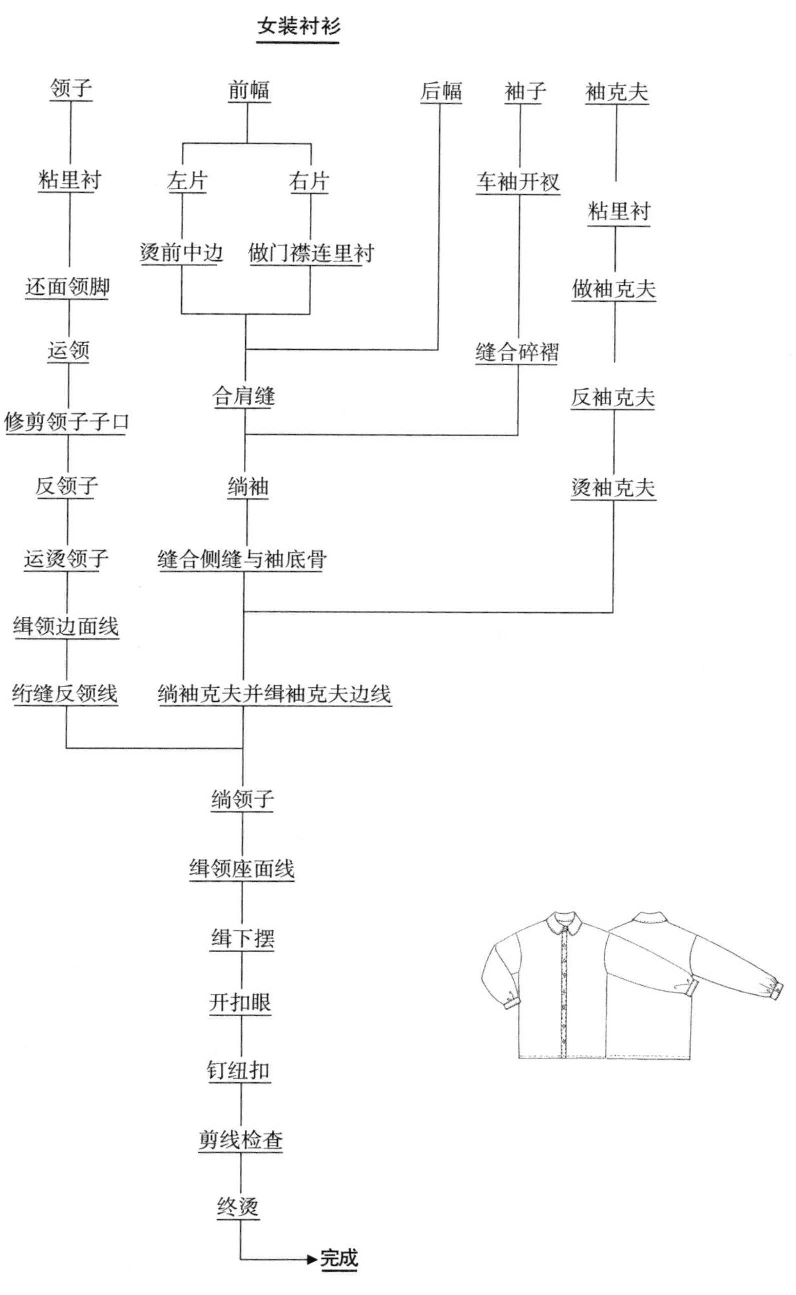


图 3-6 女装衬衫工艺流程图

### 3.5.5 牛仔裤工艺流程图（图 3-7）

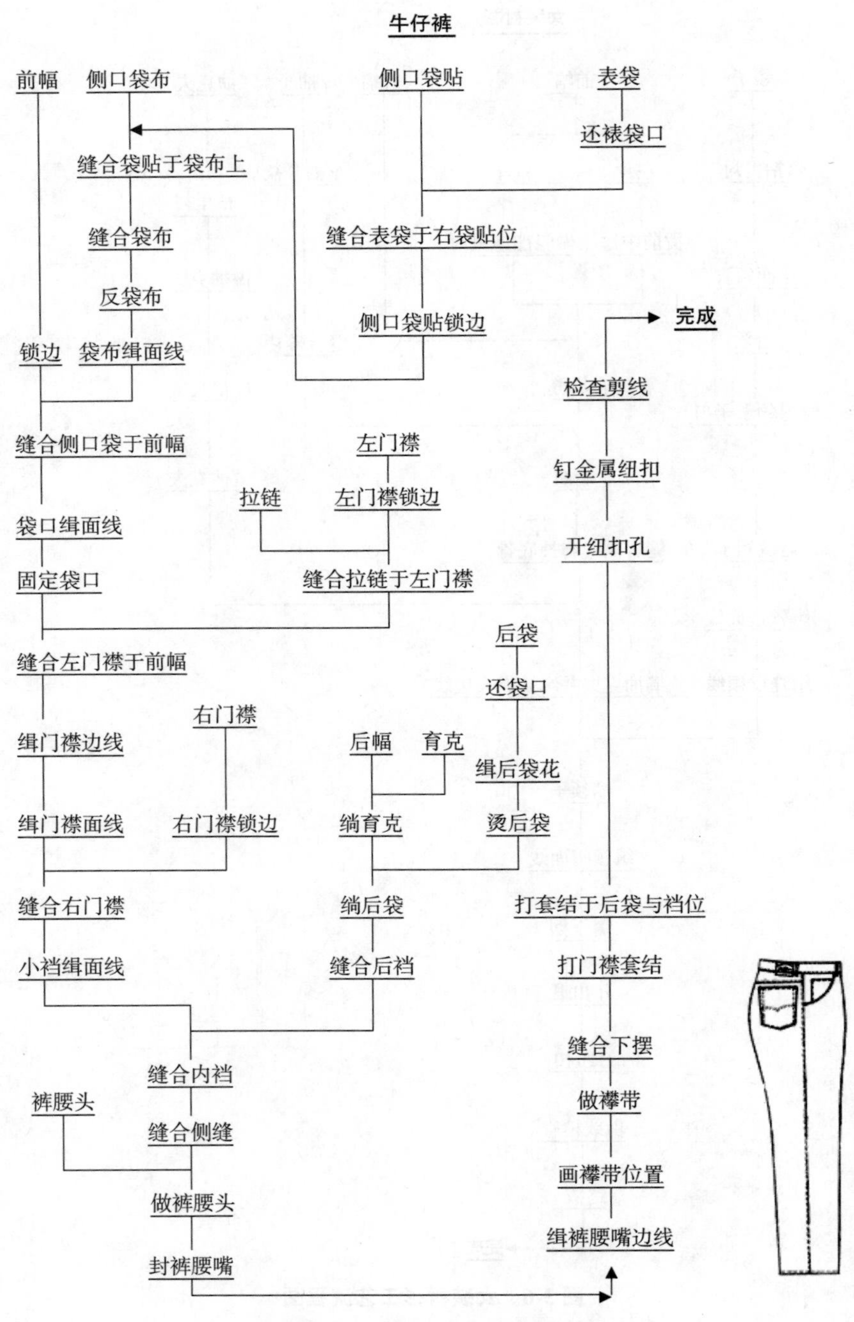


图 3-7 牛仔裤工艺流程图

## 3.6 部件解释

### 3.6.1 服装部件

（1）领子：领座立起或大衣、连衣裙、衬衫等翻开的上部。领子属于处理服装上身的部件。领子可分为几种类型，如衬衫领、西装领、立领、翻领和平领等。

（2）袖克夫：连接衣袖末端的部件，使整个袖子完整。例如，衬衫、外套、夹克等袖子底端另外附加的一块面料。

（3）领面：领子的外片或外部，但是领座是领子底面直接从衣服上升起。通常在衬衫领包含领面与领座。

（4）袋盖：覆盖口袋的部件，也叫袋盖。

（5）门襟：裤子前幅开口，由纽扣或拉链组成。

（6）暗门襟：在大衣或夹克上隐形系结物，有一下层布料扣上服装的另一边。

（7）衣袖插片：一块三角形布片嵌入袖底缝，以便调节松位和造型，通常在和服或灯笼裤可找到。

（8）三角布：一块用于插入缝合的三角形布料。

（9）小插片：裙子里的镶片，如多片裙或嵌镶片裙。

（10）下摆：采用向里折布料并缝合的边位处理方式，也叫下摆。

（11）襟贴 / 挂面：从领口位折翻的夹克或大衣部件。

（12）领边：领子外部边缘。

（13）口袋：具有功能性和装饰性。有三种基本类型，即明贴袋、缝骨袋和挖袋。

（14）明贴袋：另外一块布片直接缝在衣服上。

（15）明门襟：给予充分的空间穿衣的衬衫或夹克开口，明门襟可采用拉链或按纽进行闭合。

（16）挖袋：在衣服中间开一特殊衩位作为袋口。

（17）缝骨袋：将袋口做在缝骨位置。

（18）袖子：用于遮盖手臂到肩膀的部件，可分为绱袖型袖子与原身出袖子。

（19）衩位：在袖子、夹克或裙子处的重叠开口。

（20）裤腰头：环绕在裤子或裙子顶部。

（21）育克：将衬衫、外套或罩衫的肩部分离，或将牛仔裤或裙子的腰部形成悬挂部位。

### 3.6.2 褶的类型

（1）缝褶：采用三角形状收起多余的布料，目的在于突出体型的形状。如胸部、臀部、肩部和肘部等。

（2）装饰边：窄小的荷叶花边或边位缩碎褶的形式。

（3）裙褶：应用于服装的一条碎裥带条。

（4）宽摆：放宽下摆或衣服的底部。

（5）缩碎褶：用一条或多条针线把布料缩起。

（6）针褶：缝在衣边的一个很细的塔克。

（7）褶裥：折叠面料以便控制松位。

（8）风琴褶：窄且直的褶裥。

（9）工字褶：两个褶裥在里面相对，外呈工字型，与工字褶相反就是暗工字褶，两个褶裥在服装的外面相对。

（10）子弹褶：一种未压烫的卷褶，通常很窄小，用于装饰。

（11）踢褶：裙子上的一个小褶，是走动时的放松位置。

（12）迪奥褶：铺另外一块面料在缝合边缘翻转过来的底端。只将这块面料的顶端缝合在衣服上，两边放松不缝合。

（13）剑褶：与风琴褶相似，但所有的褶裥方向一致。

（14）太阳褶：褶裥从中间一点发出，并向衣服的边位延伸。

（15）塔克：均匀地直向折叠面料并缉面线。

## 本章小结

- 以不同案例形式介绍了制板通知单、生产指示书等服装生产文件的格式与应用，同时引出相关服装部件、成衣尺寸部位，以及服装生产与工艺结构等专业名词。
- 明确有关服装工艺结构的表达与描述。
- 通过绘制服装生产工艺流程图，进一步学习服装生产与工序名称的表达。
- 针对各种服装部件作了详细解释。
- 此章节重点培养学生对英语服装生产文件的翻译与编写能力。

# CHAPTER 4

## QUALITY INSPECTION　品质检验

**课题名称：** QUALITY INSPECTION　品质检验

**课题内容：** Quality Standards　品质标准

Inspection Report　检验报告

Expression of Garment Defects　服装疵点的表达

**课题时间：** 8 课时

**教学目的：** 让学生了解服装品质标准与规格的表达以及各种品质文件的格式与应用。理解成衣尺寸测量部位的测量描述，掌握各种服装疵点的表达与简单描述。重点掌握并熟记各品质规格与疵点名称以及尺寸测量部位等相关术语。

**教学方式：** 结合 PPT 与音频多媒体课件，以教师课堂讲述为主，学生结合相关案例分析讨论为辅。

**教学要求：** 1. 明确服装品质标准与规格的表达。

2. 熟悉成衣测量的基本描述。

3. 熟悉各种服装品质文件的格式与应用。

4. 明白各种服装疵点名称与表达方式。

**课前（后）准备：** 结合专业知识，课前预习课文内容。课后熟读课文主要部分，掌握各种服装品质文件的编写格式与应用并熟记各服装疵点名称与表达。

# CHAPTER 4

# QUALITY INSPECTION
# 品质检验

## 4.1 Quality Standards 品质标准

### 4.1.1 Introduction 简介

In garment industry terms “Quality Inspection” is looking closely and examining what is actually being produced and comparing this with what is required or what should be produced. ① The inspection activity of itself adds nothing to the product. It is only a means by which information is obtained as a basis for further action. Various method of carrying out the “inspection” function is in use in the garment industry.

①在制衣业，“品质检验”是指细心观察并检查实际的产出品并将其与需求标准作比较。

### 4.1.2 Indication of Quality Standards 品质标准的表示

For the indication of quality standards, we must refer to established firms and more standard goods. These indications are found as follows:

(1) Seam: It is the area where two or more layers of fabric are held together by stitches. A seam must be excellent to wear well, and the seam finishes and threads must also be compatible to the fabric, style and the performance required and promised by the manufacturer.

(2) Material: Fabrics and accessories must be acceptable, for example, fabric has not shaded, damaged, broken yarn, any creased mark, coarse yarn, fabric runs and pilling, etc. Accessories must be in accordance with specification.

(3) Hem: Hem-stitched design for any garment, and hems of different depths may have a particular style to achieve, which requires more or less. Generally speaking, the better work for the more excellent hem. Also, better work shows the hem to have been neatened before being stitched in place. The cheaper work, the more noticeable hem will be from the outside, and the less evenly will it hang. A perfect hem denotes a good garment.

（4）Buttonhole：Buttonholes are holes in fabric which allow buttons to pass through，securing one piece of the fabric. Buttonhole stitch should be done by machine or handwork，buttonhole must be more likely to be piped or bound.The hand-sewing stitches are always used in tailoring，embroidery，and needle lace-making.

（5）Button：Obviously，cheap work has cheap ordinary buttons，and best work has good buttons. In cheap to medium work it is possible to pull one thread and undo all the stitching. In better work of button，a safety stitch must be put on the machine，or sewn by hand.

（6）Fit：Better class work fits more closely or has more movement，and is better cut. Cheap work is shapeless and gives little freedom of movement. Sleeves especially are uncomfortable and soon distorted. Generally speaking，cheap work is skimpy in cut and better work is excellent.

（7）General Appearance：Better class work must be well cut，well made，neat inside. It fits nicely and feels comfortable. Trimmings will be of good quality and well matched，and will frequently be a feature of the garment. Collars，cuffs，pleats，pockets，waistband，and other parts will balance and hang well.

### 4.1.3 Inspection Check List 检查核对表

To inspect the garments or garment parts at selected points in the manufacturing process. All garments are through one inspection area where a team or group of specialist inspector is employed to examine garments at agreed stages. ② This approach is particularly suitable where many different styles are in production at the same time. It is convenient to consider sewing room operations under the following check list and some examples.

②所有服装都要经过由一组或一队专业人员组成的检验部门按既定标准进行的检验。

（1）Body Fabric

（a）Gauge（Fabric Contents，Construction，etc.）

（b）Weaving defects

（c）Needle chew

（d）Shrinkage

（e）Colorfast

（f）Color variation（Within a garment and between garments）

（g）Broken thread

（h）Holes / Knots

（i）Spots / Stains

（j）Streaks

（2）Accessories

（a）Lining

（b）Interlining/ Magic tape/ Elastic

（c）Button / Snap / Zipper/ Eyes & Hooks

（d）Rivet / Eyelet / Bar-tack

（e）Threads

（f）Belt

（3）Washing / Dyeing

（a）Stiff

（b）Shading

（c）Stink smell

（d）Stain / Crease mark

（e）Dye runs

（4）Cutting

（a）Stripe or edge direction

（b）Incorrect size-spec.

（5）Print / Embroidery

（a）Unclear or overprint

（b）Stained during printing

（c）Uneven shade

（d）Wrong color，design or direction

（e）Incomplete

（f）Off position

（g）Varied size

（6）Seams

（a）Twisted / Pleated

（b）Puckered

（c）Seam allowance too narrow

（d）Raw edges visible

（e）Open seam

（f）Seam strength weak

（g）Seam parts at seam not match

（7）Stitching

（a）Stitching not at straight line

（b）Stitch density too low or high

（c）Thread tension

（d）Missing / Skip / Broken / Run off stitch

（e）Overlapping stitches

（f）Wrong thread color

（8）Pressing / Trimming

（a）Wrinkle

（b）Shiny

（c）Under / Over pressed

（d）Burn marks，scorch marks，pressure marks

（e）Uncut thread

（f）Not properly folded

（9）Labeling

（a）Brand label / Size label

（b）Care label / Content label

（c）Label on poly-bag

（d）Hang tag

（e）Label position

（f）Label stitch

（10）Packing

（a）Poly-bag size / wording

（b）Folding size

（c）Missing inner box

（d）Outer carton type / size

（e）Missing / Unclear shipping marks

（f）Unequal length

（g）Twisted

（11）Pockets

（a）High-low pocket

（b）Slanting pocket mouth

（c）Pocket size

（d）Poorly shaped / Wrong placement

（e）Raw edge

（f）Twisted / Puckered / Pleated

（g）Bar-tacking or backstitching missing

（h）Pocket marks not covered

（i）Not set properly with check or stripe design material

（12）Fly

（a）Poorly shaped

（b）Twisted / Bulging

（13）Zippers

（a）Malfunction

（b）Improperly sews

（c）Wavy

（d）Size / Length and color

（14）Buttons / Snaps / Button hole

（a）Crooked button line

（b）Missing

（c）Misplaced button

（d）Faulty button

（e）Insecure button / Snap

（f）Button upside down

（g）Bubble between button

（h）Missing spare button

（i）Incorrect size / shape

（15）Belt-loop / Bar-tack / Rivet

（a）Omitted / Insecure

（b）Wrong placement

（c）Too narrow / Wide

（d）Insecure

## *Words and Expressions*

subjective [sʌb'dʒektiv] 主观的

manufacturer [mænju'fæktʃərə] 厂商，生产商

profit ['prɔfit] 利润

output ['ɑutput] 产出，产量

personnel [pə:sə'nel] 人事（部门）

style [stail] 款式
performance [pəˈfɔːməns] 性能
excellent [ˈeksələnt] 极好的
fabric [ˈfæbrik] 面料
accessory [ækˈsesəri] 配件，辅料
broken yarn [ˈbrəukən jɑːn] 断纱
creased mark [kriːs mɑːk] 折痕
coarse yarn [kɔːs jɑːn] 粗纱
pilling [ˈpiliŋ] 起球
stitched [stitʃd] 缝合
fabric runs [ˈfæbrik rʌnz] 抽丝（广东话：走纱）
specification [ˌspesifiˈkeiʃən] 规格
undo [ˈʌnˈduː] 松脱
safety stitch [ˈseifti stitʃ]（双排）保险线迹
shapeless [ˈʃeiplis] 变形
shortcoming [ˈʃɔːtkʌmiŋ] 缺点
flared [flɛəd] 展开的，喇叭的
neaten [ˈniːtn] 整理
denote [diˈnəut] 指示，表示
hem [hem] 脚围，下摆
piped [paipt] 嵌边的
bound [baund] 滚边的，包边的
comprehensive [kɔmpriˈhensiv] 全面的
skimpy [skimpi] 不足的
heading [ˈhediŋ] 标题
trimming [ˈtrimiŋ] 辅料，整理
stripe [straip] 条纹
check [tʃek] 格子
alignment [əˈlainmənt] 对齐，队列
positioning [pəˈziʃəniŋ] 定位
component [kəmˈpəunənt] 部件
collar end [ˈkɔlə end] 领口，领嘴
hanger [ˈhæŋə] 衣架
attached thread/ float thread [əˈtætʃt θred / fləut θred] 浮线
canvas [ˈkænvəs] 帆布，马尾衬
back stitching [bæk ˈstitʃiŋ] 回针
bartack [bɑːt] 套结（广东话：打枣）
stitch density [ˈstitʃ ˈdensiti] 线步密度
slipped stitch [slipid ˈstitʃ] 跳针，跳线
tension appearance [əˈpiərəns əˈpiərəns] 外形拉紧
distortion [disˈtɔːʃən] 变形
pucker [ˈpʌkəiŋ] 起皱的
soiled [sɔil] 污渍
template [ˈtemplit] 纸板
color shading[ˈkʌlə ˈʃeidiŋ] 色差
formation [fɔːˈmeiʃən] 形成
fused interlining [fjuːzd intəˈlainiŋ] 热熔衬
additional [əˈdiʃənl] 附加的
pairing left & right [ˈpɛəriŋ left & rait] 左右一对
skipped stitch [skip stitʃ]（缉线）跳针，跳线
S.A./ seam allowance [siːm əˈlauəns] 缝份，缝头（广东话：子口）
pleat [pliːt] 褶，使……打褶（广东话：活褶）
waistband [ˈweistbænd] 裤腰头，腰头
notch [nɔtʃ] 刀口，刻口（广东话：扼位）

## *Exercises*

1. Translate the following terms into Chinese.

（1）quality
（2）garment
（3）clothing
（4）thread
（5）puckering
（6）pleating
（7）pilling
（8）button
（9）pocket
（10）cuff
（11）general appearance
（12）sleeve
（13）easing
（14）collar
（15）accessory
（16）fabric runs
（17）trimming
（18）sample
（19）manufacturer
（20）cutting component
（21）stripe

(22) check
(23) notch
(24) canvas
(25) back stitching
(26) distortion
(27) template
(28) slipped stitches
(29) skipped stitches
(30) fullness
(31) seam allowance
(32) hanger
(33) stitch density
(34) material
(35) interlining
(36) color shaded

2. Suggest the definition for "Clothing Quality" by your ideas.
3. List 8 defects of fabric.

## 4.2 Inspection Report 检验报告

Quality control accumulates a very large volume of data concerning operations in a plant. This information is used to make the following types of reports: Work-in-process inspection reports, operator quality performance reports, final audit reports, describing and analyzing repairs reports, percentage defects in sewn and cut work as well as raw materials, fabric inspection reports. Not only is information regarding shading generally maintained in inventory records, but also currently computers are available which actually shade goods. Direct input of other in -process quality control data is gradually increasing.

### 4.2.1 Product Inspection Report 产品检查报告

Product inspection report is mainly for quality control function, which monitors the quality of work going through manufacturing and reports any defects discovered. This information is accumulated daily and is used to produce daily and weekly quality reports. The computer, rather than quality control auditors, processes the inspection data to produce the required reports. And the inspection process may be involved in line process and the final products inspection (Table 4-1 ~ Table 4-3).

**Table 4–1 Inspection Report（Ⅰ）**

<table>
<tr><td colspan="6">**RYKIEL TRADING CO. LTD.**<br>**IMPORT QUALITY AUDIT REPORT**<br>(　)IN-LINE　(　)FINAL</td></tr>
<tr><td colspan="6">BUYER: ______　P.O. NO.: ______　STYLE NAME: ______<br>FACTORY: ______　FTY NO. : ______　SHIPMENT: ______<br>DEL. DATE: ______　AUDIT DATE: ______　LOCATION: ______<br>TOTAL QTY IN LOT: ______　AUDIT SAMPLE SIZE: ______<br>LOT NO.: ______　ACCEPT/REJECT LEVEL: ______<br>COLOUR: ______　AUDITOR: ______<br>DESCRIPTION: ______</td></tr>
<tr><td></td><td colspan="5">WAIST SIZE（MORE THAN 1/2" SMALL OR LARGE）</td></tr>
<tr><td></td><td colspan="5">INSEAM（MORE THAN 1/2" SHORT OR LONG）</td></tr>
<tr><td></td><td colspan="5">INSIDE LABELS（PLACEMENT WRONG OR NOT SECURE）</td></tr>
<tr><td></td><td colspan="5">OUTSIDE LABELS（PLACEMENT WRONG OR NOT SECURE）</td></tr>
<tr><td></td><td colspan="5">SKIPPED STITCHES（OPEN SEAM OR TOPSTITCH）</td></tr>
<tr><td></td><td colspan="5">BUTTON-HOLES（MISSING OR NOT CUT OPEN PROPERLY）</td></tr>
<tr><td></td><td colspan="5">BUTTONS（MISSING OR BROKEN）</td></tr>
<tr><td></td><td colspan="5">FABRIC FLAWS（HOLES, SHADE）</td></tr>
<tr><td></td><td colspan="5">SPOTS（DIRT, OIL）</td></tr>
<tr><td></td><td colspan="5">GENERAL APPERANCE（STRINGS, WASH, PRESSING）</td></tr>
<tr><td colspan="2">TOTAL DEFECT UNITS</td><td colspan="4" rowspan="2">ACCEPT（　）　　REJECT（　）</td></tr>
<tr><td></td><td></td></tr>
<tr><td></td><td>ACCEPT</td><td>REJECT</td><td></td><td>ACCEPT</td><td>REJECT</td></tr>
<tr><td>STYLE DETAILS</td><td>(　)</td><td>(　)</td><td>COLOUR SHADE</td><td>(　)</td><td>(　)</td></tr>
<tr><td>WASHING</td><td>(　)</td><td>(　)</td><td>FABRIC</td><td>(　)</td><td>(　)</td></tr>
<tr><td>PRESSING</td><td>(　)</td><td>(　)</td><td>TWISTED LEG</td><td>(　)</td><td>(　)</td></tr>
<tr><td>HAND FEEL</td><td>(　)</td><td>(　)</td><td>PACKING</td><td>(　)</td><td>(　)</td></tr>
<tr><td>MEASUREMENT</td><td>(　)</td><td>(　)</td><td>ACCESSORIES</td><td>(　)</td><td>(　)</td></tr>
<tr><td>SHIPPING MARKS</td><td>(　)</td><td>(　)</td><td>WORKMANSHIP</td><td>(　)</td><td>(　)</td></tr>
<tr><td colspan="6">NO. OF FTY PERSONS TO TACK MEASUREMENT: ______<br>CARTON NO. INSPECTED: ______<br>REMARKS: ______<br><br>______ RYK REPRESENTATIVE　　______ FTY REPRESENTATIVE</td></tr>
</table>

Continued

<table>
<tr><td colspan="7">RYKIEL TRADING CO. LTD.<br>MEASUREMENT CHART（TOPS）<br>VENDOR: ____________ DATE: ____________<br>BUYER ORDER NO.: ____________ STYLE: ____________<br>CONTRACT CONFIRMATION NO. : QUANTITY: ____________</td></tr>
<tr><td>SIZE / MEASURED POINTS</td><td>SPEC</td><td>ACT</td><td>ACT</td><td>SPEC</td><td>ACT</td><td>ACT</td></tr>
<tr><td>CHEST（1" UNDER ARMHOLE）</td><td></td><td></td><td></td><td></td><td></td><td></td></tr>
<tr><td>WAIST</td><td></td><td></td><td></td><td></td><td></td><td></td></tr>
<tr><td>BOTTOM</td><td></td><td></td><td></td><td></td><td></td><td></td></tr>
<tr><td>SHOULDER</td><td></td><td></td><td></td><td></td><td></td><td></td></tr>
<tr><td>F. MID ARMHOLE</td><td></td><td></td><td></td><td></td><td></td><td></td></tr>
<tr><td>SLEEVE LENGTH</td><td></td><td></td><td></td><td></td><td></td><td></td></tr>
<tr><td>ARMHOLE</td><td></td><td></td><td></td><td></td><td></td><td></td></tr>
<tr><td>SLEEVE WIDTH</td><td></td><td></td><td></td><td></td><td></td><td></td></tr>
<tr><td>C.B. LENGTH</td><td></td><td></td><td></td><td></td><td></td><td></td></tr>
<tr><td>B. MID ARMHOLE</td><td></td><td></td><td></td><td></td><td></td><td></td></tr>
<tr><td>NECK DROP</td><td></td><td></td><td></td><td></td><td></td><td></td></tr>
<tr><td>COLLAR/NECK BAND</td><td></td><td></td><td></td><td></td><td></td><td></td></tr>
<tr><td>POCKET</td><td></td><td></td><td></td><td></td><td></td><td></td></tr>
<tr><td>HOOD</td><td></td><td></td><td></td><td></td><td></td><td></td></tr>
<tr><td></td><td></td><td></td><td></td><td></td><td></td><td></td></tr>
<tr><td colspan="7">REMARK:</td></tr>
</table>

**Continued**

**RYKIEL TRADING CO. LTD.**

**MEASUREMENT CHART ( BOTTOMS )**

VENDOR: ______________ DATE: ______________

BUYER ORDER NO.: ______________ STYLE: ______________

CONTRACT CONFIRMATION NO. : ______________ QUANTITY: ______________

| SIZE / MEASURED POINTS | SPEC | ACT | ACT | SPEC | ACT | ACT |
|---|---|---|---|---|---|---|
| WAIST ( Relax ) | | | | | | |
| WAIST ( Stretch ) | | | | | | |
| HIGH HIP | | | | | | |
| SEAT | | | | | | |
| THIGH | | | | | | |
| KNEE 1 | | | | | | |
| KNEE 2 | | | | | | |
| BOTTOM | | | | | | |
| FRONT RISE | | | | | | |
| BACK RISE | | | | | | |
| INSEAM | | | | | | |
| OUT-LEG | | | | | | |
| ZIPPER | | | | | | |
| | | | | | | |

REMARK:

**Table 4–2 Inspection Report（Ⅱ）**

<table>
<tr><td colspan="6"><u>QUALITY CONTROL DEPARTMENT</u><br><u>INSPECTION REPORT</u><br>(　) Intermediate　(　) Final　(　) Re-Inspection</td></tr>
<tr><td colspan="6">Customer:　P.O. No.:　Date:　Time:</td></tr>
<tr><td colspan="6">MFTR:　Style:　QTY:　Del. Date:</td></tr>
<tr><td colspan="6">Garment:　AQL:　FTY Code:</td></tr>
<tr><td colspan="6">Note: Inspection Method is relative to Production Order QC Manual</td></tr>
<tr><th>Check Points</th><th>Right</th><th>Wrong</th><th>Check Points</th><th>Right</th><th>Wrong</th></tr>
<tr><td>A. Label</td><td></td><td></td><td>F. Fabric Color</td><td></td><td></td></tr>
<tr><td>B. Style / Color</td><td></td><td></td><td>G. Belt Color</td><td></td><td></td></tr>
<tr><td>C. Shipping / Side Marks</td><td></td><td></td><td>H. Thread Color</td><td></td><td></td></tr>
<tr><td>D. Hangtag Description</td><td></td><td></td><td>I. Zipper Color</td><td></td><td></td></tr>
<tr><td>E. Folding & Packing</td><td></td><td></td><td>J. Button Color</td><td></td><td></td></tr>
<tr><th>Defects</th><th>Major</th><th>Minor</th><th>Defects</th><th>Major</th><th>Minor</th></tr>
<tr><td>01. Material Defect</td><td></td><td></td><td>10. Closures</td><td></td><td></td></tr>
<tr><td>02. Shading Problem</td><td></td><td></td><td>11. Damage</td><td></td><td></td></tr>
<tr><td>03. Dyeing Problem</td><td></td><td></td><td>12. Threads</td><td></td><td></td></tr>
<tr><td>04. Printing Problem</td><td></td><td></td><td>13. Handicraft</td><td></td><td></td></tr>
<tr><td>05. Cleanliness</td><td></td><td></td><td>14. Fit & Balance</td><td></td><td></td></tr>
<tr><td>06. Component & Assembly</td><td></td><td></td><td>15. Creased Marks</td><td></td><td></td></tr>
<tr><td>07. Seam & Stitching</td><td></td><td></td><td>16. Finishing & Hand</td><td></td><td></td></tr>
<tr><td>08. Pressing</td><td></td><td></td><td>17. Packing</td><td></td><td></td></tr>
<tr><td>09. Measurement</td><td></td><td></td><td colspan="3">Total Defective Pieces:</td></tr>
<tr><td colspan="3">Inspected:________PCS</td><td colspan="3">From Carton No.:</td></tr>
</table>

<table>
<tr><th>Detailing</th><th>%</th><th>Suggestion & Remarks</th></tr>
<tr><td>A.</td><td></td><td></td></tr>
<tr><td>B.</td><td></td><td></td></tr>
<tr><td>C.</td><td></td><td></td></tr>
<tr><td>D.</td><td></td><td></td></tr>
<tr><td>E.</td><td></td><td></td></tr>
</table>

<table>
<tr><td rowspan="2">Garment Size<br>& Quantity</td><td></td><td></td><td></td><td></td><td></td><td></td><td></td><td></td></tr>
<tr><td></td><td></td><td></td><td></td><td></td><td></td><td></td><td></td></tr>
<tr><td colspan="5">(　) Good　(　) Satisfactory　(　) Fair<br>(　) Unacceptable　(　) L/G</td><td colspan="4">Inspector: ____________<br>Packing In-charge: ________<br>QC Manager: __________</td></tr>
</table>

Table 4-3 Inspection Report（Ⅲ）

| RYKIEL APPAREL LTD.<br>CUTTING QUALITY INSPECTION REPORT<br>Inspection Date: | | | | |
|---|---|---|---|---|
| P.O. No.: | Vendor: | Goods Description: | | |
| Style No.: | Quantity: | Color: | | |
| Spreading Method: | Layout No.: | Spreading Quantity: | | |
| **Ⅰ. Marker** | | | Yes | No |
| A. Are the marker and the fabric width the same? | | | | |
| B. Does the marker use the original pattern? | | | | |
| C. Do the cutting pieces need the notch? | | | | |
| D. Is the marker line smooth? | | | | |
| E. Are the pieces of pattern not missing in the marker? | | | | |
| F. Is the join fabric position correct ? | | | | |
| G. Are the size of marker cut pieces and size of pattern the same? | | | | |
| Inspection Date: | | | | |
| **Ⅱ. Cutting Pieces** | | | Yes | No |
| A. Are the size of cutting pieces top and bottom the same? | | | | |
| B. Is the cutting piece line smooth? | | | | |
| C. Is the Cutting pieces marker line correct? | | | | |
| D. Is the number of cutting pieces correct? | | | | |
| E. Are the laying fabric start and end allowance in the standard? | | | | |
| F. Is the marker put in the correct position before cutting? | | | | |
| G. Do the fabric in side & up side lay? | | | | |
| H. Are the fabric sample card & fabric saved? | | | | |
| Inspection Date: | | | | |
| **Ⅲ. Bundling** | | | Yes | No |
| A. Are the details of the sewing tickets and the cutting pieces the same? | | | | |
| B. Are the details of the sewing tickets and the bundle quantity the same? | | | | |
| C. Are the fabric ticket quantity and the bundle quantity the same? | | | | |
| Inspection Date: | | | | |
| Comments:<br><br>Cutting Room Supervisor:__________ Inspector:__________ QC Supervisor:__________ | | | | |

## 4.2.2 Defects Inspection Report 疵点检查报告

Inspection report forms are used to record the information about the checked garment, such as garment defects inspection. The quality controllers or checkers would bring this form while they are inspecting goods. All defects would be found and marked. This report would be submitted to the merchandiser or merchandising manager to take further remedy action（Table 4-4、Table 4-5）.

**Table 4–4 Defect Inspection Report（Ⅰ）**

**IN–PROCESS INSPECTION REPORT**

**DEFECT CHECKLIST**

Style No.________ Page No.________

| **Fabric Defects** | Major | Minor | **Sewing Defects** | Major | Minor |
|---|---|---|---|---|---|
| Holes | | | Open Seams | | |
| Soiling | | | Weak Seams | | |
| Flaws | | | Raw Edges | | |
| Pilling | | | Wavy Stitch | | |
| Uneven Dyeing | | | Skip Stitch | | |
| Burn Marks | | | Broken Stitch | | |
| Barre | | | Incorrect Link | | |
| | | | | | |

| **Garment Defects** | Major | Minor | **Pressing Defects** | Major | Minor |
|---|---|---|---|---|---|
| Fabric Color Mismatch | | | Uneven Hem | | |
| Component Color Mismatch | | | Misaligned Parts | | |
| Defective Snaps | | | Missing Parts | | |
| Defective Zippers | | | Uneven Plaids | | |
| Exposed Zippers | | | Puckering | | |
| Excessive Thread Ends | | | Other | | |
| Loose Buttons | | | | | |
| Defective Buttonholes | | | | | |
| | | | | | |

Pieces available for inspection:________________________

No. pieces inspected:________________ Acceptance level:________________

No. pieces rejected:________________ Minor defects:________________

| Date | Comments<br>Production Status: | Action to be Taken by QC | Vendor's Signature |
|---|---|---|---|
| | | | |
| | | | |

**Continued**

| The vendor is responsible for correcting all defects found during the inspection and summarized in this report. However, the inspection does not relieve the vendor from its responsibility for defects found in the merchandise shipped. |
|---|
| Vendor's Signature:____________ Date:____________<br>Inspector's Signature:____________ Date:____________ |

**Table 4–5 Defect Inspection Report ( Ⅱ )**

| CLOTHING INDUSTRY TRAINING CENTER<br>DEFECT INSPECTION REPORT<br>BUNDLE CODE: | | | | | | | |
|---|---|---|---|---|---|---|---|
| DEFECTS ITEM | 1 | 2 | 3 | 4 | 5 | 6 | TOTAL |
| ( 1 ) Garment Parts Asymmetric | | | | | | | |
| ( 2 ) Garment Parts Bubbling | | | | | | | |
| ( 3 ) Garment Parts Missing | | | | | | | |
| ( 4 ) Mismatched Meet Point of Seam | | | | | | | |
| ( 5 ) Incorrectly Seaming | | | | | | | |
| ( 6 ) Pleated Seam | | | | | | | |
| ( 7 ) Seam Pucker | | | | | | | |
| ( 8 ) Garment Parts Misplaced | | | | | | | |
| ( 9 ) Skipped Stitching | | | | | | | |
| ( 10 ) Broken Stitching | | | | | | | |
| ( 11 ) Poorly Back-Stitching | | | | | | | |
| ( 12 ) Out-Seam Run off Stitching | | | | | | | |
| ( 13 ) Uneven Tension Stitching | | | | | | | |
| ( 14 ) Improperly Pressing | | | | | | | |
| ( 15 ) Uneven Top-Stitching | | | | | | | |
| ( 16 ) Missing Operation | | | | | | | |
| ( 17 ) Seam Broken | | | | | | | |
| ( 18 ) Missing Trimmings | | | | | | | |
| ( 19 ) Fabric Defects | | | | | | | |
| ( 20 ) Wrong Accessory | | | | | | | |
| ( 21 ) Others | | | | | | | |
| TOTAL | | | | | | | |
| REMARK: | | | | | | | |

### 4.2.3 Measurement Report 尺寸检查报告

In the clothing industry, we have a measurement report to record garment measurement in final line or in-process line, and control the garment measurements to satisfy customer's requirements（Table 4-6）.

**Table 4-6 Measurement Report**

| CLOTHING INDUSTRY TRAINING CENTER<br>MEASUREMENT REPORT<br>CHECK BY:________ DATE:________ MARKS:________<br>BUNDLE CODE:________ SIZE:________ TIME:____TO____ | | | | | | |
|---|---|---|---|---|---|---|
| JACKET, SHIRT, ETC. | MEAS.SPEC. | TOLERANCE | 1 | 2 | 3 | 4 |
| MEASUREMENT POINTS | | | | | | |
| COLLAR | | | | | | |
| BUST/CHEST | | | | | | |
| HEM | | | | | | |
| RELEASED HEM | | | | | | |
| C/B LENGTH | | | | | | |
| ACRYLIC HEM WIDTH | | | | | | |
| SHOULDER WIDTH | | | | | | |
| SLEEVE LENGTH | | | | | | |
| CUFF | | | | | | |
| ACRYLIC CUFF WIDTH | | | | | | |
| SCYE | | | | | | |
| | | | | | | |
| REMARK: | | | | | | |

## *Words and Expressions*

inspection report [in'spekʃən ri'pɔ:t] 验货报告

audit report ['ɔ:dit ri'pɔ:t] 检查报告

fabric defects ['fæbrik di'fekts] 布疵

color shade ['kʌlə ʃeid] 色差

classification [klæsifi'keiʃən] 分类

composition [kɔmpə'ziʃən] 结合，构成

quality report ['kwɔliti ri'pɔ:t] 品质报告

audit level/ Aud. L ['ɔːdit 'lev(ə)l]. 稽查水平
remedy action ['remidi 'ækʃən] 修补行为
in-line audit [in lain 'ɔːdit] 在线检查
final audit ['fainl 'ɔːdit] 最后检查
Lot No. [lɔt nəu] 批量编号
accept/ acc. [ək'sept] 接受
reject/ rej. [ri'dʒekt] 拒绝
fabric flaws ['fæbrik flɔːz] 布料瑕疵
water spots ['wɔːtə spɔt] 水渍
fabric streak ['fæbrik striːk] 织物条花疵
general appearance ['dʒenərəl ə'piərəns] 总体外形
workmanship ['wəːkmənʃip] 手艺，技艺
hand feel [hænd fiːl] 手感
representative [repri'zentətiv] 代表，描述
b. mid. arm-hole [biː 'mid ɑːm-həul] 后背宽
f. mid. arm-hole [ef 'mid ɑːm həul] 前胸宽
intermediate [intə'miːdjət] 中间
re-inspection [riː-in'spekʃən] 翻查
mftr. / manufacturer [mænju'fæktʃərə] 制造商，厂商
QC manual [QC 'mænjuəl] 质检手册
shipping marks ['ʃipiŋ mɑːks] 正箱唛
side marks [said mɑːks] 侧箱唛
cleanliness ['kliːnlinis] 清洁
component & assembly [kəm'pəunənt ə'sembli] 服装组件
seam & stitching [siːm'stitʃiŋ] 缝型线迹
closures ['kləuʒə] 闭合件
handicraft ['hændikrɑːft] 手工
fit & balance [fit 'bæləns] 合身
spreading method ['srediŋ 'meθəd] 铺布方法
layout No. ['leiaut nəu] 排料图编号
marker ['mɑːkə] 排料（广东话：唛架，马克）
original pattern [ə'ridʒənəl 'pætən] 原版纸样
cutting piece ['kʌtiŋ piːs] 衣片（广东话：裁片）
allowance [ə'lauəns] 放缝，余量
bundling ['bʌndliŋ] 捆扎
spec./ specification [ˌspesifi'keiʃən] 规格，说明书
act./ actual [ækt 'æktjuəl] 实际
confirmation [kɔnfə'meiʃən] 批准，确定
acknowledged [ək'nɔlidʒd] 被认可的
evaluation report [iˌvælju'eiʃən ri'pɔːt] 评定报告
disposition [dispə'ziʃən] 处理，布置
laboratory test [lə'bɔrətri test] 实验室测试
content label ['kɔntentˌ 'leibl] 成分商标
flammability test [ˌflæmə'biləti test] 防火测试
soil [sɔil] 污迹
pilling ['piliŋ] 起球
uneven dyeing ['ʌniːvən 'daiiŋ] 染色不匀
wavy stitch ['weivi stitʃ] 波浪形线迹
incorrect link [inkə'rekt liŋk] 错误接缝
thread ends [θred endz] 线头尾
major defect ['meidʒə di'fekt] 主要疵点
minor defect ['mainə di'fekt] 微小疵点
uneven hem ['ʌniːvən hem] 不匀边脚
missing parts ['misiŋ pɑːts] 部件遗失
uneven plaids ['ʌniːvən plæd] 不均匀格子
misaligned [ˌmisə'laind] 排列不准齐
tolerance ['tɔlərəns] 公差，宽松度
jeans [dʒeinz] 牛仔裤
improperly [im'prɔpəli] 不适当地
run-off stitching [rʌn-ɔːf 'stitʃiŋ] 面缝线迹不平直（落坑）
broken stitching ['brəukən 'stitʃiŋ] 断线
bubbling ['bʌbl] 起泡
bundle code ['bʌndl kəud] 捆扎号
fullness ['fulnis] 不平服的 / 松动的
asymmetric [æsi'metrik] 不对称
mat./ material [mə'tiəriəl] 原材料
off-pressing [ɔːf 'presiŋ] 大烫，终烫
under-pressing ['ʌndə'presiŋ] 中烫，缝制

前熨烫
oil stain [ɔil stein] 油渍，油污
label misplaced ['leibl 'mis'pleis] 商标错位
shoulder slope ['ʃəuldə sləup] 肩斜
seam broken [si:m 'brəukən] 接缝爆裂（广东话：爆子口）
burns [bə:nz] 烫煳 *
shiny ['ʃaini] 发亮的（广东话：起镜面）
grinning stitch [grin stitʃ] 露齿线迹
seam twist [si:m twist] 缝口扭曲，扭缝口
uneven tension ['ʌni:vən 'tenʃən] 张力不均
slv. placket ['plækit] 袖衩
twisted placket [twist 'plækit] 扭门襟（广东话：明筒扭曲）
collar stand ['kɔlə stænd] 领座
collar fall ['kɔlə fɔ:l] 翻领
seam slippage [si:m 'slipidʒ] 脱缝，缝口脱开（广东话：散口）
rough yarn [rʌf jɑ:n] 粗纱
wrinkle ['riŋkl] 皱褶
missing stitch ['misiŋ stitʃ] 漏针
broken needle ['brəukən 'ni:dl] 断针
barre [bɑ:'rei] 横裆疵
open seam ['əupən si:m] 开缝
weak seam [wi:k si:m] 不牢的缝口
raw edge [rɔ: edʒ] 毛边
puckering ['pʌkəiŋ] 起皱
broken pick / broken weft ['brəukən pik / 'brəukən weft] 断纬 *
crease mark [kri:s mɑ:k] 折痕
knot [nɔt] 纱结 *
selvage ['selvidʒ] 布边 *
fuzz balls [fʌz bɔ:lz] 起毛球 *
broken end/ broken warp ['brəukən end / 'brəukən wɔ:p] 断经 *
workshop ['wə:kʃɔp] 车间 *

## *Exercises*

1. Translate the following terms into Chinese.

(1) run-off stitching
(2) side seam
(3) stitch
(4) audit report
(5) fabric defect
(6) thread ends
(7) color shading
(8) major defect
(9) uneven hem
(10) textile fibre
(11) trade dept.
(12) uneven plaids
(13) quality report
(14) audit level
(15) skipped stitching
(16) in-line audit
(17) final audit
(18) sweep
(19) jeans
(20) broken stitching
(21) seam broken
(22) fabric flaw
(23) bubbling
(24) water spots
(25) bundle code
(26) workmanship
(27) easing
(28) grinning stitch
(29) hand feel
(30) asymmetric
(31) twisted leg
(32) off-pressing
(33) under-pressing
(34) armhole
(35) burns mark
(36) shiny
(37) nape to waist
(38) oil stain
(39) size spec.
(40) seam twist
(41) placket
(42) collar
(43) rough yarn
(44) content label
(45) wrinkles
(46) care label
(47) missing stitch
(48) soiling
(49) broken needle
(50) pilling
(51) dyeing
(52) warp
(53) knot
(54) raw edge

（55）puckering
（56）overlocking
（57）size label
（58）crotch point
（59）enterprise
（60）counter sample
（61）main label
（62）fit sample
（63）intermediate
（64）re-inspection
（65）manufacture
（66）QC Manual
（67）shipping mark
（68）cleanliness
（69）seam & stitching
（70）closures
（71）handicraft
（72）fit & balance
（73）spreading method
（74）Layout No.
（75）marker
（76）original pattern
（77）notch
（78）cutting pieces
（79）allowance
（80）sewing ticket

2. List 10 types of garment defects.

Example：uneven stitching

3. Point out the following indications（Jacket with Hood）.

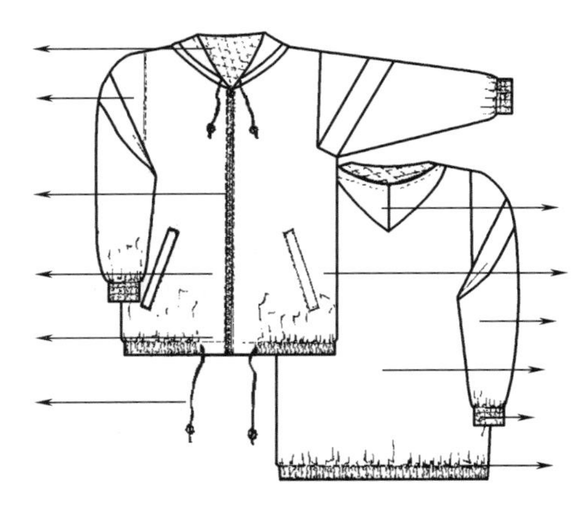

**Jacket with Hood**

4. List the major garment measurements for the following styles, and then explain how to measure them.

Example: a. Shoulder: Lay the waistcoat flat with the back facing to you, and measure straight across from shoulder point to shoulder point.（For Men's Waistcoat）

b. Hips: Lay the skirt flat, and eliminate all wrinkles at the hips-part, and then measure straight across from side-seam to side-seam at right angle with the tape.（For Jeans Skirt）

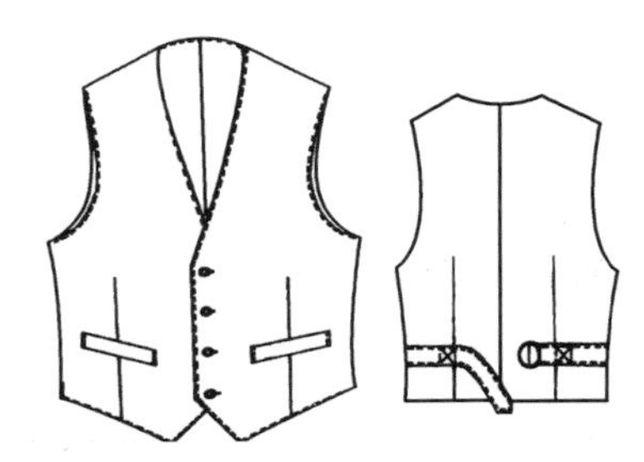

**Men's Waistcoat**

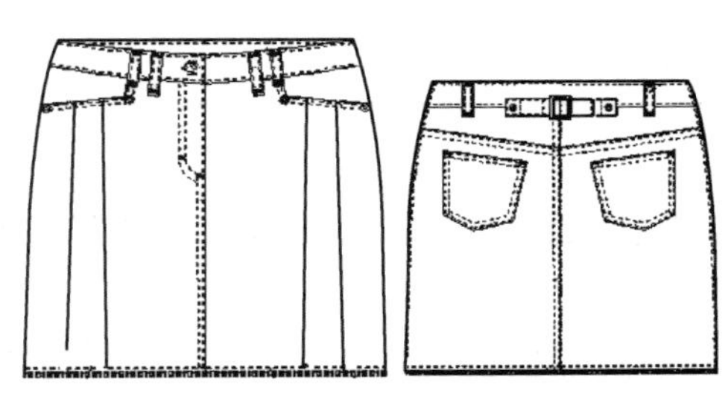

**Jeans Skirt**

## 4.3 Expression of Garment Defects 服装疵点的表述

### 4.3.1 Defective Item of Garment 服装疵点术语（Table 4–7）

Table 4–7 General Defects

| No. | Description | No. | Description |
|---|---|---|---|
| 1 | Skipped stitch | 22 | Insecure back stitching |
| 2 | Broken stitch | 23 | Oil stain |
| 3 | Uneven thread tension | 24 | Omitted or uncut buttonhole |
| 4 | Grinning stitch | 25 | Button/ buttonhole misplace |
| 5 | Wrong stitch density | 26 | Missing or insecure buttons |
| 6 | Run off stitch | 27 | Label part misplace |
| 7 | Thread not trimmed | 28 | Wrong placement of pocket flap |
| 8 | Loose thread | 29 | Fabric dirty / soiling |
| 9 | Tight thread | 30 | Poorly shaped of pocket |
| 10 | Uneven zipper facing | 31 | Zipper show out |
| 11 | Stitching not uniform | 32 | Collar points not uniform |
| 12 | Seam pucker | 33 | Collar fullness or puckering |
| 13 | Mismatched seam | 34 | Asymmetrical garment part |
| 14 | Raw edge shows | 35 | Garment parts sewing bubbles |
| 15 | Seam pleat | 36 | Garment parts missing |
| 16 | Improper seam setting | 37 | Garment part improper position |
| 17 | Uneven width of inlay | 38 | Uneven edge / seam on showing |
| 18 | Seam twisted & puckered | 39 | Incorrect polybay |
| 19 | Wrong shade of thread | 40 | Unsightly over-lapping stitch |
| 20 | Label up side down | 41 | Misaligned at crotch / under-seam |
| 21 | Fabric defect | 42 | Size label wrongly placement |

### 4.3.2 Defect of Jacket 夹克疵点（Table 4–8）

Table 4–8 Jacket Defects

| No. | Description | No. | Description |
|---|---|---|---|
| 1 | High / low collar ends | 3 | Missing size label |
| 2 | Uneven pocket jet | 4 | Broken knitted at cuff |

**Continued**

| No. | Description | No. | Description |
|---|---|---|---|
| 5 | Uneven pocket length | 20 | Pleated seam on zipper facing |
| 6 | High/ low hem facing | 21 | Sleeve misplaced |
| 7 | Irregular top stitch on zipper facing | 22 | Fabric flaw on lining |
| 8 | Wrongly used zipper | 23 | Pleated seam on front facing |
| 9 | Uneven ply on yoke | 24 | Uneven length of slv. |
| 10 | Irregular top stitch on back panel | 25 | Twisted sleeve |
| 11 | Pleated seam on slv. | 26 | No stitch setting on sleeve side seam |
| 12 | Seam broken on collar tip | 27 | Broken stitch on front yoke |
| 13 | Visible hided jet pkt. facing | 28 | Broken stitch on front pkt. |
| 14 | Edge of center front not closing | 29 | Shorter lining on front facing |
| 15 | No stitch setting between outer and inter sleeve | 30 | Broken stitch on shoulder of lining |
| 16 | Loose stitch seriously affecting appearance | 31 | Inter pocket wrongly placed |
| 17 | High/ low yoke at side seam | 32 | Seam broken on lining and hem facing |
| 18 | Broken stitch on under collar | 33 | Broken stitch on zip. facing |
| 19 | Stitch broken on inter pocket | 34 | Bubbling on front facing |

### 4.3.3 Defects of Dress 长裙疵点 (Table 4-9)

**Table 4-9 Dress Defects**

| No. | Description | No. | Description |
|---|---|---|---|
| 1 | Color shading with front & back | 11 | Material damaged at top of zipper setting |
| 2 | Uneven zipper jet | 12 | Side seam belt-loop missing |
| 3 | Unbalanced front gathering | 13 | Center front belt-loop too large / small |
| 4 | Uneven gathering waist | 14 | Misplaced instruction label |
| 5 | Wrong size zipper | 15 | Bottom lining hemming fraying |
| 6 | Uneven bottom hemming | 16 | Missing size label / washing label |
| 7 | Dressing lining too tight | 17 | Pressing marks / discoloration |
| 8 | Poor pressing | 18 | Incorrectly folded edge |
| 9 | Stain / Dirt on body | 19 | Wrong grain line cutting |
| 10 | Wrong packing method | | |

### 4.3.4 Defect of Jeans 牛仔服疵点（Table 4–10）

Table 4–10 Jeans Defects

| No. | Description | No. | Description |
|---|---|---|---|
| 1 | Rough yarn | 23 | Stud insecure |
| 2 | Uneven dye | 24 | Uneven front fly width |
| 3 | Shaded parts | 25 | Front fly length too long |
| 4 | Waistband stitch skipped | 26 | Waistband twisted |
| 5 | High / low waistband ends | 27 | Wrong waistband height |
| 6 | Waistband end not close to front fly | 28 | Stitch per inch less than specified |
| 7 | Raw edge at body | 29 | Incorrect shape of back pocket |
| 8 | Uneven belt-loop size | 30 | Twisted bottom |
| 9 | Missing belt-loop | 31 | Stud misplaced |
| 10 | Pocket lining caught in bar-tacking | 32 | Exposed drill holes will be not closed |
| 11 | Too short pocket bag | 33 | Uneven top-stitch at left fly |
| 12 | Inseam skipped stitch | 34 | Belt-loop not as sample |
| 13 | Under layer exposed fullness | 35 | Missing bartack |
| 14 | Pocket setting misplaced | 36 | Measurement out of tolerance |
| 15 | Pocket edge not straight | 37 | Wrong care label |
| 16 | Too much seam allowance in setting pocket | 38 | The under crotch meet point do not matched |
| 17 | Uneven size of back yoke | 39 | Zipper cannot reach at top |
| 18 | Improper layer shown out | 40 | Out-seam run off stitch |
| 19 | Mismatched meet point of back yoke | 41 | Fraying the end of catching facing |
| 20 | Pocket-bag not extend to front fly | 42 | Fold fabric by improper bar-tack setting |
| 21 | Uneven tension at front fly top stitch | 43 | Uneven tension at pkt. top stitch |
| 22 | Inseam grinning stitch | | |

## 4.3.5 Defect of Shirt 衬衫疵点（Table 4-11）

Table 4-11 Shirt Defects

| No. | Description | No. | Description |
|---|---|---|---|
| 1 | Collar edge asymmetric | 27 | Uneven distance of slv. pleat |
| 2 | Uneven collar stand | 28 | Cuff assembles to slv. not closed |
| 3 | Collar tip grinning stitch | 29 | Collar assemble to body not close |
| 4 | Under layer exposed | 30 | Seam slippage at bottom hemming |
| 5 | Uneven tension at top-stitch collar | 31 | Collar bubbling due to incorrect handling |
| 6 | Placket bubbling | 32 | Seam broken at collar stand end |
| 7 | Collar bubbling due to uneven pressing | 33 | Pleats or puckers outside seriously affecting appearance |
| 8 | Twisted placket | 34 | Button not sewn through all eyes |
| 9 | Incorrect pocket size | 35 | Button sewn on face down |
| 10 | Under ply of collar appeared after collar top stitch | 36 | Buttonholes cut in opposite direction to specification |
| 11 | Defective buttons | 37 | One or both collar stays missing |
| 12 | Collar size incorrect | 38 | Wrong type seam |
| 13 | Seam broken of cuff edge | 39 | High / Low pkts. & flaps |
| 14 | Uneven of collar tip | 40 | Seam broken in sleeve placket |
| 15 | Uneven length of front placket | 41 | Fullness on the under ply of the cuff |
| 16 | Pleated at sleeve joining | 42 | Incorrectly folded edge |
| 17 | Improperly pressed | 43 | Fullness or twist of cuff |
| 18 | Collar tip seam broken | 44 | Mismatched checks or stripes |
| 19 | Wrong edge margin | 45 | Uneven collar fall shape |
| 20 | Uneven height in both side of cuff | 46 | Woven label 1/2" or more off center |
| 21 | Thread impression fused into collar | 47 | Uneven tension of top stitch cuff |
| 22 | Side edge of pkt. exposed beyond side edge of flap | 48 | Shirt boards & collar boards missing where specified |
| 23 | Buttonholes stitching too narrow or not catching fabric | 49 | Neck buttonhole more than 1/8" off center on collar stand |
| 24 | Raw edge safety-stitch seam | 50 | Glaring shine mark on outside |
| 25 | Vent bar-tack causing pleat | 51 | Pkt. edge not in straight line |
| 26 | Pressed in pleats | | |

## Words and Expressions

affect appearance [ə'fekt ə'piərəns] 影响外形
uniform ['ju:nifɔ:m] 一致的，相同的
thread tension [θred 'tenʃən] 缝线张力
crotch meet point [krɔtʃ mi:t pɔint] 十字裤裆
improper [im'prɔpə] 错误的，不一致的
inlay ['in'lei] 镶嵌
insecure [,insi'kjuə] 不牢固
over-lapping stitch ['əuvə-læpiŋ stitʃ] 驳缝线迹
misaligned [,misə'laind] 不成一直线
omitted [əu'mit] 省略，遗漏
misplaced ['mis'pleis] 错位
missing ['misiŋ] 错失
up side down [ʌp said daun] 倒转
opposite direction ['ɔpəzit di'rekʃən] 相反方向
collar point/ tip ['kɔlə pɔint / tip] 领尖
asymmetric [æsi'metrik] 不对称
glaring shine mark ['glεəriŋ ʃain ma:k] 反光（广东话：起镜面）
stitch setting [stitʃ setiŋ] 合缝
hided jet pkt. [haidid dʒet 'pɔkit] 暗袋
ends not closing [endz nɔt 'kləuziŋ] 不贴紧（广东话：凸嘴）
front facing [frʌnt 'feisiŋ] 前襟贴
hem facing [hem 'feisiŋ] 下摆折边
twist [twist] 扭曲，歪曲
irregular [i'regjulə] 不规则的，不整齐的
back panel [bæk 'pænl] 后衣片（广东话：后幅）
inter pocket [in'tə: 'pɔkit] 内袋，暗袋
press marks [pres ma:ks] 烫痕
poor pressing [puə 'presiŋ] 烫工差
out of tolerance [aut ɔv, 'tɔlərəns] 超出松量
front fly [frʌnt flai] 前门襟
specify ['spesifai] 说明，要求
run off stitch [rʌn ɔ:f stitʃ] 面缝线迹不平直（落坑）
stud [stʌd] 装饰纽扣
exposed [iks'pəuzd] 暴露的
under layer ['ʌndə 'leiə] 下层
reverse [ri'və:s] 反面，倒转
mismatch ['mis'mætʃ] 不配合
catch facing [kætʃ 'feisiŋ] 贴边，挂面
front placket [frʌnt 'plækit] 前开口，开襟
seam slippage [si:m 'slipidʒ] 脱缝，缝口脱开（广东话：散口）
face down [feis daun] 向下
fraying [freiiŋ] 散边，毛边（广东话：散口）
collar stay ['kɔlə stei] 领插骨片（广东话：领插竹）
placement ['pleismənt] 位置
flap [flæp] 袋盖
sleeve joining [sli:v 'dʒɔiniŋ] 上袖
vent [vent] 衩位
edge margin / S.A. [edʒ 'ma:dʒin] 子口

## Exercises

1. Translate the following terms into Chinese.

(1) glaring shine mark
(2) thread
(3) crotch point
(4) jet pkt
(5) S.A.
(6) front facing
(7) over-lapping stitch
(8) hem
(9) back panel
(10) inter pocket
(11) press mark
(12) collar point
(13) pressing
(14) tolerance
(15) front fly

（16）seam slippage
（17）run off stitch
（18）collar stay
（19）press stud
（20）flap
（21）sleeve vent
（22）catch facing
（23）seam
（24）front placket

2. List 10 types of jeans defects.
3. List 10 kinds of garment parts for the Tailored Jacket.
   Example: collar, etc.

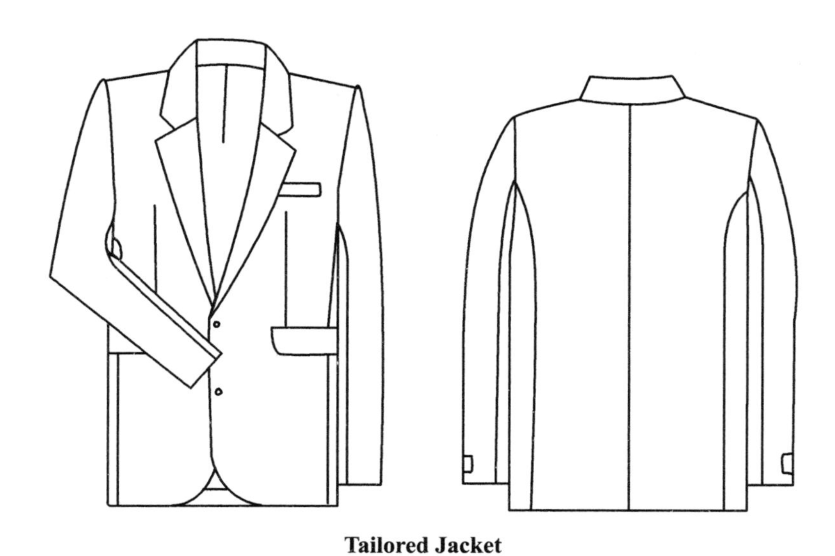

**Tailored Jacket**

# 译文

## 第四章　品质检验

### 4.1　品质标准

#### 4.1.1　简介

在制衣业，“品质检查”是指细心观察并检查实际的产出品并将其与需求标准作比较。检验活动本身并没有为产品增加任何东西。这只是一个手段，获得的信息可以为下一行动作基础。由各种各样的方法组成的“检查”功能被用于制衣业中。

#### 4.1.2　品质标准的表示

对于质量标准的表示，我们必须参考已建公司规范和更多的标准样品。从这些表示中可发现如下信息：

（1）缝口：缝口是指能把两层或多层面料缝合在一起的区域。衣服要穿着完美必须有极好的缝口，且缝口处理与缝纫线选择必须与面料、款式、所需性能以及制造商所承诺的标准一致。

（2）原材料：面料和辅料必须是令人满意的。例如，面料不能有色差、损坏、断纱、任何折痕、粗纱、抽丝和起球等。辅料必须符合规格要求。

（3）下摆：任何服装都有下摆缝合设计，不同的下摆大小能实现特殊的款式效果。总体来说，越好的工艺会有越好的下摆。更好的工艺也显示出下摆在缝合定位之前已经经过边位整理。较粗糙的工艺则更易在衣服外面显露不良下摆，甚至会吊起不均。完美的下摆可代表一件高质服装。

（4）孔眼：孔眼是在面料上容许纽扣穿过并固定在另一块面料的小孔，孔眼缝合可由用机器或手工进行制作，孔眼处理后必须进行嵌边或滚边处理。手缝线迹一般用在洋服定制、刺绣以及花边制作工艺。

（5）纽扣：很明显，粗糙的工艺会制作出廉价的纽扣，精湛的工艺能制作出优质的纽扣。在廉价与中等工艺技术中，可能只用一根缝线进行钉扣工序，从而可能松脱线迹。在精湛的纽扣缝纫工艺中，必须通过机器或手工缝制安全线迹。

（6）合体：更好的工艺与良好的裁剪会使衣服更加舒适合体或有更多的活

动范围。廉价的工艺会使衣服容易变形，给予的活动空间也很小。尤其是袖子会让人特别不舒适且很快扭曲变形。总的来说，廉价的工艺会使服装裁剪不足，好的工艺会使衣服更加完美。

（7）总体外观：好的工艺是指良好的裁剪、好的制作、整洁的衣里。它穿起来非常的合体、舒适。配料必须是高质量的而且能很好地匹配，它通常可作为衣服的一个特征。领子、袖级、褶裥、口袋、腰头和其他部件必须能很均衡且方便悬挂。

### 4.1.3 检查核对表

在指定的生产环节中要检查服装和零部件。所有的服装都要经过由一组或者一队专业人员组成的检验部门按既定标准进行的检验。这种方法特别适合检验在同一时间生产的许多不同款式的服装。以下检查表和例子为车间运作提供参考。

（1）衣身面料

（a）规格（面料成份、结构等）

（b）织物疵点

（c）缝针痕迹

（d）缩水

（e）色牢度

（f）色调变化（在同件服装或不同服装之间的色差）

（g）断线

（h）破洞 / 纱结

（i）斑点 / 污点

（j）洗水痕

（2）辅料

（a）里料

（b）里衬 / 魔术贴 / 松紧带

（c）纽扣 / 按纽 / 拉链 / 钩眼扣

（d）铆钉 / 孔眼 / 打套结

（e）缝线

（f）皮带

（3）成衣洗水 / 染色

（a）硬度

（b）色差

（c）异味

（d）污点 / 折痕

（e）染色痕迹

（4）裁剪

（a）条纹或布边方向

（b）不正确的尺码规格

（5）印花 / 刺绣

（a）不清晰或叠印

（b）印花时沾污

（c）颜色不均匀

（d）错色、设计错误或者方向错误

（e）不完整

（f）位置不符

（g）尺码不同

（6）缝口

（a）扭曲 / 打褶裥

（b）起皱

（c）子口不足

（d）毛边外露

（e）开缝

（f）缝纫强力弱

（g）缝口位置与缝口不对位

（7）线迹

（a）线迹不在同一直线上

（b）针密太低 / 太高

（c）缝线张力

（d）漏针 / 跳针 / 断针 / 线迹落坑

（e）线迹重叠 / 接线

（f）缝线颜色不对

（8）熨烫 / 剪线

（a）皱褶

（b）起镜面

（c）熨烫时间不够 / 熨烫时间太长

（d）烫坏烧焦

（e）未修剪缝线

（f）折叠不适当

（9）商标

（a）主唛商标 / 尺寸商标

（b）洗水商标 / 成分商标

（c）包装袋商标

（d）吊牌

（e）商标位置

（f）商标线迹

（10）包装

（a）包装袋尺寸 / 用词描述

（b）折衣尺寸

（c）缺少内包装盒

（d）外箱类型 / 尺寸

（e）缺少箱唛 / 箱唛不清晰

（f）长度不等

（g）扭曲

（11）口袋

（a）高低袋

（b）斜袋口

（c）袋子尺寸

（d）形状欠佳 / 位置错误

（e）毛边

（f）扭曲 / 起皱 / 起褶

（g）套结或回针遗漏

（h）点袋位外露（未遮盖）

（i）条格不对

（12）门襟

（a）形状欠佳

（b）扭曲 / 凸起

（13）拉链

（a）故障

（b）缝法不恰当

（c）起伏不平

（d）尺寸、长度和颜色

（14）纽扣 / 按纽 / 纽孔

（a）纽扣歪斜

（b）遗漏

（c）纽扣错位

（d）劣质纽扣

（e）纽扣 / 按纽不牢固

（f）纽扣倒置

（g）纽扣之间起伏不平

（h）缺少备用纽扣

（i）不正确的尺码 / 形状

（15）裤耳 / 打套结 / 铆钉

（a）遗漏 / 不牢固

（b）位置错误

（c）太窄 / 宽

（d）不牢固

## 4.2 检验报告

质量控制积存了大量有关车间操作的数据。这些数据被用来制作以下报告：在制品检验报告、操作人员质量表现报告、最终检查报告、描述与分析修补报告、缝纫和裁剪造成的缺损及织物原材料疵点占百分比的检验报告。检验报告不仅记录了货存商品的色差信息，而且现在可以使用电脑对货物进行色差检查。直接输入其他在制品的质量控制数据的操作在逐渐增加。

### 4.2.1 产品检查报告

产品检查报告主要起品质控制的作用，目的是在制造和发现疵点的过程中对产品质量进行监控。这些每日积累来的信息被用于编制每日和每周的质量报告。电脑相当于一个质量控制的稽查员，对检查数据进行加工并编制所需报告。检查过程会包括在制品检查和最终产品检查（表4-1 ~ 表4-3）。

**表 4-1 检查报告（Ⅰ）**

**RYKIEL 贸易有限公司**

**内部质量稽查报告**

（ ）在制品 （ ）成品

买家：______ 采购订单编号：______ 款式名称：______

加工厂：______ 工厂编号：______ 运送方式：______

交货期：______ 稽核日期：______ 口岸：______

总数量：______ 稽查样板码：______

批量编号：______ 接受 / 拒接水平：______

颜色：______ 稽查人：______

产品描述：______

| | |
|---|---|
| | 腰围（大于或小于 1/2"） |
| | 内长（大于或小于 1/2"） |
| | 内商标（位置错误或不牢） |
| | 外商标（位置错误或不牢） |
| | 跳线（分开缝或面缝线迹） |
| | 扣眼（漏开或没有剪开口） |
| | 纽扣（漏钉或损坏） |
| | 面料瑕疵（孔洞、色差） |
| | 污点（污垢、油污） |
| | 整体外观（车间尾期、洗水后、熨烫后） |

| 总疵点数 | | 接受（ ） 拒接（ ） |
|---|---|---|
| | | |

| | 接受 | 拒接 | | 接受 | 拒接 |
|---|---|---|---|---|---|
| 款式细节 | （ ） | （ ） | 色差 | （ ） | （ ） |
| 洗水 | （ ） | （ ） | 面料 | （ ） | （ ） |
| 熨烫 | （ ） | （ ） | 裤腿扭曲 | （ ） | （ ） |
| 手感 | （ ） | （ ） | 包装 | （ ） | （ ） |
| 尺寸 | （ ） | （ ） | 辅料 | （ ） | （ ） |
| 箱唛 | （ ） | （ ） | 手工艺 | （ ） | （ ） |

工厂负责量尺寸的工人数量：______

已检查的纸箱编号：______

备注：______

______ RYK 代表 ______ 加工厂代表

续表

**RYKIEL贸易有限公司**

**尺寸测量表(上装)**

客商：________　　日期：________

订单编号：________　　款式：________

合同编号：________　　数量：________

| 尺码 / 测量部位 | 规格 | 实际 | 实际 | 规格 | 实际 | 实际 |
|---|---|---|---|---|---|---|
| 胸围（袖窿下1英寸测量） | | | | | | |
| 腰围 | | | | | | |
| 下摆 | | | | | | |
| 肩宽 | | | | | | |
| 前胸宽 | | | | | | |
| 袖长 | | | | | | |
| 袖窿 | | | | | | |
| 袖臂宽 | | | | | | |
| 后中长 | | | | | | |
| 后背宽 | | | | | | |
| 领圈深 | | | | | | |
| 领围 | | | | | | |
| 口袋 | | | | | | |
| 帽高 | | | | | | |
| | | | | | | |

备注：

**续表**

**RYKIEL 贸易有限公司**

**尺寸测量表（下装）**

客商：________________　　　　日期：________________

订单编号：________________　　　　款式：________________

合同编号：________________　　　　数量：________________

| 尺码 / 测量部位 | 规格 | 实际 | 实际 | 规格 | 实际 | 实际 |
|---|---|---|---|---|---|---|
| 腰围（放松） | | | | | | |
| 腰围（拉紧） | | | | | | |
| 上臀围 | | | | | | |
| 臀围 | | | | | | |
| 股上围 | | | | | | |
| 膝围 1 | | | | | | |
| 膝围 2 | | | | | | |
| 下摆 | | | | | | |
| 前裆 | | | | | | |
| 后裆 | | | | | | |
| 内长 | | | | | | |
| 外长 | | | | | | |
| 拉链 | | | | | | |
| | | | | | | |

备注：

表 4-2　检查报告（Ⅱ）

| 质控部检查报告<br>（　）中期　（　）尾期　（　）翻查 | | | | | |
|---|---|---|---|---|---|
| 顾客： | 订单编号： | | 日期： | 时间： | |
| 制造商： | 款式： | | 数量： | 交付期： | |
| 服装： | 质量可接受水平： | | 工厂编号： | | |
| 注意：检查方法与生产制造通知单的 QC 手册相符 | | | | | |
| **检查要点** | 正确 | 错误 | **检查要点** | 正确 | 错误 |
| A. 商标 | | | F. 面料颜色 | | |
| B. 款式 / 颜色 | | | G. 皮带颜色 | | |
| C. 箱唛 / 侧箱唛 | | | H. 缝纫线颜色 | | |
| D. 吊牌描述 | | | I. 拉链颜色 | | |
| E. 折叠与包装 | | | J. 纽扣颜色 | | |
| **疵点** | 主要 | 微小 | **疵点** | 主要 | 微小 |
| 01. 面料疵点 | | | 10. 开口 | | |
| 02. 色差问题 | | | 11. 损坏 | | |
| 03. 染色问题 | | | 12. 缝纫线 | | |
| 04. 印花问题 | | | 13. 手工艺 | | |
| 05. 清洁程度 | | | 14. 合身与平衡 | | |
| 06. 部件与合并工序 | | | 15. 折痕 | | |
| 07. 缝骨与线迹 | | | 16. 后整理与手工 | | |
| 08. 熨烫 | | | 17. 包装 | | |
| 09. 尺寸 | | | 总疵点件数： | | |
| 已检查：______件 | | | 检查的纸箱编号：________ | | |

| **细节** | **%** | **建议与备注** |
|---|---|---|
| A. | | |
| B. | | |
| C. | | |
| D. | | |
| E. | | |

| 服装尺码与数量 | | | | | | | | | | |
|---|---|---|---|---|---|---|---|---|---|---|
| | | | | | | | | | | |

| | |
|---|---|
| （　）好　（　）满意　（　）一般<br>（　）不接受　（　）担保走货 | 检查人员：____________<br>包装主管：____________<br>QC 经理：____________ |

**表 4–3 检查报告（Ⅲ）**

| **RYKIEL 成衣有限公司**<br>**裁剪质量检查报告**<br>检查日期：__________ | | |
|---|---|---|
| 订单编号： 客商： 产品描述： | | |
| 款式编号： 数量： 颜色： | | |
| 铺料方法： 床号： 铺料数量： | | |
| **Ⅰ. 铺料图** | 是 | 否 |
| A. 铺料图宽度与布幅是否一致？ | | |
| B. 铺料图是否采用原版纸样？ | | |
| C. 裁片是否需要打剪口？ | | |
| D. 铺料图划线是否顺滑？ | | |
| E. 铺料图的纸样数量是否遗漏？ | | |
| F. 衔接布料位置是否正确？ | | |
| G. 铺料图的裁片尺码是否与纸样相同？ | | |
| 检查日期： | | |
| **Ⅱ. 裁片** | 是 | 否 |
| A. 上下层裁片的尺码是否一致？ | | |
| B. 裁片外形是否顺滑？ | | |
| C. 裁片铺料图是否正确？ | | |
| D. 裁片数量是否正确？ | | |
| E. 摆放布料的起始宽余位是否标准？ | | |
| F. 裁剪前铺料图是否放在正确的位置？ | | |
| G. 布料是否合掌铺放？ | | |
| H. 是否预存布板卡或布板？ | | |
| 检查日期： | | |
| **Ⅲ. 捆扎** | 是 | 否 |
| A. 工票细节是否与裁片一致？ | | |
| B. 工票细节与捆扎数量是否一致？ | | |
| C. 工票标示的数量是否与捆扎数量一致？ | | |
| 检查日期： | | |
| 备注：<br><br>裁床主管：__________ 检查员：__________ QC 主管：__________ | | |

### 4.2.2 疵点检查报告

检查报告表用来记录检查服装的信息，比如服装疵点检查。质量控制员或者检查员会在检查产品时带上表格。所有被发现的疵点会被记录。该报告会呈交给跟单员或跟单部经理，用以进一步的修补行动（表 4–4、表 4–5）。

**表 4–4 疵点检查报告（Ⅰ）**

| 在制品检查报告<br>疵点列表<br>款式编号_______页码_________ | | | | | |
|---|---|---|---|---|---|
| 面料疵点 | 主要 | 微小 | 缝纫疵点 | 主要 | 微小 |
| 孔洞 | | | 分开缝 | | |
| 污渍 | | | 缝骨不牢 | | |
| 瑕疵 | | | 毛边 | | |
| 起球 | | | 波浪纹线迹 | | |
| 染色不匀 | | | 跳线 | | |
| 烧焦痕迹 | | | 断线 | | |
| 横裆疵点 | | | 连接不正确 | | |
| | | | | | |
| 服装疵点 | 主要 | 微小 | 熨烫疵点 | 主要 | 微小 |
| 面料颜色不匹配 | | | 下摆不均匀 | | |
| 部件颜色不匹配 | | | 部件不直 | | |
| 有缺陷的按扣 | | | 部件遗漏 | | |
| 有缺陷的拉链 | | | 不均匀格子 | | |
| 拉链暴口 | | | 起皱 | | |
| 线尾过长 | | | 其他 | | |
| 纽扣松脱 | | | | | |
| 有缺陷的纽孔 | | | | | |
| | | | | | |
| 需检查件数：<br>已检查件数：________　可接受水平：________<br>拒接件数：________　微小疵点：________ | | | | | |
| 日期 | 备注<br>生产环节： | | QC 采取的行动 | 供应商签名 | |

续表

| | | | |
|---|---|---|---|
| | | | |
| | | | |
| 供应商有责任改正在检查中发现或在此报告表总结的疵点。不管怎样，供应商都要对装船出货时发现的疵点负责 | | | |
| 供应商签名：________________ 日期：________________<br>检查人签名：________________ 日期：________________ | | | |

**表 4–5 疵点检查报告（II）**

| 制衣工业培训中心<br>疵点检查报告<br>捆扎号： | | | | | | | |
|---|---|---|---|---|---|---|---|
| 疵点项目 | 1 | 2 | 3 | 4 | 5 | 6 | 总共 |
| （1）服装部件不对称 | | | | | | | |
| （2）服装部件起泡 | | | | | | | |
| （3）服装部件遗失 | | | | | | | |
| （4）十字缝骨不对位 | | | | | | | |
| （5）缝合不正确 | | | | | | | |
| （6）缝骨打褶裥 | | | | | | | |
| （7）缝骨起皱 | | | | | | | |
| （8）服装部件错位 | | | | | | | |
| （9）跳线 | | | | | | | |
| （10）断线 | | | | | | | |
| （11）回针欠佳 | | | | | | | |
| （12）侧缝缉线落坑 | | | | | | | |
| （13）缉线张力不均 | | | | | | | |
| （14）熨烫不恰当 | | | | | | | |
| （15）面缝线迹不均匀 | | | | | | | |
| （16）工序遗漏 | | | | | | | |
| （17）缝骨暴口 | | | | | | | |
| （18）配件遗漏 | | | | | | | |
| （19）面料疵点 | | | | | | | |
| （20）辅料错误 | | | | | | | |
| （21）其他 | | | | | | | |
| 总共 | | | | | | | |
| 备注： | | | | | | | |

### 4.2.3 尺寸检查报告

在制衣工业中，我们采用尺寸检查报告记录在最后环节或生产线上的衣服尺寸，并限定服装尺寸以便满足顾客的要求（表 4–6）。

表 4–6 尺寸测量报告

制衣工业培训中心

尺寸测量报告

检查:________ 日期:________ 评分:________

捆扎号:________ 尺码:________ 时间______至______

| 夹克、衬衫等 | 尺码规格 | 宽容位 | 1 | 2 | 3 | 4 |
| --- | --- | --- | --- | --- | --- | --- |
| 测量部位 | | | | | | |
| 领围 | | | | | | |
| 胸围 | | | | | | |
| 下摆 | | | | | | |
| 放松的下摆 | | | | | | |
| 后中长 | | | | | | |
| 下摆宽 | | | | | | |
| 肩宽 | | | | | | |
| 袖长 | | | | | | |
| 袖级 | | | | | | |
| 袖级宽 | | | | | | |
| 袖窿 | | | | | | |
| | | | | | | |

备注:

# 4.3 服装疵点的表述

## 4.3.1 服装疵点术语（表 4-7）

表 4-7 常规疵点

| 编号 | 疵点描述 | 编号 | 疵点描述 |
|---|---|---|---|
| 1 | 跳线 | 22 | 回针不牢固 |
| 2 | 断线 | 23 | 油污 |
| 3 | 缝线张力不均匀 | 24 | 扣眼漏开 |
| 4 | 齿牙状线迹 | 25 | 纽扣 / 扣眼错位 |
| 5 | 线迹密度错误 | 26 | 纽扣漏钉或不牢 |
| 6 | 缉线落坑 | 27 | 商标错位 |
| 7 | 没有剪线 | 28 | 袋盖错位 |
| 8 | 缝线太松 | 29 | 布料油污 |
| 9 | 缝线太紧 | 30 | 袋形欠佳 |
| 10 | 链贴不均匀 | 31 | 拉链外露 |
| 11 | 线迹不均匀 | 32 | 领尖不均匀 |
| 12 | 缝型起皱 | 33 | 领子起皱 |
| 13 | 缝骨不对位 | 34 | 服装部件不对称 |
| 14 | 毛边外露 | 35 | 服装部件起泡 |
| 15 | 缝骨打褶 | 36 | 服装部件遗漏 |
| 16 | 缝骨做法不恰当 | 37 | 服装部件位置不对 |
| 17 | 子口不均匀 | 38 | 外露边位 / 缝骨不均匀 |
| 18 | 缝骨扭曲或起皱 | 39 | 包装袋不正确 |
| 19 | 缝纫线颜色错误 | 40 | 重线 / 驳线不美观 |
| 20 | 商标倒置 | 41 | 裤裆 / 袖底缝不对位 |
| 21 | 布料疵点 | 42 | 尺码商标错位 |

## 4.3.2 夹克疵点（表 4-8）

表 4-8 夹克疵点

| 编号 | 疵点描述 | 编号 | 疵点描述 |
|---|---|---|---|
| 1 | 高低领嘴 | 3 | 遗漏尺码商标 |
| 2 | 袋嵌边不均匀 | 4 | 针织袖口暴口 |

续表

| 编号 | 疵点描述 | 编号 | 疵点描述 |
|---|---|---|---|
| 5 | 袋长不均 | 20 | 拉链贴缝骨打褶 |
| 6 | 高低下摆贴 | 21 | 袖子错位 |
| 7 | 链贴面缝线迹不规范 | 22 | 里布疵点 |
| 8 | 用错拉链 | 23 | 前襟贴缝骨打褶 |
| 9 | 育克底面层不均 | 24 | 袖长不均匀 |
| 10 | 后片面缝线迹不规范 | 25 | 袖子扭曲 |
| 11 | 袖缝打褶 | 26 | 袖侧缝未缝合 |
| 12 | 领肩暴口 | 27 | 前幅育克断线 |
| 13 | 袋贴外露 | 28 | 前袋断线 |
| 14 | 前中边位不齐 | 29 | 前幅襟贴位里布太短 |
| 15 | 面里袖之间未缝线迹 | 30 | 里布肩缝断线 |
| 16 | 缝线太松严重影响外形 | 31 | 内袋错位 |
| 17 | 在侧缝处育克高低 | 32 | 下摆贴和里布之间暴口 |
| 18 | 底领断线 | 33 | 链贴断线 |
| 19 | 内袋断线 | 34 | 前襟贴起泡 |

## 4.3.3 长裙疵点（表 4–9）

表 4–9 长裙疵点

| 编号 | 疵点描述 | 编号 | 疵点描述 |
|---|---|---|---|
| 1 | 前后片色差 | 11 | 拉链顶端面料破损 |
| 2 | 链贴不均匀 | 12 | 侧缝襻带遗漏 |
| 3 | 前幅碎褶不均匀 | 13 | 前中襻带太大 / 小 |
| 4 | 腰头碎褶不均 | 14 | 指示商标错位 |
| 5 | 拉链错码 | 15 | 里子下摆散口 |
| 6 | 下摆缝边不均 | 16 | 遗漏尺码商标 / 洗水商标 |
| 7 | 裙里太短 | 17 | 烫痕 / 变色 |
| 8 | 熨烫欠佳 | 18 | 折边不正确 |
| 9 | 裙身有油污 | 19 | 布纹错误裁剪 |
| 10 | 包装方法错误 | | |

### 4.3.4 牛仔服疵点（表 4–10）

表 4–10 牛仔裤疵点

| 编号 | 疵点描述 | 编号 | 疵点描述 |
| --- | --- | --- | --- |
| 1 | 粗纱 | 23 | 纽扣不牢 |
| 2 | 染色不匀 | 24 | 前门襟宽度不均 |
| 3 | 部件色差 | 25 | 前门襟太长 |
| 4 | 腰头缉线跳线 | 26 | 腰头扭曲 |
| 5 | 高低裤腰嘴 | 27 | 腰头高度错误 |
| 6 | 腰头嘴凸出 | 28 | 每英寸线迹比规定少 |
| 7 | 裤身暴口 | 29 | 后袋形状错误 |
| 8 | 襻带大小不均 | 30 | 下摆扭曲 |
| 9 | 襻带遗漏 | 31 | 纽扣错位 |
| 10 | 套结钉住袋布 | 32 | 钻孔位外露 |
| 11 | 袋布太短 | 33 | 左门襟面缝线迹不均 |
| 12 | 内缝跳线 | 34 | 襻带与样板不一致 |
| 13 | 下层太松 | 35 | 套结遗漏 |
| 14 | 口袋缝合错位 | 36 | 尺寸超出宽余位 |
| 15 | 袋边不直 | 37 | 洗水商标错误 |
| 16 | 褙袋子口太大 | 38 | 裤裆缝不对位 |
| 17 | 后育克尺寸不均 | 39 | 拉链未伸到顶位 |
| 18 | 布层外露不恰当 | 40 | 侧缝线迹落坑 |
| 19 | 后育克缝不对位 | 41 | 门襟贴底端散口 |
| 20 | 袋布未延伸到门襟位 | 42 | 打套结不恰当使布料折叠 |
| 21 | 前门襟面缝线迹张力不均 | 43 | 口袋面缝线迹张力不均 |
| 22 | 内裆缉线齿牙状 | | |

### 4.3.5 衬衫疵点（表 4–11）

表 4–11 衬衫疵点

| 编号 | 疵点描述 | 编号 | 疵点描述 |
| --- | --- | --- | --- |
| 1 | 领边不对称 | 7 | 熨烫不均匀使领子起泡 |
| 2 | 领座不均 | 8 | 门襟开口扭曲 |
| 3 | 领尖缉线齿牙状 | 9 | 口袋尺寸不正确 |
| 4 | 下层外露 | 10 | 领子缉线后下层外露 |
| 5 | 领子面缝线迹张力不均 | 11 | 有疵点的纽扣 |
| 6 | 门襟开口起泡 | 12 | 领子尺寸不正确 |

续表

| 编号 | 疵点描述 | 编号 | 疵点描述 |
| --- | --- | --- | --- |
| 13 | 袖克夫边位暴口 | 33 | 打褶或起皱严重影响外形 |
| 14 | 领尖不对称 | 34 | 纽扣未钉缝所有孔位 |
| 15 | 前门襟开口长短不一 | 35 | 纽扣面反置 |
| 16 | 绱袖打褶 | 36 | 扣眼开口与规格方向相反 |
| 17 | 熨烫不恰当 | 37 | 一个或两个领子插竹遗漏 |
| 18 | 领尖暴口 | 38 | 缝型错误 |
| 19 | 子口错误 | 39 | 高低口袋或袋盖 |
| 20 | 袖克夫两端高度不均 | 40 | 袖衩位暴口 |
| 21 | 浮线粘进领子里面 | 41 | 袖克夫下层太松 |
| 22 | 口袋边大于袋盖边位 | 42 | 折边不正确 |
| 23 | 扣眼线迹太窄或未能缝合到布料 | 43 | 袖克夫太松或扭曲 |
| 24 | 安全保险线迹散边 | 44 | 条格不对 |
| 25 | 衩位打套结引起打褶 | 45 | 领型不均 |
| 26 | 熨烫打褶 | 46 | 机织商标偏离中位 1/2” 或大于 1/2” |
| 27 | 袖口褶裥距离不均 | 47 | 袖克夫面缝线迹张力不均 |
| 28 | 袖克夫凸嘴 | 48 | 规定的衬衫纸板或领子纸板遗漏 |
| 28 | 领子凸嘴 | 49 | 领圈扣眼偏离领座中位大于 1/8” |
| 30 | 下摆还口暴口 | 50 | 熨烫起镜 |
| 31 | 操作不正确使领子起泡 | 51 | 袋边不直 |
| 32 | 领座嘴暴口 | | |

## 本章小结

- 系统地介绍了各种服装的品质标准与规格表达，同时引出大量的相关专业术语。
- 进一步对成衣尺寸测量部位作了更详细的解释。
- 结合企业的实际应用与操作，对各种品质文件作了详细的绘编。
- 针对服装疵点的不同表达作了详细列表说明。
- 此章节重点培养学生对服装QC岗位的认识，提高学生对各种品质文件的阅读与编写能力。

# CHAPTER 5

## APPAREL MARKETING AND MERCHANDISING　服装市场营销与采购

**课题名称：** APPAREL MARKETING AND MERCHANDISING
服装市场营销与采购

**课题内容：** Introduction　简介
Leisure Wear Market in China　中国的休闲服市场
Case Analyzing—Baleno　案例分析——班尼路
The Apparel Merchandising Cycle　成衣采购循环
Relative Glossary　相关术语

**课题时间：** 6 课时

**教学目的：** 让学生了解服装市场与采购的基本知识。通过对中国休闲服市场的简单描述以及学习相关的专业术语与品牌名称。掌握成衣采购的基本流程以及服装工艺单、成本报价单等相关跟单文件的认识与编写。重点掌握并熟记市场营销与贸易等相关术语的表达与描述。

**教学方式：** 结合 PPT 与音频多媒体课件，以教师课堂讲述为主，学生结合相关案例分析讨论为辅。

**教学要求：**
1. 明确服装市场与采购的基本知识，掌握中国休闲服市场的简单描述。
2. 熟悉成衣采购的基本流程，以及商品采购涉及的文件应用与编写。
3. 熟悉各种服装工艺单、成本报价单的形式与应用。
4. 熟悉市场营销与贸易等相关术语的表达与描述。

**课前（后）准备：** 结合专业知识，课前预习课文内容。课后熟读课文主要部分，掌握成衣采购的文件应用与编写，并熟记市场营销与贸易等相关术语。

# CHAPTER 5

# APPAREL MARKETING AND MERCHANDISING
# 服装市场营销与采购

## 5.1 Introduction 简介

①服装生意是指将全球性服装从设计师的陈列室转移到零售店并卖给消费者。

Fashion marketing is the business that moves the global fashions from designers' showrooms to the retail floor and into the hands of consumers. ① On the other hand, most apparel companies take emphasis on planning new strategies to sales monitor so as to forecast markets successfully (Figure 5-1).

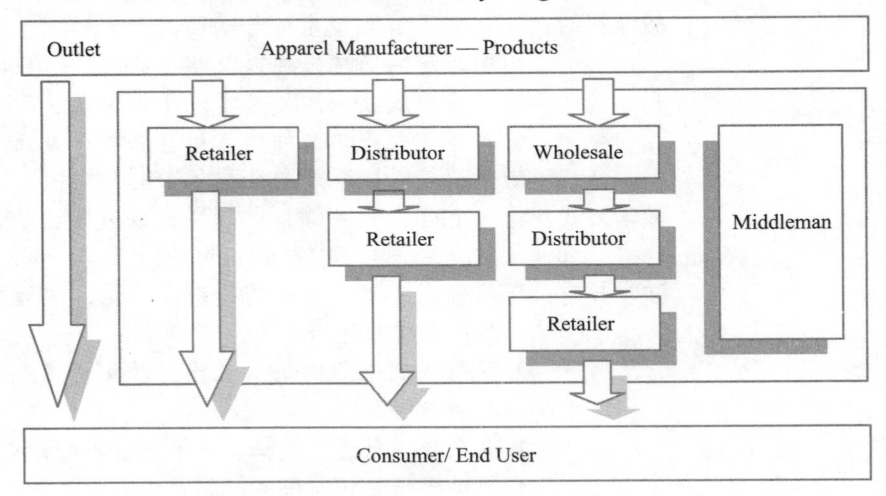


**Figure 5-1 Products Supply Chain**

②成衣营销主要涉及如成衣产品设计、确定生产规格，以及成衣供应的各种技术问题。

Apparel merchandising mostly relates to such areas as designing apparel merchandise, setting production specifications, and various technical issues of apparel supply. ② It further involves determining the market need, selecting the supplier, arriving at a proper price, specifying transaction terms and conditions, issuing orders and following up progress. The function of apparel merchandising in practice covers these activities:

(1) Anticipating the fashion trend and the market need.

(2) Understanding the role of each party in the product supply chain.

(3) Preparing the merchandising plans, e.g. design features, manufacturing specification, quality standard and delivery date, etc.

(4) Conducting procurement negotiation with suppliers.

(5) Placing orders and expediting progress to ensure timely production and distribution.

(6) Monitoring garment production and quality.

(7) Arranging transportation and documentation.

## 5.2 Leisure Wear Market in China 中国的休闲服市场

### 5.2.1 Market Background 市场背景

The improvement of consumers' purchasing power in China, the trend towards a completely opened retail market, together with the government-supported franchising policy, the fashion trend, and the consumer preference, all of these are the major drivers of China's fashion market.

In apparel retailing industry, fashion trends and consumers' preferences are crucial factors that make the buying decision process.[①] However, the fashion trends and consumers' preferences appear to be different in different regional markets. Described as follows:

①成衣零售行业中，时装流行趋势与消费者的偏好是决定购买过程的关键要素。

(1) Southern China: Consumers generate purchase impulse for design or color of clothes and particularly prefer the taste of Hong Kong or Taiwan.

(2) Central China: Consumers regard Shanghai as the center of fashion trend and consider brand image the most important factor.

(3) Northern China: Consumers are more affected by Japanese or Korean tastes and consider brand image and quality equally important factors.

(4) Southwestern and Northwestern region: Consumers follow the trend of Shanghai and Beijing but lay 1 ~ 2 years behind. They prefer buying low-to-medium priced clothing.

### 5.2.2 Strategic Planning 策略性计划

When clothing apparel enters into a relative market, it must set the strategic planning for the marketing operation, and the firm should take the advantage of opportunities and its constantly changing environment, which involves the following details:

（1）Clear Company Mission: A modern apparel company must develop a mission statement to guide staff working with organization goal, e.g. market oriented mission statement, which defines the retailing service for different customer needs especially after garment was sold. ②

②一家现代的成衣公司必须制订一套任务目标以指引全体员工，并配合企业的机构组织目标。例如，任务目标是市场为主导，则要求公司注重在服装售出以后对不同客户需求的服务。

（2）Supporting Objective: In order to achieve the marketing objective, the other departments must set up the supporting objective, and the objective should be as specific as possible. For example, production, merchandising, design and finance, etc.

（3）Co-coordinated Functional Strategies: Set up the detailed plan for each marketing unit, e.g. pricing, position, distribution, promotion, advertising, retailing or franchising, etc.

### 5.2.3 Position of Leisure Wear 休闲装的定位

In China, the number of foreign investors in the garment industry has increased, and recently strong local competitors adopt western quality management and marketing techniques to promote their brand. ③ However, Baleno still considers those Hong Kong based fashion brands as major competitors, which include Bossini, Giordano, U2, Jeanswest, Esprit, Theme, Crocodile, and Goldion. And the positioning map below shows the market position of some of Baleno's major competitors（Figure 5-2）.

③在中国服装业中，国外投资者的数量大量增加，目前，国内品牌企业也开始采用西方的质量管理与营销技术推广他们的品牌。

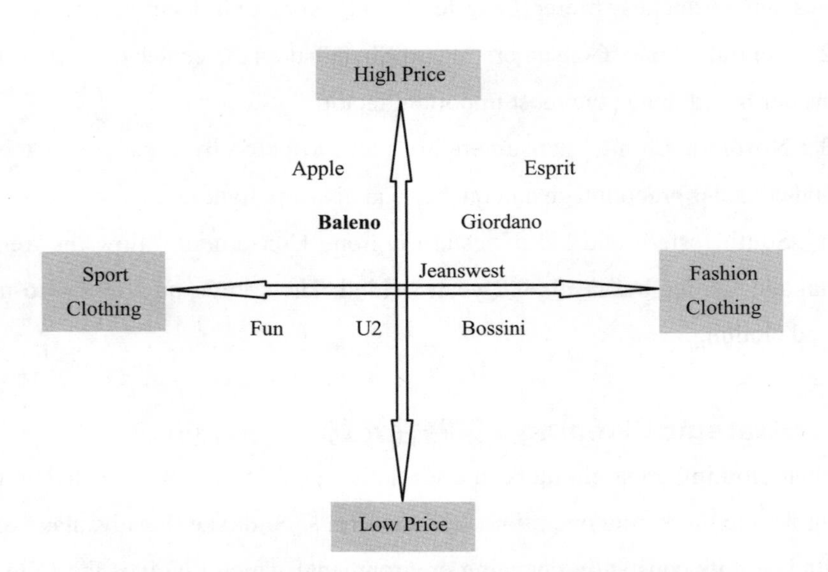


**Figure 5-2 Positioning Map of Leisure Wear Brands in China**

### 5.2.4 Franchising Strategy 特许经营策略

Garment companies rely on franchising as the major expansion strategy, or have joint ventures with local enterprises in Mainland China, because the franchising strategy could reduce overhead, and improve production efficiency through business process restructuring.④ Franchising is the most common type of marketing and distribution strategy for most retail business. Franchising involves the licensing of independent franchisees to reduce and sell branded products under tightly specified and controlled procedures. Franchising involves strong contracts and unique business format that the franchisees are required to operate in the same way and to deliver the services according to their specifications.

④服装公司一般将特许经营作为主要扩展策略，或与中国内地的本地企业进行合资经营，因为特许经营可通过业务重组将成本费用减少，并且提高生产效率。

## 5.3 Case Analyzing—Baleno 案例分析——班尼路

### 5.3.1 About Baleno 关于班尼路

Baleno first entered into the Chinese market in 1992, it promotes a full collection of casual wear to Chinese teenagers and youth. At first, the senior management replicated the business model of Giordano to launch Baleno in China. In particular, the management realized that one of the key success factors of Giordano was its ability to pick up good sites for its shops, and adopting the "me-too" approach. Baleno responds to open shops right next to Giordano's shops. This "follower strategy" enable the company to enjoy competitive advantages like low risk, cost efficient, and high traffic flow to visit Baleno shops.① Today there are over 350 Baleno outlets in China, which include self-owned shops, franchisees, or a combination of franchising and self-owned shops. And Baleno advises the corporate mission which is to provide excellent retailing service and high quality casual wears for enhancing the youth's quality of life.

①这个"跟随者策略"能够使公司享受到类似低风险、成本效率，高客流量光顾班尼路店铺的竞争优势。

### 5.3.2 Positioning and Promotion 定位与推广

Baleno adopts price, products and quality customer services as major means to support its positioning strategy, also to treat franchisees as business partners and provide satisfactory rewards as incentive.

Baleno positions itself as "value for money" in terms of mass apparel design, good quality, competitive price, superior customer service, and comfortable shopping environment. The management of Baleno believes that the fashion trend of the next decade would be a strong preference towards simple, clean and natural design.

②班尼路把有竞争力的定价与优质的产品作为主要的支持性定位策略。

Baleno adopts competitive pricing and quality products as a major means to support its positioning strategy. [②] Customers also demand quality customer service before or after sales, such as enquiry service during shopping, fitting service and the return of goods after purchasing. Thus, provision of superior customer service is also a major component of image of "value for money".

Baleno intends to build up a strong brand image which enable customers to associate the benefits of high quality with value, stylish and status. To launch a series of marketing communication programs to teach Chinese youths how to achieve a perfect image of casual wears to enhance personal style.

### 5.3.3 Franchising 特许经营

The marketing objectives of expanding Baleno in China through franchising cover several areas. Baleno wants to be the market leader in the youth segment. Baleno aims to strengthen its competitive position in China by building numbers of outlets. [③] Baleno also wants to make full use of franchising outlets to promote brand image of good quality, good price and excellent service in the most efficient way. And there are several reasons for Baleno to make full advantages of franchising.

③班尼路通过建立无数的销售点来加强在中国的竞争地位。

（1）To be the market leader in the youth segment.

（2）To promote and enhance Baleno's image of good quality, good price and excellent service.

（3）Provide information systems including sales information and inventory management.

（4）Training provided to franchisees on Baleno's operating system.

（5）Site selection for best location and floor plan design.

（6）Provide "Quality merchandise" to franchisees.

（7）Nation-wide marketing communication program.

（8）Advice on merchandise selection.

（9）Customer service monitoring program.

Otherwise, a franchisee may become a future competitor to Baleno. Once they have acquired the management skills, technology, and the management information system from Baleno in operating retailing business, they can open their own shops under another trademark and run the business with the same model.

## *Words and Expressions*

showroom ['ʃəurum] 陈列室
transaction [træn'zækʃən] 交易
anticipate [æn'tisipeit] 预期，期望
conduct ['kɔndʌkt] 引导
procurement [prə'kjuəmənt] 采购，获得
negotiation [ni,gəuʃi'eiʃən] 商议，谈判
expedite ['ekspədait] 加速，派出
merchandising ['mə:tʃəndaiziŋ] 商品跟进，商品采购跟单
transportation [,trænspɔ:'teiʃən] 运输
marketing ['mɑ:kitiŋ] 市场营销
casual wear ['kæʒjuəl wiə] 便服
franchising ['fræntʃaiziŋ] 特许经营
fashion trend ['fæʃən trend] 流行趋势
consumer [kən'sju:mə] 消费者
preference ['prefərəns] 偏好
crucial ['kru:ʃiəl] 关键的
brand [brænd] 品牌
taste [teist] 品位
strategy ['strætidʒi] 策略
opportunity [ɔpə'tju:niti] 时机，机会
mission ['miʃən] 使命
organization goal[ɔ:gənɑi'zeiʃən gəul] 组织目标
pricing ['praisiŋ] 定价
position [pə'ziʃən] 定位
distribution [distri'bju:ʃən] 分销
promotion [prə'məuʃən] 推广
advertising ['ædvətaiziŋ] 广告
retailing ['ri:teiliŋ] 零售
overhead ['əuvə'hed,] 企业一般管理费用（广东话：厂皮开支）
joint ventures [dʒɔint 'ventʃə] 合资
business process restructuring/ BPR ['biznis 'prəuses ri'strʌktʃə] 业务重组
franchisees [,fræntʃai'zi:] 特许经营商
licensing ['laisəns] 许可证
contract ['kɔntrækt,] 合同
teenager ['ti:n,eidʒə] 青少年
launch [lɔ:ntʃ] 开办，投入，发动
approach [ə'prəutʃ] 方法
respond [ris'pɔnd] 向应，反应
self-owned shop [self -əund ʃɔp] 自营店
satisfactory [,sætis'fæktəri] 满意
incentive [in'sentiv] 激励，动机
environment [in'vɑiərənmənt] 环境
provision [prə'viʒən] 供应
associate with [ə'səuʃieit wið ] 联合
benefit ['benifit] 利益
stylish [stailiʃ] 时髦的
personal style ['pə:sənl stail] 个人风格
outlet ['ɑut-let] 经销商，零售商
trademark [treidmɑ:k] 注册商标
quality merchandise ['kwɔliti 'mə:tʃəndɑiz] 优质商品
inventory management ['invəntri 'mænidʒmənt] 存货管理
management information system/ MIS ['mænidʒmənt infə'meiʃən 'sistəm] 信息管理系统

## *Exercises*

1. Translate the following terms into Chinese.

（1）CIF
（2）quota
（3）CMT
（4）marketing
（5）vendor
（6）wholesales price
（7）consumer
（8）casual wear
（9）stylish
（10）outlet
（11）trademark
（12）franchising
（13）fashion trend
（14）brand
（15）promotion
（16）advertising
（17）retailing
（18）BPR
（19）contract
（20）self-owned shop
（21）MIS

2. List 10 brands of casual wear, which promote products in the Chinese market.

## 5.4 The Apparel Merchandising Cycle 成衣采购循环

①在实际操作中，商品采购职能远远超过商品跟单部门的活动范围。

In practice, the scope of merchandising functions is far wider than the activities of the merchandising department. [①] There are many other functions within a company actively engaged in the merchandising process. Figure 5-3 schematically illustrates the major steps of a generalized merchandising business process. It shows that these activities are closely interrelated and yet exposed to different environmental influences. The fundamental objectives of apparel merchandising process can be summarized as follows:

(1) To plan and procure consistent apparel products.

(2) To organize continuity of supply by maintaining effective relationships with suppliers and by developing other supply source opportunities.

(3) To maintain co-operative relationships with other departments, providing information and advice as necessary to ensure the effective operation of the organization as a whole.

(4) To maintain updated information and knowledge for proper and responsive decision-making.

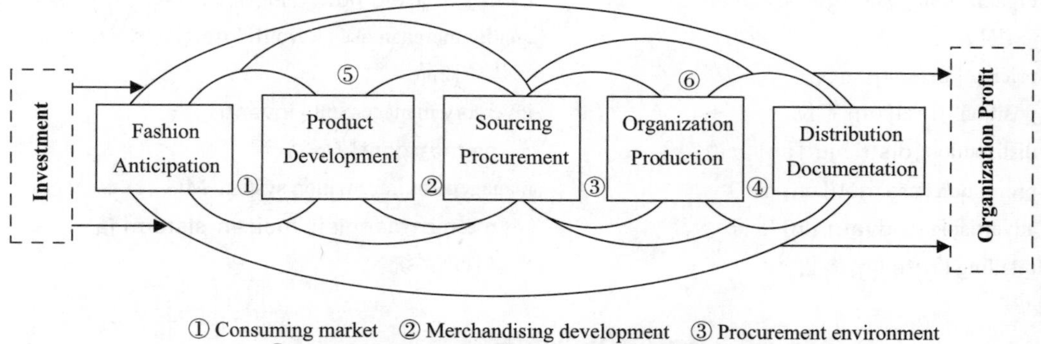


① Consuming market ② Merchandising development ③ Procurement environment
④ Logistic environment ⑤ Consuming market ⑥ Supply market

**Figure 5-3 Apparel Merchandising Process**

### 5.4.1 Anticipation of Fashion Trend 时装潮流的预测

Today fashion begins and ends with the consumer. Consumers are people who buy and use merchandise. Marketing strategies have become more important, manufacturers and retailers have had to consider consumers' wants and needs. This understanding is to foresee what apparel product will be accepted in coming seasons. The anticipation includes observation of general market-wide development and the environmental opportunities for apparel merchandising.② Furthermore, the process focuses on the characteristics of existing merchandises and the strengths and weaknesses of the merchandise on the basis of market popularity.

②该内容的理解是预见下一季节将流行什么样的成衣产品。预测包括对成衣市场整体发展与服装营销环境机会的观察。

## 5.4.2 Development of Product 产品的开发

Following the identification of market trends and preferences in the previous phase, development teams, comprising designers, buyers and merchandisers attempt to define criteria to develop the portfolio program in terms of fashion features that appeal to customers' choice. ③ It comprises the choice of colors, cutting, silhouette, texture, trimming adoption and workmanship. A new seasonal design collection is sometimes compiled as a portfolio that is established to extend the existing product line, as shown in Figure 5-4. In practice, varieties of portfolio are developed for one season and screened until the final one is consistent with the company's objectives. ④

③遵循早期确定的市场趋势与偏好，由设计师、买手和销售经理构成的产品开发团队按照顾客选择的时装特征共同确定开发产品的设计图稿的标准。

④在实际操作中，在同一季节会开发多种设计图稿进行筛选，直到开发出符合公司目标的设计图稿为止。

And the functional department may require some information to plan a schedule, such as stock & location information, traffic assignments, production information, etc, and has an ability to balance the production costs with the costing department, as shown in Table 5-1、Table 5-2.

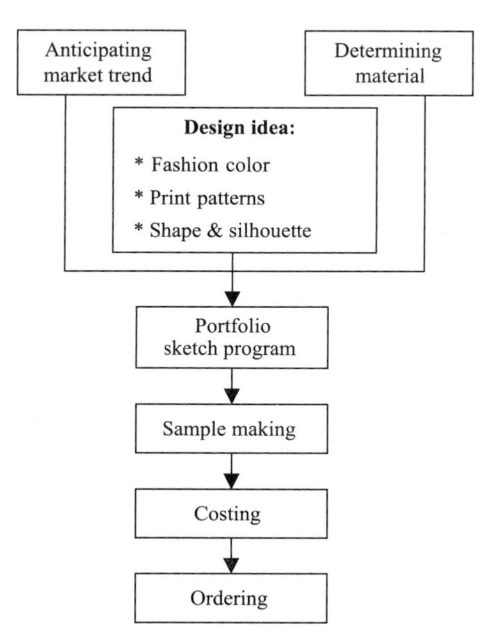


**Figure 5-4 Process of Product Development**

**Table 5-1 Design Specification of Women's Pants**

**Design Specification**

Date: 2-3-2020

| Season | Description | Quantity | Size | Designer | Pattern |
| --- | --- | --- | --- | --- | --- |
| Winter | Women's Pants | 240 | S ~ XL | Sandy | Ben |

**Style Details**

| Production Sketch | |
| --- | --- |
|  | 1. A coin pocket and two curved pockets are top stitched with double needles.<br>2. Two patterned hips pockets in back part.<br>3. Rivets are set in the all pocket corners.<br>4. Triangular back yoke with double needles.<br>5. 3 cm Cuffed bottom with single needle.<br>6. Five belt-loops with double needles on the waistband.<br>7. Flower embroidery for the leather logo. |

**Fabrics Details**

| Types | Suppliers | Contents | Construction | Color | Width |
| --- | --- | --- | --- | --- | --- |
| Shell 1 | CK Textile Corporation | 80%Cotton<br>15%Polyester<br>5%Lycra | 1/3Twill<br>"Z" Direction<br>54 × 35/18S × 18S | Grey-446C | 150cm |
| Pocketing | CK Textile Corporation | T/C | Plain | White-1200 | 150cm |
| Interlining for fly | ALBB Co. | 100% Cotton | Plain/ Fusible | White | 90cm |

**Accessories Details**

| Types | Supplier | Size | Content & Color | Position |
| --- | --- | --- | --- | --- |
| Press Stud | YKNK | 25L | Brass | C/F Opening |
| Zipper | YKK | # 4YG/ 12 ~ 14cm | Brass | C/F Opening |
| Thread 1 | DALI | 40S/ 3 | Grey/ Wash Color | Body for RS |
| Thread 2 | DALI | 30S/ 3 | Grey/ Wash Color | Body for WS |
| Rivet | YKNK | VR-511 | Brass | Corner for pockets |
| Leather logo | YKNK | 5cm × 4cm | Leather | Right Back WB |
| Care label/ Content label | YKNK | Woven label | Polyester | Right side inner WB |
| Size label | YKNK | Woven label | Polyester | Right side inner WB |
| PO No. label | YKNK | Woven label | Polyester | Right side inner WB |

**Finished: Stone Wash**

**Special Instruction:** Prepare the Size specification with "S—M—L—X—XL"

Design Group: Alice Tam 2-3-2020    Merchandising Group: Ben Liu 2-3-2020

**Table 5-2 Cost Sheet of Women's Pants**

**Cost Sheet**

Ref. No.:CT-010 Date: 5-3-2020

| Buyer: Tommy | Item: Women's Pants | Quantity: 240Pcs | Style No.: RD-018 |
|---|---|---|---|
| Size Range: S ~ XL | Country: Fm. China To USA | Target FOB: $12USD | Retail Price:$120USD |

**Fabric Cost** ***Exchange rate: 1USD =6.517CNY***

| Types | Source | Unit Price | Width | Amount | Total Amount |
|---|---|---|---|---|---|
| Shell 1 | CK Textile Corporation | ¥30/m | 150cm | 1.4m/ Pcs | ¥42/Unit |
| Pocketing | CK Textile Corporation | ¥6/m | 150cm | 1m/8 Pcs | ¥0.76/ Unit |
| Interlining for fly | ALBB Co. | ¥12/m | 90cm | 1m/120 Pcs | ¥0.1/ Unit |

**① Fabric Sub Cost=RMB ¥42.86/Unit**

**Accessory Cost**

| Types | Source | Unit Price | Amount/ Unit | Total Spared QTY（10%） | Total Amount |
|---|---|---|---|---|---|
| Press Stud | YKNK | ¥68/1Pkg（100Pcs） | 1 Pc | 24Pcs/ ¥0.068 | ¥0.68/Unit |
| Zipper | YKK | ¥150/1Pkg（100Pcs） | 1 Pc | 12Pcs/ ¥0.075 | ¥1.5/ Unit |
| Thread 1 | DALI | ¥4.6/ Thread | Thread/ 12Pcs | 2 Pcs/ ¥0.038 | ¥0.38/ Unit |
| Thread 2 | DALI | ¥3.2/ Thread | Thread/ 12Pcs | 2 Pcs/ ¥0.027 | ¥0.27/Unit |
| Rivet | YKNK | ¥58/ 1Pkg（100Pcs） | 10 Pcs | 24Pcs/ ¥0.058 | ¥5.8/ Unit |
| Leather logo | YKNK | ¥36/100Pcs | 1 Pc | 12Pcs/ ¥0.018 | ¥0.36/ Unit |
| Care label/ Content label | YKNK | ¥22/100Pcs | 1 Pc | 12Pcs/ ¥0.011 | ¥0.22/Unit |
| Size label | YKNK | ¥12/100Pcs | 1 Pc | 12Pcs/ ¥0.006 | ¥0.12/ Unit |
| PO No. label | YKNK | ¥22/100Pcs | 1 Pc | 12Pcs/ ¥0.011 | ¥0.22/ Unit |
| Polybag | ALB | ¥16/100Pcs | 1 Pc | 24 Pcs/ ¥0.016 | ¥0.16/ Unit |
| Carton Box | ALB | ¥4.8/1Pc | Box/ 24Pcs | 1 Pc/ ¥0.02 | ¥0.2/ Unit |
| Sub Total | | | | ¥0.348/Unit | ¥9.91/Unit |

**② Accessory Sub Cost= ¥0.348+ ¥9.91 = RMB 10.26/Unit（Included spared QTY）**

**Finish：Stone Wash** **Packing: Flat** **Making Cost & Other Charges**

**Continued**

| Production Sketch | Stone Wash | ¥12/Doz |
|---|---|---|
| | CMT | ¥216.8/Doz |
| | Package | ¥2.4/Doz |
| | Managing Charge | ¥5/Doz |
| | Transportation | ¥10.5/Doz |
| | Trimmings | ¥6/Doz |
| | Warehousing | ¥3/Doz |
| | Others | ¥20/Doz |
| | **Total Making Cost** | **¥275.8/Doz** |
| | **③ Making Sub Cost=¥22.9/Unit** | |
| Total Unit Cost= ① + ② + ③ = ¥76.12 | | |
| | | **④ Profit=Total Cost × 10%= ¥7.612** |
| | | **⑤ Target FOB Price=（¥76.12+ ¥7.612）/ $6.517=12.85USD/Unit** |
| Prepared By: May He 5-3-2020 | Check By: Ben Liu 5-3-2020 | |

## 5.4.3 Material Sourcing and Procurement 原材料搜集与采购

The merchandiser searches for sourcing possibilities in which could optimize the profitability, assures the least uncertainty in the supply market, schedules timely delivery and arranges the financing aspect. Procurement refers to activities of purchasing, storing, handling traffic, receiving and inspecting incoming material. The merchandiser should be alert to various sourcing restrictions according to predetermined business criteria. [⑤] In most cases, merchandising functions in this phase involve material management and order placement to achieve a timely production schedule.

An important activity of the purchasing department is to monitor the performance of suppliers and track new orders, also to ensure timely delivery of the proper quantity and quality of goods or services at reasonable prices. [⑥] There are many activities in purchasing.

（1）The evaluation of alternative suppliers, and the use of a small but effective number of these suppliers for the more important goods and materials. [⑦]

（2）To provide the chance to buy effective materials with competitive prices.

⑤根据预定的经营标准，跟单员须注意到各种有关资源搜集的限制。

⑥采购部的一项重要任务是监督供应商及跟踪新订单的进展，确保对方及时以合理的价格交付规定数量和质量的商品或服务。

⑦对不同供应商进行评估，较为重要的产品和原材料应选用较少但较有效的供应商。

（3）The rating or assessment of suppliers on delivery, service and quality, as well as price.

（4）The enhancement of the company reputation in the eyes of suppliers and competitors.

（5）Have an ability to select new materials by constant communication with the marketing department.

（6）The maintenance of adequate stock levels throughout the company, in conjunction with other departments.

It will be evident from these activities that the purchasing function could have an effect on the profitability of the company.

### 5.4.4 Organizing Production 组织生产

This phase emphasizes how to arrange production, define quality level, make schedule of material requirement and delivery, and allocate control personnel. The process involves detailed expedition of material ordering and receiving, inspection, resolving technical problems and arrangement of finished product packaging and dispatch. [8] Merchandisers here should be clear on the technical specifications and their tolerance levels for errors.

⑧该程序包括原材料的订购与接收、检验，技术问题的分解以及成品包装与付运的详细进程。

### 5.4.5 Shipment Arrangement 船运安排

In this stage, the merchandiser evaluates shipping-ordering systems, to decide transportation modes and routes, studies packaging and handling methods and finally advises the best arrangement of delivery. Merchandising would assist shipping and finance operation in proceeding adequate and accurate documents for order settlement.

In addition to the activities that are distinctly within the responsibility of merchandising, there are a number of duties that are typically shared with other departments. Among these duties are:

（1）Laboratory testing and inspection.

（2）Material searching and budget.

（3）Development of standards and specifications.

（4）Supervision of production progress.

（5）Inventory control.

## Words and Expressions

apparel [ə'pærəl fə:m] 成衣
portfolio [pɔ:t'fəuljəu] 设计图稿
distribution channel [distri'bju:ʃən 'tʃænl] 分销渠道 *
product supply chain ['prɔdəkt sə'plai tʃein] 产品供应链
industry marketing ['indəstri 'mɑ:kitiŋ] 工业营销 *
schematically [ski'mætikəli] 示意性的
generalized ['dʒenərəlaizd] 无显著特点的
interrelate [ˌintə(:)ri'leit] 相互关连
fundamental [ˌfʌndə'mentl] 基本的，基本原则
co-operative [kəu-'ɔpərətiv] 合作的
anticipation [ˌæntisi'peiʃən] 预期，预料
sourcing [əsɔ:siŋ] 搜集资源
procurement [prə'kjuəmənt] 获取，采购
complex ['kɔmpleks] 合成的，联合体
silhouette [ˌsilu(:)'et] 轮廓
texture ['tekstʃə] 织品的质地
trimming ['trimiŋ] 辅料
workmanship ['wə:kmənʃip] 手工、技艺
consistent [kən'sistənt] 一致的，相容的
screened [skri:nid] 筛过的
cost sheet [kɔst ʃi:t] 成本单
inventory ['invəntri] 存货，物品清单
investment [in'vestmənt] 投资
distribution [distri'bju:ʃən] 分派，分配
decision-making [di'siʒən-meikiŋ] 决策
evaluating cost [i'væljueitiŋ kɔst] 成本核算
traffic ['træfik] 交通，运输
retail price ['ri:teil prɑis] 零售价格
shell fabric [ʃel 'fæbrik] 面布
interlining ['intə'lainiŋ] 里衬
source mill [sɔ:s mil] 原材料加工厂
amount [ə'mɑunt] 总数，量额
lining ['lɑiniŋ] 里料
pocketing ['pɔkitiŋ] 袋布
exchange [iks'tʃeindʒ] 兑换
snap [snæp] 按纽
thread [θred] 缝纫线
hanger ['hæŋə] 衣架
eyelet ['ailit] 孔眼（广东话：凤眼）
style [stail] 款式
transportation [ˌtrænspɔ:'teiʃən] 运输
charge [tʃɑ:dʒ] 费用
sales contract [seil 'kɔntrækt,] 销售合同
target market ['tɑ:git 'mɑ:kit] 目标市场
stand up packing [stænd ʌp 'pækiŋ] 直立式包装
hanger ['hæŋə] 衣架
flat packing [flæt 'pækiŋ] 扁平式包装
forecasting ['fɔ:kɑ:stiŋ] 预测
merchandising ['mə:tʃəndaiziŋ] 营销
delivery dates [di'livəri 'deits] 交货期
shipping instructions [ʃipiŋ in'strʌkʃənz] 货运指示 *
production sketch [prə'dʌkʃən sketʃ] 生产图
category ['kætigəri] 类目，部门
unit price ['ju:nit prɑis] 单价
payment ['peimənt] 付款方式
optimize ['ɔptimaiz] 使最优化
storing ['stɔ:riŋ] 保管，存储
brief [bri:f] 摘要，大纲
quota ['kwəutə] 配额
tariff ['tærif] 关税，价格表
requisition [ˌrekwi'ziʃən] 申请表，通知单
quotation [kwəu'teiʃən] 报价单
verify ['verifai] 校验，核实
invoice ['invɔis] 发票，发货单
correspond [kɔris'pɔnd] 符合，协调
route [ru:t] 路线，路程
settlement ['setlmənt] 解决，结算
shipping order['ʃipiŋ 'ɔ:də] 装货单
client ['klaiənt] 顾客
shoulder splice ['ʃəuldə splais] 肩缝
splits [split] 裂口，开衩

twin needle stitch [twin 'ni:dl stitʃ] 双针线迹

micro-fibre [ˌmaɪkrəu 'faibə] 微型纤维

pin-stitching [pin 'stitʃiŋ] 缉边线

## Exercises

1. List content points of purchase order.

   Example: Order date.

2. Translate the following terms into Chinese.

(1) pattern design
(2) size label
(3) clothing engineering
(4) overhead
(5) R & D
(6) retail price
(7) unit price
(8) size assortment
(9) size chart
(10) payment
(11) shell fabric
(12) interlining
(13) source mill
(14) pocketing
(15) eyelet
(16) FOB
(17) target market
(18) hanger
(19) merchandising
(20) delivery date
(21) production sketch
(22) linen
(23) apparel firm
(24) core business
(25) closed-fit
(26) fully open
(27) magyar sleeve
(28) button closure
(29) poly-bag
(30) logo
(31) manufacturing specification
(32) yardage
(33) supplier
(34) letter of credit
(35) shipping
(36) financial control
(37) apparel firm
(38) goods
(39) approved
(40) fabric swatch
(41) delivery date
(42) enterprise
(43) sample swatch
(44) signature
(45) marketing
(46) purchase order
(47) specification
(48) style
(49) warehouse
(50) vendor
(51) unit price
(52) poly-bag
(53) payment
(54) inventory
(55) initial sample
(56) approval sample
(57) bulk-production
(58) applicant
(59) manufacturing
(60) material
(61) accessory
(62) shell fabric
(63) chambray

3. Write corresponding abbreviations for the following special terms.

   Example: Pocket—Pkt.

(1) Zipper
(2) Free on Board
(3) Cut, Make & Trim
(4) Cost, Insurance & Freight
(5) Measurement
(6) Quantity
(7) Size Specification
(8) Company Limited
(9) Department
(10) Style
(11) Letter of Credit
(12) Piece
(13) Yard
(14) Letter of Guarantee
(15) Size
(16) Package
(17) Button

4. Prepare a cost sheet according to the following style (Elastic Shorts).

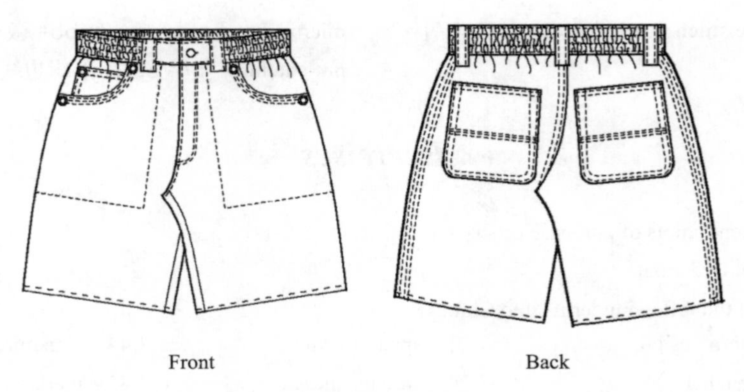


**Elastie Shorts**

## 5.5 Relative Glossary 相关术语

### 5.5.1 Merchandising Terms 市场营销词汇

（1）Advertising：paid promotional message from an identified sponsor.

（2）Apparel outlets：stores owned by apparel producers that sell seconds and over-runs to the public at low prices.

（3）Buyer：a market participant who takes a long futures position or buys an option. An option buyer is also called a taker，holder or owner.

（4）Branch Stores：When a well-established department store opens a store in another location，the new one is called a branch.① It operates from the original flagship store.

①当一家经营很好的百货公司在另外一个场所开店时，新的一家就叫分店。

（5）Chain Stores：a chain is a group of stores owned，managed，and controlled by a central office. All of a company's chain stores look alike. No store is considered to be the main store.

（6）Consumers：those who buy and wear the garments. Consumer purchases and use apparel as well as other economic goods. Consumer is very important in determining what fashions will or will not become popular.

（7）Department Stores：retail establishments that offer large varieties of many types of merchandise placed in appropriate departments. Almost all clothing and household needs are sold in a wide variety of colors，sizes，and styles.

（8）Discount Stores：sell clothing and other merchandise in large，simple buildings with low overhead. Large amounts of garments are sold on racks and shelves.

（9）Franchises：individually owned businesses that use the name and merchandise of an established firm. The franchiser provides a franchisee with exclusive use of the name and goods in a specified city or area. ②

②使用一家公司的名称与商品进行独立经营。特许经营商在指定的城市或地区给代销商提供专用的名号与商品。

（10）Factory Outlets：store owned by a manufacturer who sells company products to the public at reduced prices.

（11）Indirect Selling：non-personal promotion aimed at a large general audience.

（12）Mail Order Houses：sell to consumers through catalogs. They offer shopping at home for customers who cannot，or prefer not to go out.

（13）Merchandising：the process through which products are designed，developed，and promoted to the point of sale.

（14）Promotions：to sell particular fashions that conducted nationally by the manufacturer，and locally by the retail store. The promotional activities are the advertising and merchandising efforts to improve sales.

（15）Purchase Order：document written by a buyer that authorizes a seller to deliver certain goods at specified prices.

（16）Retail stores：sell to consumers. They advertise and sell their items directly to the general public. Retail stores include department stores，chain and discount stores.

（17）Specialty Stores：Might handle only apparel，or they might specialize even further into a specific kind of apparel. Such as maternity shops，shoe stores，bridal boutiques and children apparel stores.

（18）Visual Merchandising：presenting goods in an attractive and understandable manner. Displays and exhibits are ways that clothing items are visually promoted. Stores will have special events，and have models display merchandise.

（19）Video Merchandising：uses video in retail stores to show new fashion trends，promote merchandise. Videos are set up in retail store departments near the merchandise they are showing. They attract the attention of passing customers with sound and movement.

（20）Wholesale：sell goods in large quantities to retailers. Wholesalers usually distribute their goods from large warehouses，and each goods must be purchased by dozens. ③

③批发商通常从大仓库配送产品且每种产品必须以打为数量单位进行采购。

### 5.5.2 Trading Terms 贸易词汇

（1）C & FRT：（Cost and Freight）：paid to a point of destination and included in the price quoted.

（2）CIF（Costs，Insurance and Freight）：an international commercial term used in international sales contracts，meaning that the selling price includes all cost，insurance and freight for any goods sold. The seller arranges and pays for all relevant expenses involved in shipping goods from their point of exportation to a given point of importation.

（3）CMT（Cut，Make & Trim）：the buyer provides the fabric，and the supplier makes the garments.

（4）Commission：the charge made by a commission house for buying and selling commodities.

（5）Commercial Invoice：a document prepared by the exporter or freight forwarder，and required by the importer to prove ownership and arrange for payment to the exporter. It should include basic information about the transaction such as a description of the goods，address of the shipper and seller as well as delivery and payment terms.

（6）Contract：an agreement to buy or sell a specified commodity，detailing the amount and grade of the product and the date on which the contract will mature and become deliverable.

（7）Delivery Date：the date on which the commodity or instrument of delivery must be delivered to fulfill the terms of a contract.

（8）Export Quota：a specific restrictions imposed by an exporting country on the value or volume of certain exports to protect domestic producers and consumers.④ The quota is also often applied in orderly marketing agreements and voluntary restraint agreements，and to promote domestic processing of raw materials in countries that produce them.

④为了保护本国生产商与消费者的利益，由出口国对出口商品的价格与数量进行一种特定的限制。

（9）Exchange Rate：the price of one currency stated in terms of another currency.

（10）FOB（Free On Board）：this price is all costs of the goods loaded on board a ship whose destination is stated in the commercial contract. It is same as FOA（FOB Airport）.

（11）L/C（Letter of Credit）：a document issued by a bank in which guarantees the payment of a customer's drafts for a specified period and up to a specified amount.

(12) Packing List: It gives details of the contents of all the packages making up the consignments and is required by customer authorities if the packing information is not shown on the invoice.

(13) Quotation: a quotation represents the price and conditions at which a seller is prepared to supply goods.

(14) Trademark: a mark or symbol secured by legal registration used by a manufacturer or trader to distinguish his or her goods from competing goods.

## *Words and Expressions*

advertising ['ædvətaiziŋ] 广告，广告的
sponsor ['spɔnsə] 赞助商，主办人
apparel outlet [ə'pærəl 'autlet] 成衣零售店
flagship store ['flægʃip stɔː] 旗舰店
over-runs [ˌəuvə-'rʌn] 超过，泛滥
participant [pɑː'tisipənt] 参与者，参与的
holder ['həuldə] 持有者，占有者
branch store [brɑːntʃ stɔː] 分店
department store [di'pɑːtmənt stɔː] 百货公司
chain store [tʃein stɔː] 连锁商店
household ['haushəuld] 家庭，家属的
discount store ['diskaunt stɔː] 折扣店
franchiser ['fræntʃai'zə] 特权经营商
franchisee [ˌfræntʃai'ziː] 有特许经营代销权的人
exclusive [iks'kluːsiv] 独占的
factory outlet ['fæktəri 'autlet] 工厂直销店
indirect selling [ˌindi'rekt 'seliŋ] 间接销售
audience ['ɔːdiəns] 听众，观众
mail order [meil 'ɔːdə] 邮购
promotion [prə'məuʃən] 推广
retail store ['riːteil stɔː] 零售店
specialty store ['speʃəlti stɔː] 专卖店
maternity shop [mə'təːniti ʃɔp] 孕妇商场
bridal boutique ['braidl buː'tiːk] 专卖新婚衣服的小商店
visual merchandising ['vizjuəl 'məːtʃəndaiziŋ] 视觉营销
video merchandising ['vidiəu 'məːtʃəndaiziŋ] 视频营销
wholesale ['həulseil] 批发
destination [ˌdesti'neiʃən] 目的地
relevant ['relivənt] 有关的
commission [kə'miʃən] 佣金
commercial invoice [kə'məːʃəl 'invɔis] 商务发票
forwarder ['fɔːwədə] 代运人，运输业者，转运公司
rests with [rest wið] 取决于，由…负责
delivery date [di'livəri deit] 交货日期
fulfill [ful'fil] 履行，实现
export quota ['ekspɔːt 'kwəutə] 出口配额
restrictions [ris'trikʃən] 限制，约束
temporary ['tempərəri] 暂时的
exchange rate [iks'tʃeindʒ reit] 汇率，换率
currency ['kʌrənsi] 流通货币
C & FRT / Cost and Freight [kɔst ænd, freit] 成本加运费
CIF / Cost, Insurance & Freight [kɔst in'ʃuərəns freit] 到岸价
export license [eks'pɔːt, 'laisəns] 出口许可证
CMT / Cut, Make & Trim [kʌt meik & trim] 来料加工
L/C / Letter of Credit ['letə ɔv 'kredit] 信用证

guarantee [ˌgærən'tiː] 担保，保证
packing list ['pækiŋ list] 装箱单
quotation [kwəu'teiʃən] 报价单
trademark ['treidmɑːk] 商标
distinguish [dis'tiŋgwiʃ] 区别，辨别
export taxes [eks'pɔːtˌ tæks] 出口税款
consumer market [kən'sjuːmə 'mɑːkit] 消费者市场
business market ['biznis 'mɑːkit] 商业市场
vendor ['vendɔː] 客商，买主

## *Exercises*

1. Translate the following terms into Chinese.

(1) advertising
(2) sponsor
(3) apparel outlet
(4) department store
(5) chain store
(6) discount store
(7) apparel
(8) design portfolio
(9) sourcing
(10) fashion silhouette
(11) texture
(12) decision-making
(13) overhead
(14) tariff
(15) quotation
(16) overhead
(17) franchise
(18) exclusive
(19) retail store
(20) specialty store
(21) boutique
(22) merchandising
(23) wholesale
(24) commission
(25) CIF
(26) CMT
(27) L/C

2. Explaining the flowing glossaries.

(1) Chain Stores
(2) Franchises
(3) Merchandising
(4) Wholesale
(5) CIF
(6) Export Quota
(7) FOB
(8) Trademark
(9) Department Stores

# 译文

# 第五章 服装市场营销与采购

## 5.1 简介

成衣生意是指将全球性服装从设计师的陈列室转移到零售店并卖给消费者。另一方面，大部分服装公司把重点放在规划新策略对销售进行监控，以便成功地做好市场预测（图 5-1）。

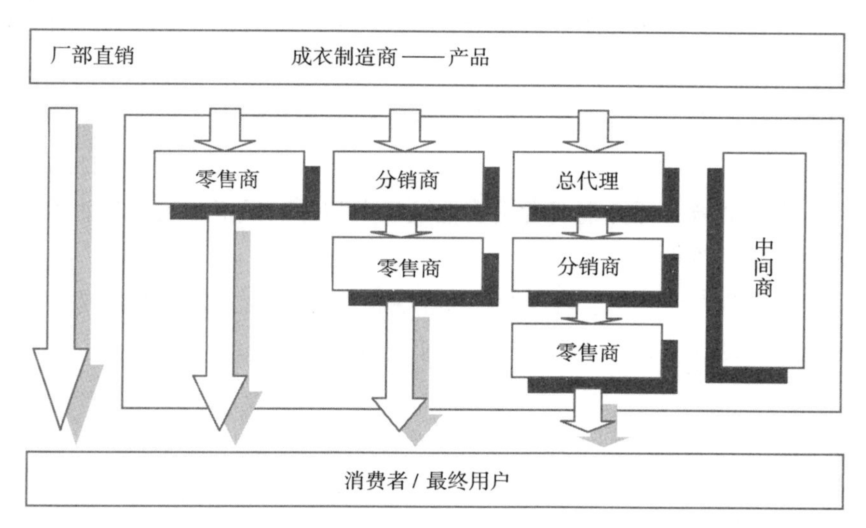


**图 5-1 产品供应链**

成衣营销主要涉及诸如成衣设计、确定生产规格以及成衣供应的各种技术问题。它还进一步涉及确定市场需求、选择供应商、合理定价、指定交易方式和条款、发出订单及跟进采购等工作。成衣营销在实践中的应用包括以下几个活动：

（1）预测时装趋势和市场需求。

（2）了解产品供应链中每一个角色的作用。

（3）准备商品采购计划，如设计特点、制作规格、质量标准和交货期等。

（4）准备与供应商谈判的采购文件。

（5）给加工厂下订单并预测生产进度，保证准时生产和交货。

（6）监测服装生产和质量。

（7）安排运输和单证处理。

## 5.2 中国的休闲服市场

### 5.2.1 市场背景

中国消费者购买力的增长、零售市场朝着完全开放的趋势发展、政府支持特许经营的政策、时尚趋势以及消费者偏好等所有因素，都成为中国服装市场发展的主要动力。

成衣零售行业中，时装流行趋势与消费者的偏好是决定购买过程的关键要素。然而，时装趋势与消费者的偏好在不同区域的市场似乎不一样，描述如下：

（1）中国南方：南方消费者的购买冲动源于服装的设计、颜色，且特别喜欢中国香港或中国台湾的品位。

（2）中国中部：中部消费者把上海看成时装流行中心，且认为品牌形象是最重要的因素。

（3）中国北方：北方消费者更多地被日本和韩国风格影响，认为品牌形象和产品质量是同等重要的因素。

（4）西北和东北地区：该地区消费者跟随北京和上海的潮流趋势，但比上海和北京迟一两年。他们更倾向于买中低档服装。

### 5.2.2 策略性计划

当成衣时装进入相关的市场时，必须为市场操作调整战略性计划，且公司应该利用机会优势和不断变化的环境进行操作，其中涉及以下细节：

（1）清晰的公司使命：一家现代的服装企业必须制订一套任务目标以指引全体员工，并配合企业的机构组织目标。例如，任务目标是市场为主导，则要求公司注重在服装售出以后对不同客户需求的服务。

（2）辅助的目标：为了达到市场的目标，其他部门必须制订一个辅助目标，且该目标必须尽可能具体，如生产、商品跟进、设计和财政等。

（3）协同作用的策略：对每一个市场单元制订详尽的计划，如定价、定位、分销、推广、广告以及零售或特许经营等。

### 5.2.3 休闲装的定位

在中国服装行业中，国外投资者的数量大量增加，目前，国内品牌企业也开始采用西方的质量管理与营销技术推广他们的品牌。然而，班尼路

（Baleno）总是把中国香港本地时尚品牌作为主要竞争对手，其中包括堡狮龙（Bossini）、佐丹奴（Giordano）、U2、真维斯（Jeanswest）、艾斯普瑞特（Esprit）、掂（Theme）、鳄鱼（Crocodile）和金利来（Goldion）。以下的位置图显示出班尼路的主要竞争对手的市场定位（图 5-2）。

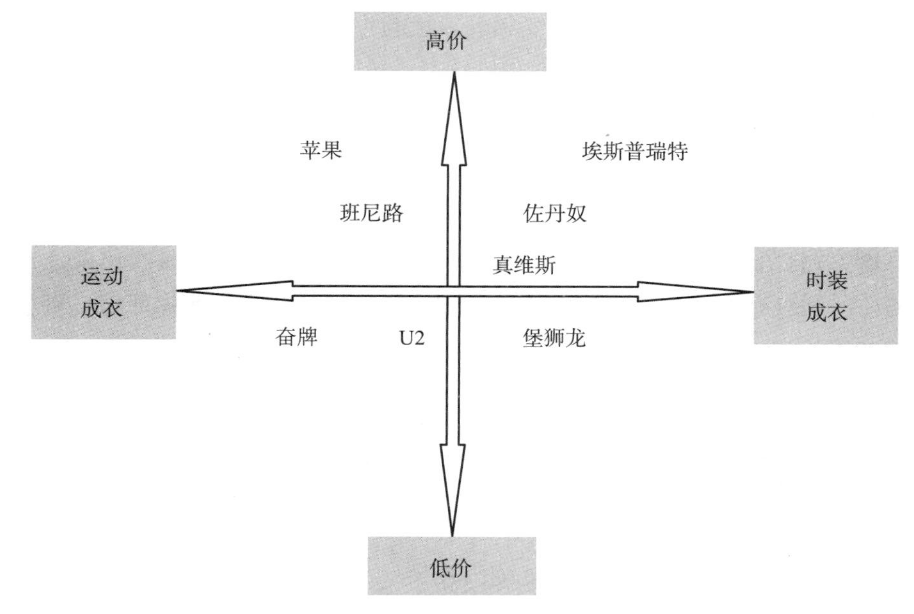


**图 5-2 中国休闲服品牌定位图**

### 5.2.4 特许经营策略

服装公司一般将特许经营作为主要扩展策略，或与中国内地的本地企业进行合资经营，因为特许经营可通过业务重组将成本费用减少，并且可以提高生产效率。对于大部分的零售业务来说，特许经营是最普遍的营销和分销策略。特许经营涉及许可的独立经销商在严格的规定和控制下减少和出售品牌产品。特许经营涉及强烈的对比和独特的商业形式，因此特许经营需要用同样的处理方式，根据提供商的标准提供服务。

## 5.3 案例分析——班尼路

### 5.3.1 关于班尼路

班尼路首次进驻中国市场是在 1992 年，它向中国年轻一代推出休闲服所有系列产品。起初，高层管理人员复制佐丹奴的商业模式开始经营班尼路。他们尤其意识到佐丹奴成功的关键因素是选择最佳地点作为店铺的能力以及“模仿”的策略。班尼路把店铺开在佐丹奴的旁边作为回应。这个“跟随者策略”能够使公司享受到类似低风险、成本效率，高客流量光顾班尼路店铺的竞争优势。如今，中国有超过 350 个班尼路零售商，其中包括直营店、特许经营店，或者是特许跟直营模式结合的店铺。班尼路践行的企业使命是通过提供优秀的零售服务和高品质的休闲服来改变年轻一代的生活质量。

### 5.3.2 定位与推广

班尼路采用价格、产品和优质的顾客服务作为主要的支持性定位策略，同时把特许经营商当成是商业伙伴，并对他们提供丰厚的回报作为激励。

班尼路把自身定位为“物超所值”，这主要基于其自身的成衣化设计、良好的品质、具有竞争力的价格、优质的顾客服务以及舒适的购物环境。班尼路的管理层相信未来十年的潮流趋势将会是偏向朴素、整洁和自然的设计。

班尼路把有竞争力的定价与优质的产品作为主要的支持性定位策略。顾客在购物之前或之后都会要求高质量的顾客服务，比如提供购物时的咨询服务、试穿服务和购买后的退货服务。因此，提供优良的顾客服务也是“物超所值”的一个主要部分。

班尼路想要建立一个能够让顾客把高质量的好处、价值、时尚和地位联系到一起的强大品牌形象。它推出一系列营销宣传项目，教会中国年轻人怎样获得一个完美的休闲服印象来改变个人风格。

### 5.3.3 特许经营

班尼路在中国的市场扩张目标通过覆盖众多地区的特许经营点来达到。班尼路想在年轻消费群里成为市场领袖。班尼路通过建立无数的销售点来加强在中国的竞争地位。班尼路还想充分利用特许经销商以最有效的方法推广高质量、合理价格和优质服务的品牌形象，以下是班尼路要充分利用特许经营的若干理由。

（1）成为年轻消费群的市场领袖。

（2）推广和树立班尼路高质量、合理价格和优质服务的品牌形象。

（3）提供销售和存货管理的信息系统。

（4）向特许经营商提供班尼路经营管理系统的培训。

（5）最佳店铺位置的选择和门面设计。

（6）向特许经营商提供“优质产品”。

（7）全国营销沟通系统。

（8）在商品选择上提供建议。

（9）顾客服务的监控系统。

否则，特许经营商就会在将来成为班尼路的竞争对手。一旦他们拥有管理技巧、工艺技术和班尼路零售管理运作的信息管理系统，他们就能够采用相同的模式经营自己其他品牌的店铺。

## 5.4 成衣采购循环

在实际操作中，商品采购职能远远超过商品跟单部门的活动范围。一家公司的商品采购程序还有很多其他职能。如图 5-3 所示，以图解形式说明了一般商品采购程序的主要步骤。该图显示了各种活动的紧密联系，同时也暴露了采购程序在不同环境所受的影响。成衣采购的基本目标可以总结如下：

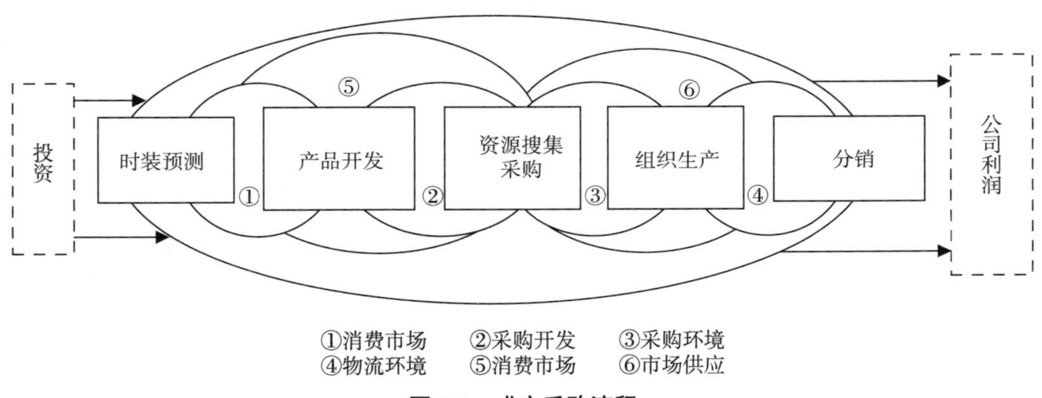


**图 5-3 成衣采购流程**

（1）计划或采购一致的成衣产品。

（2）与供应商保持有效关系并开发其他有效资源，以便组织连续的产品供应。

（3）与其他部门保持合作关系，提供所需的信息与建议，确保整个公司的有效运作。

（4）保证及时更新信息与知识，以便决策之用。

### 5.4.1 时装潮流的预测

今天的时装以消费者开始，也以消费者而结束。消费者是购买和使用商品的人群。市场策略变得更为重要，制造商与零售商必须考虑消费者的各种需求。该内容的理解是预见下一季节将流行什么样的成衣产品。预测包括对成衣市场整体发展与服装营销环境机会的观察。此外，这种程序以现有的商品和市场流行为基础的商品优劣为焦点。

### 5.4.2 产品的开发

遵循早期鉴定的市场趋势与偏好，由设计师、买手和销售经理构成的产品开发团队按照顾客选择的时装特征共同确定开发产品的设计图稿的标准。其中包括颜色、裁剪、外形轮廓、面料质地、采用的辅料与手工的选择。新季节的设计产品有时会编成一个设计组合，以便延长现有的产品线，如图 5-4 所示。在实际操作中，在同一季节会开发多种设计图稿进行筛选，直到开发出符合公司目标的设计图稿为止。

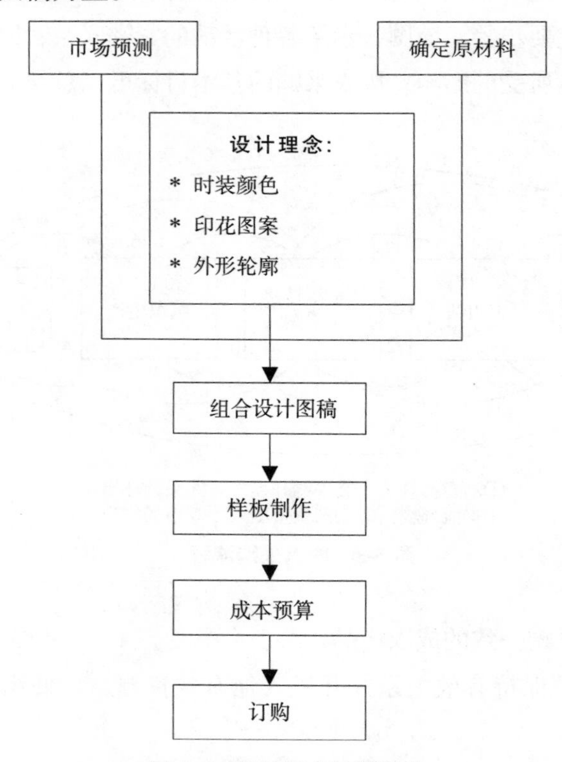


**图 5-4 设计组合计划程序**

职能部门需要一些信息用于计划生产排期，如存仓与生产场所信息、交通运输分配与生产信息等。职能部门还需要有与成本控制部门一起平衡生产成本的能力，见表 5-1、表 5-2。

**表 5-1 女装裤子规格表**

**设计规格**

日期：2-3-2020

| 季节 | 款式描述 | 数量 | 尺码 | 设计师 | 纸样师 |
| --- | --- | --- | --- | --- | --- |
| 冬季 | 女装裤 | 240 件 | S~XL | Sandy | Ben |

**款式细节**

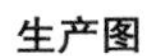

1. 表袋和 2 个弯袋双针面线。
2. 后袋缉袋花。
3. 所有袋角位上铆钉。
4. 三角形育克双针面线。
5. 3cm 单针反裤脚。
6. 腰头 5 个裤串带双针裤串带。
7. 皮牌花样绣花。

**面料细节**

| 种类 | 供应商 | 成分 | 组织 | 颜色 | 布封 |
| --- | --- | --- | --- | --- | --- |
| 面料 1 | CK 纺织公司 | 80% 棉<br>15% 涤纶<br>5% 莱卡 | $\frac{1}{3}$ 斜纹<br>“Z” 向<br>54 × 35/18S × 18S | 灰色—446C | 150cm |
| 袋布 | CK 纺织公司 | 涤 / 棉 | 平纹布 | 白色—1200 | 150cm |
| 门襟里衬 | ALBB 公司 | 100% 棉 | 平纹 / 黏性 | 白色 | 90cm |

**辅料细节**

| 种类 | 供应商 | 尺码 | 成分 & 颜色 | 位置 |
| --- | --- | --- | --- | --- |
| 掀纽 | YKNK | 25L | 黄铜 | 前中开口 |
| 拉链 | YKK | # 4YG/ 12~14cm | 黄铜 | 前中开口 |
| 缝纫线 1 | DALI | 40S/ 3 | 灰色 / 洗水色 | 裤身正面线 |
| 缝纫线 2 | DALI | 30S/ 3 | 灰色 / 洗水色 | 裤身底面线 |
| 铆钉 | YKNK | VR-511 | 黄铜 | 袋角位 |
| 皮牌商标 | YKNK | 5cm × 4cm | 皮 | 右后腰头 |
| 洗水标 / 成分标 | YKNK | 机织商标 | 涤纶 | 右后腰头内侧 |
| 尺码标 | YKNK | 机织商标 | 涤纶 | 右后腰头内侧 |
| 合同款号标 | YKNK | 机织商标 | 涤纶 | 右后腰头内侧 |

**后整：石洗**

**特殊指示：准备“S—M—L—X—XL”尺码规格**

设计组：Alice Tam 2-3-2020　　跟单组：Ben Liu 2-3-2020

**表 5-2　女装裤成本单**

<table>
<tr><td colspan="7">成本单</td></tr>
<tr><td colspan="4">参考编号：CT-010</td><td colspan="3">日期：5-3-2020</td></tr>
<tr><td colspan="2">买家：Tommy</td><td colspan="2">项目：女装裤</td><td colspan="2">数量：240 条</td><td>款号：RD-018</td></tr>
<tr><td colspan="2">尺码范围：S ~ XL</td><td colspan="2">国家：从中国到美国</td><td colspan="2">FOB 价格：12 美元</td><td>零售价格：120 美元</td></tr>
<tr><td colspan="3">面料成本</td><td colspan="4">汇率：1 美元 =6.517 人民币</td></tr>
<tr><td>种类</td><td colspan="2">供应商</td><td>单价</td><td>布幅宽</td><td>损耗数量（10%）</td><td>总金额</td></tr>
<tr><td>面料 1</td><td colspan="2">CK 纺织公司</td><td>¥30/m</td><td>150cm</td><td>1.4m / 件</td><td>¥42/ 条</td></tr>
<tr><td>袋布</td><td colspan="2">CK 纺织公司</td><td>¥6/m</td><td>150cm</td><td>1m /8 件</td><td>¥0.76/ 条</td></tr>
<tr><td>门襟里衬</td><td colspan="2">ALBB 公司</td><td>¥12/m</td><td>90cm</td><td>1m /120 件</td><td>¥0.1/ 条</td></tr>
<tr><td colspan="7">① 面料总成本 = ¥42.86/ 条</td></tr>
<tr><td colspan="7">辅料成本</td></tr>
<tr><td>种类</td><td>供应商</td><td colspan="2">单价</td><td>数量</td><td>总损耗数（10%）</td><td>总金额</td></tr>
<tr><td>揿纽</td><td>YKNK</td><td colspan="2">¥68/1 包（100 件）</td><td>1 个 / 条</td><td>24 个</td><td>¥0.68/ 条</td></tr>
<tr><td>拉链</td><td>YKK</td><td colspan="2">¥150/1 包（100 件）</td><td>1 条 / 条</td><td>12 条</td><td>¥1.5/ 条</td></tr>
<tr><td>缝纫线 1</td><td>DALI</td><td colspan="2">¥4.6/ 件</td><td>1 个线 / 12 条</td><td>2 个</td><td>¥0.38/ 条</td></tr>
<tr><td>缝纫线 2</td><td>DALI</td><td colspan="2">¥3.2/ 件</td><td>1 个线 / 12 条</td><td>2 个</td><td>¥0.27/ 条</td></tr>
<tr><td>铆钉</td><td>YKNK</td><td colspan="2">¥68/ 1 包（100 件）</td><td>10 个 / 条</td><td>24 个</td><td>¥5.8/ 条</td></tr>
<tr><td>皮牌商标</td><td>YKNK</td><td colspan="2">¥36/100 件</td><td>1 个 / 条</td><td>12 个</td><td>¥0.36/ 条</td></tr>
<tr><td>洗水标 / 成分标</td><td>YKNK</td><td colspan="2">¥22/100 件</td><td>1 个 / 条</td><td>12 个</td><td>¥0.22/ 条</td></tr>
<tr><td>尺码标</td><td>YKNK</td><td colspan="2">¥12/100 件</td><td>1 个 / 条</td><td>12 个</td><td>¥0.12/ 条</td></tr>
<tr><td>合同款号标</td><td>YKNK</td><td colspan="2">¥22/100 件</td><td>1 个 / 条</td><td>12 个</td><td>¥0.22/ 条</td></tr>
<tr><td>包装袋</td><td>ALB</td><td colspan="2">¥16/100 件</td><td>1 个 / 条</td><td>24 个</td><td>¥0.16/ 条</td></tr>
<tr><td>纸箱</td><td>ALB</td><td colspan="2">¥4.8/ 件</td><td>箱 / 24 件</td><td>1 个</td><td>¥0.2/ 条</td></tr>
<tr><td colspan="4">小计</td><td>¥0.348/ 条</td><td></td><td>¥9.91/ 条</td></tr>
<tr><td colspan="7">② 辅料总成本 = ¥0.348+ ¥9.91= ¥10.26/ 条（包括损耗数）</td></tr>
<tr><td colspan="4">后整：石洗　包装：平装</td><td colspan="3">制作成本 & 其他费用</td></tr>
</table>

续表

| 生产图 | | |
|---|---|---|
| | 石洗 | ¥12/ 打 |
| | CMT | ¥216.8/ 打 |
| | 包装 | ¥2.4/ 打 |
| | 管理费用 | ¥5/ 打 |
| | 运输 | ¥10.5/ 打 |
| | 修整 | ¥6/ 打 |
| | 存仓 | ¥3/ 打 |
| | 其他 | ¥20/ 打 |
| | **制作成本** | **¥275.8/ 打** |
| | **③ 制作总成本 =¥22.9/ 条** | |
| 单件总成本 = ① + ② + ③ = ¥76.12 | | |
| **④ 利润 = 总成本 × 10%= ¥7.612** | | |
| **⑤ 目标 FOB 价 =（¥76.12+ ¥7.612）/ $6.517=12.85 美元 / 条** | | |
| 制表：May He 5-3-2020 | 检查：Ben Liu 5-3-2020 | |

## 5.4.3 原材料搜集与采购

为了优化企业利润，跟单员尽可能地搜集各种资源，保证市场供应、准时交货、财务安排在最少不确定因素下进行。采购是指购买、存储、运输处理以及接受与检查进料等活动。根据预定的经营标准，跟单员须注意到各种有关资源搜集的限制。在很多案例中，在这个阶段的跟单采购职能涉及物料管理与订单生产安排，以便准时完成生产计划。

采购部的一项重要任务是监督供应商及跟踪新订单的进展，确保对方及时以合理的价格交付规定数量和质量的商品或服务。采购活动包括如下过程。

（1）对不同供应商进行评估，较为重要的产品和原材料应选用少量但效率高的供应商。

（2）以有竞争力的价格提供购买有效物料的机会。

（3）对供应商的交货期、服务、质量和价格进行分级和评估。

（4）提高公司在供应商和竞争者心中的声誉。

（5）通过与市场部的不断交流而获得选择新材料的机会。

（6）通过与其他部门的联系，确保全公司有足够的库存。

从以上活动可清楚看到采购对公司的利润会产生影响。

### 5.4.4 组织生产

这一阶段强调如何安排生产、确定质量标准、安排所需物料与落货排期、配备控制人员。该程序包括原材料的订购与接收、检验，技术问题的分解以及成品包装与付运的详细进程。在此，跟单员必须明白技术规格与疵点的容许偏差。

### 5.4.5 船运安排

在此阶段，跟单员评估装货单系统，决定运输方式与路线，研究包装与搬运方法，最后建议用最好的方式安排付货。跟单工作能为订单结算准备适当且准确的文件，以便辅助出货装船与财务运作。

另外，各项跟单活动的职责是清晰的，与其他部门共同承担一定责任。这些职责如下：

（1）实验室测试与检查。

（2）物料搜集与预算。

（3）标准与规范的制定。

（4）生产进度监督。

（5）存货控制。

## 5.5 相关术语

### 5.5.1 市场营销词汇

（1）广告：由某个确定的赞助商支付促销信息费。

（2）成衣直销商：成衣生产商拥有的专门销售处理品或以低价向公众大甩卖的商店。

（3）买手：一个能接受期货产品或有购买选择权的市场参与者。一个有选择权的买手也叫购买者、持有者或所有者。

（4）分店：当一家经营很好的百货公司在另外一个场所开店时，新的一家就叫分店。分店的经营源于原来的旗舰店。

（5）连锁店：一组由中心机构所拥有、管理和控制的专营店。所有公司的连锁店都如此。没有店铺被认为是主店。

（6）消费者：购买并穿着服装的人群。消费者像购买其他经济产品一样采购并使用成衣。消费者对决定什么时装流行与什么时装不流行来说是非常重要的。

（7）百货公司：在适当的部门提供各种不同的商品的零售机构。各种颜色、尺码和款式的大量服装与家居产品在百货公司里销售。

（8）折扣店：在宽大、简朴的建筑物中以低成本销售服装与其他商品。大量的服装在货架上进行销售。

（9）特许经营：使用一家公司的名称与商品进行独立经营。特许经营商在指定的城市或地区给代销商提供专用的名号与商品。

（10）工厂直销：制造商自身拥有的商店以减价形式向公众销售本公司的产品。

（11）直销：没有个性化推广，只针对大量普通观众进行销售。

（12）邮购机构：通过产品目录销售产品给消费者。他们为那些不能或不喜欢外出的消费者提供在家购物的服务。

（13）商品采购：贯穿产品设计、开发与推广到销售环节的营销程序。

（14）促销：由制造商引导在全国范围内的零售店销售特殊的时装。促销活动指通过广告和推销等方式促进销量。

（15）采购订单：由买家编写的授权卖家以指定价格交付产品的文件。

（16）零售店：将产品直接销售给消费者。他们直接向普通公众做广告并销售产品。

（17）专卖店：只销售成衣或者专门销售一种特定的成衣。例如，孕妇商店、鞋店、婚宴商店和童装成衣店。

（18）视觉营销：以一种具有吸引力、易于理解的方式展示产品。通过陈列与展示从视觉上推广服装。商店会举办特殊的活动，甚至让模特展示商品。

（19）视频营销：在零售店采用视频展示最新的服装流行趋势以推广商品。视频放置在零售店内其展示商品的附近。它们用声音与动作吸引过客的注意力。

（20）批发商：出售大量的产品给零售商。批发商通常从大仓库配送产品且每种产品必须以打为数量单位进行采购。

### 5.5.2 贸易词汇

（1）成本和运费：指定交货地点，报价包括支付“成本和运费”。

（2）到岸价（成本、保险和运费）：一种使用国际销售合同的国际商业付款方式，销售价包括销售产品的所有成本、保险和运费。卖家安排从出口国到进口国的所有出货环节并支付涉及的所有费用。

（3）来料加工（裁剪、制作和后整理）：买家提供面料，供应商负责制作服装。

（4）佣金：支付给委托机构进行商品买卖的费用。

（5）商业发票：由出口商或货物转运公司准备的付款证明文件，必需在进口商证明所有权后安排付款给出口商。它应该包括产品描述、发货人、发货地址以及交付期与付款方式等有关交易的基本信息。

（6）合同：一份买卖协议，规定了具体商品、产品数量、产品等级和合同到期与交付使用的日期。

（7）交付期：商品或仪器的交付日期必须履行合同条款进行交付。

（8）出口配额：为了保护本国生产商与消费者的利益，由进口国对出口商品的价格与数量进行一种特定的限制。出口配额也通常应用在有秩序的买卖协议或非官方限制协议中以便促进国内企业使用本国生产的原材料。

（9）汇率：用一种货币购买另一种付款货币的价格。

（10）离岸价（船上交货）：价格包括在商业合同中规定的产品装船付运所有成本。它与FOA性质相同（FOB空运）。

（11）信用证：由银行发行的信用文件，保证顾客在规定期间支取规定金额作为付款。

（12）包装表：该表给出托运包裹内容的详细资料，如果包装信息在发票上没有显示，必须由顾客授权确认。

（13）报价单：报价单包含了卖家准备提供的产品价格和条件。

（14）商标：制造商或交易商使用的合法注册的保护标识和符号，以此区分其他竞争产品。

## 本章小结

- 简单介绍了服装市场与采购的基本知识，并简单描述中国休闲服市场的相关情况。以班尼路品牌为案例，介绍了特许经营的市场策略与营运模式，从而导出相关专业术语与品牌名称。
- 描述了成衣采购的基本流程，以服装工艺单与成本报价单为案例说明了采购文件的编写与跟进工作。
- 熟悉面料采购基本理论与跟单文件的认识与应用。
- 对市场营销与贸易等相关术语进行详细的描述。
- 此章节重点培养学生对成衣采购与跟单的认识，提高学生对商品采购文件的阅读与编写能力。

**CHAPTER 6**

# BUSINESS ENGLISH　商务英语

**课题名称：**BUSINESS ENGLISH　商务英语

课题内容：Practical English in Chain Store　连锁店实用英语

Telephone Conversation　电话交谈

Business Demonstration　商务洽谈示范

**课题时间：**4课时

**教学目的：**让学生了解服装连锁店内不同情况的英语沟通，掌握办公室不同情形的电话沟通方式与技巧。掌握服装企业有关业务洽谈与沟通的基本对话方法。重点掌握并熟记各种典型的句型与对话并能熟练应用与表达。

**教学方式：**结合PPT与音频多媒体课件，以教师课堂讲述为引导，学生分组进行相关对话训练为主。

**教学要求：**1. 熟悉服装连锁店内不同情况的英语沟通与对话。

2. 熟悉不同情形的电话沟通方式与技巧。

3. 熟悉服装企业业务洽谈与沟通方式。

4. 主要引导学生课后进行自发的交流与训练。

**课前（后）准备：**结合专业知识，课前预习课文内容。课后熟读课文主要部分，掌握各种典型的句型与对话，并能熟练应用与表达。

# CHAPTER 6

# BUSINESS ENGLISH
# 商务英语

## 6.1 Practical English in Chain-Store 连锁店实用英语

### 6.1.1 Showing Visitors Round the Shop 陪客巡看

（Notes：S for Salesman，and C for Customer）

S：Welcome to our shop. You seem to be in a hurry. What's the matter？

C：It's my first time to Guangdong. I want to buy something for my mother. But I have only an hour for shopping before leaving for the airport. Would you mind if I ask for your help？①

①这是我第一次来广东。我想为我母亲买些东西。但我去机场之前仅有一小时时间购物。我可以请你帮忙吗？

S：Not at all. What can I do for you？

C：I want to buy a blouse and a scarf，but I'm afraid I may miss the plane.

S：Don't worry. Please follow me. I'm sure you can take what you need，and you won't be late for your plane.

C：It's very kind of you. I will be most grateful to you.

S：It's my pleasure to help you. The clothing and knitted-wears counter is on the second floor. This way，please.（They go upstairs.）By the way，where are you from？

C：I'm from America.

S：What's your job？

C：I'm a teacher.

S：Oh，it is a nice occupation. What color does your mother like？

C：Any color except blue.

S：What's her bust measurement？

C：About thirty eight inches.

S：This blouse is 100% cotton，and the quality is excellent and the style is smart.② What do you think of it？

②这款罩衫是纯棉的，质量好且款式漂亮。

C：Very nice. I like it. What time is it now？

S：It took us only five minutes. Don't worry. Let's go over there.

C：I want to buy a yellow and a red scarf.

S：There is still more than ten minutes to go. Please contact the service room downstairs for a taxi to take you to the airport. ③

③还有十多分钟的时间。请联系楼下的服务部叫出租车带你到机场。

C：Thank you very much. Good-bye.

S：Good journey.

### 6.1.2 At "JW" Chain Store 在"JW"连锁店

（Notes：Afor Salesman，Bfor Customer）

A：Good morning，sir. Can I help you？

B：Yes. I am looking for something to give as a gift.

A：What about some smart ties？ We have some quality ties from France，Italy，the United States，and Switzerland. Their designs are extra-ordinary. None of them are commonplace. The color is well blended with taste. You can get them in contrasting color. Every tie is a piece of art.

B：How much does such a smashing tie cost？

A：RMB Ninety-five，sir.

B：That is shocking.

A：But sir，look at the quality，the design，and the coloring. Tremendous work and effort have gone into every tie. It is a piece of art，not mass-produced. ④

④但是先生，看一下它的质量、设计和颜色，每条领带都花费了大量的心血和努力。它是一件艺术品，而不是批量生产的普通货。

B：Yes. They do look nice，quite unusual. All right，give me four pieces. I'll have the red one with splotches of blue and yellow，the pink one with blue and green circles.

A：Sir，you have excellent taste. They are really outstanding.

B：Could you please gift wrap them？

A：Certainly，sir.

### 6.1.3 Purchasing Clothing 购买服装

（Notes：A for Salesman，B and C for Customer）

A：Good afternoon. Have you been taken care of？

B：No，not yet. I'm looking for light summer suits. Do you have anything ready-made to fit me？

A：Let me take your measurements，sir. I'm afraid our ready-made sizes are just a little too small for you. We could fit you in trousers，but the shirt would be a little narrow in the shoulder. ⑤

⑤让我帮你量一下尺寸，先生。我想我们的成衣尺码对你太小了。我们可以给你挑选合身的裤子，但衬衫肩宽有点窄。

B：Well，may I try a suit on？ Perhaps it could be altered.

A：Certainly，sir. Please step this way. This is our largest size. Please try the shirt on for this size.

B：I see what you mean. It's the same，too. I like this material very much.

A：Well，sir，how about having our tailoring department makes one up for you？ There's not much difference in the price between a ready-wear and a tailored one.⑥

⑥好的，先生，我们让裁剪部给你做一件怎么样？成衣和量身定制之间的价格没有多大的差异。

B：How much is this ready-wear one？

A：It's two hundred and fifty. And I believe we could tailor you one for about two hundred.

B：Well，that's not so bad. Let me talk to your tailor.

A：Certainly，sir. If you'll just step this way，I'll introduce you to our tailoring clerks.

B：Thank you.

A：Not at all.

C：Excuse me，is this color in trend？

A：Yes，it is quite in fashion.

C：Thanks for your suggestion. I'll take it.

A：Anything else？

C：I want to buy some bath towels.

A：What color do you wish？

C：Two white and three red.

A：These are made of fine material.

C：What is the price？

A：White is seven Yuan per piece；red is eight Yuan fifty cents.

C：Do you have anything cheaper？

A：Here is another one at four Yuan and fifty cents for white；red is five Yuan one piece.

C：Well，I will take the cheaper one.

A：Thanks. Can I show you anything else？

C：Nothing else，thank you. What is the total amount？

A：The total amount is one hundred and twenty- four Yuan.

C：Here is one hundred and twenty- five Yuan.

A：Thank you. Here is your change，one Yuan.

### 6.1.4 About Bargaining 议价

（Notes：C for Customer，and S for Salesmen）

C：Have you got any 100% cotton trousers like that？

S：That's the last one.

C：But there is a little defect with it. Can you lower the price？

S：Wait a moment，please. I'll ask my boss.

C：O.K.

S：I'm sorry to have kept you waiting. My boss has agreed to give you a five percent discount.

C：Five percent？ It's still too much. Can't you make it any cheaper than that？

S：I'm sorry，I can't.

C：By the way，these trousers are unfashionable.

S：But these trousers own good quality fabric，and they are very suitable for you.

C：But I don't have confidence in workmanship.

S：This being the case，you can enjoy another five percent discount. The rule of our department store，the customer can get a five percent discount when he finds a fault in goods. ⑦

⑦对于这种情况，你可以另外享受5％折扣。我们百货公司规定，顾客发现疵品可得到5%的折扣。

C：All right. I'll take it then.

### 6.1.5 Agreement with Customer 认同顾客

（Notes：S for Salesmen，and C for Customer/ Mrs. Wu）

S：Mrs. Wu，do you prefer 100% wool or wool-polyester suits？

C：I prefer wool-polyester suits. They seem to hold their shape and not wrinkle as much as 100% wool suits. I'm looking for a suit for working；I think wool-polyester holds up better during the work days. ⑧

⑧我喜欢毛涤套装。它们的保形性好，不像纯毛产品易起皱。我想买工作套装，我想毛涤面料更适合日常穿着。

S：I think you are quite right about the wool-polyesters. The suits you are looking at are really nice. That navy suit has a real classic look，doesn't it？

C：Yes，it is really nice.

S：Did you say to want a suit with a classic look？

C：Yes. I think that is best for business clothes.

S：And you said that you wear a size L？

C：Well，it depends on the cut of the suit. But I think the L is the right size.

S：Here are two navy suits with a classic look in size L. Which one do you like better？

C：I think the one on the right is better.

S：Don't you like the edge-stitching around the collar？

C：That's what attracted me to the suit.

S：Why don't you just try it on？ It will take a second look. The dressing room is right there.

## *Words and Expressions*

customer ['kʌstəmə] 顾客
information desk [infə'meiʃən desk] 咨询处
furnishing ['fə:niʃiŋ] 服饰品
escalator ['eskəleitə] 自动扶梯
pure silk [pjuə silk] 真丝
tie counter [tai 'kauntə] 领带柜台
obliged to [ə'blaidʒd tu:] 使感激
salesman ['seilzmən] 售货员
airport ['eəpɔ:t] 机场
shopping ['ʃɔpiŋ] 购物
blouse [blauz] 女装罩衫
scarf [ska:f] 围巾
knitted-wear ['nitid wiə] 针织服装
store [stɔ:] 店铺
clothing ['kləuðiŋ] 衣服
upstairs ['ʌp'stɛəz] 楼上的
bust [bʌst] 胸围
measurement ['meʒəmənt] 尺寸
smart [sma:t] 漂亮的，时髦的
occupation [ɔkju'peiʃən] 职业
downstairs ['daun'steəz] 楼下的
contact ['kɔntækt,] 接触
service ['sə:vis] 服务
journey ['dʒə:ni] 旅行
gift/ present [gift / 'preznt] 礼物
quality ['kwɔliti] 质量
extra-ordinary ['ekstrə 'ɔ:dinəri] 异常的
contrasting ['kɔntræstiŋ] 对比鲜明的
smashing ['smæʃiŋ] 绝妙的
commonplace ['kɔmənpleis] 陈腔滥调
shocking ['ʃɔkiŋ] 震惊的，可怕的
tremendous [tri'mendəs] 惊人的
mass-produced [mæs prə'dju:st] 批量生产
splotch [splɔtʃ] 污迹，油污
outstanding [aut'stændiŋ] 醒目的，杰出的
taste [teist] 品味
knitted garment ['nitid 'ga:mənt] 针织服装
perfect ['pə:fikt,] 完美的
tight [tait] 紧身的
summer suit ['sʌmə sju:t] 夏装
ready-made ['redi meid] 成衣制作
shoulder ['ʃəuldə] 肩宽
alter ['ɔ:ltə] 改变
material [mə'tiəriəl] 原材料
tailor ['teilə] 裁缝师
ready-wear ['redi wiə] 成衣
recommend [rekə'mend] 推荐
commodity [kə'mɔditi] 商品
cotton ['kɔtn] 棉
fashion ['fæʃən] 时装
bath towel [ba:θ 'tauəl] 毛巾
total amount ['təutl ə'maunt] 总数
bargain ['ba:gin] 议价

fabric ['fæbrik] 布料
five percent / 5% [faiv pə'sent] 百分之五
discount/ sales off ['diskaunt / seilz ɔ:f] 折扣
workmanship ['wə:kmənʃip] 手工艺
confidence ['kɔnfidəns] 信心
fault [fɔ:lt] 缺点
presentation [prezen'teiʃən] 献礼，展示
wool-polyester [wul ' pɔliestə] 羊毛 / 涤纶混纺
wrinkle ['riŋkl] 皱褶
classic look ['klæsik luk] 传统样式
business clothes ['biznis kləuðz] 职业装
edge stitching [edʒ 'stitʃiŋ] 缝边线迹（广东话：缉边线）
attract [ə'trækt] 吸引
dressing room ['dresiŋ ru:m] 试衣间
deliver [di'livə] 传送

## *Exercises*

1. Translate the following terms into Chinese.

（1）customer
（2）furnishing
（3）pure silk
（4）salesman
（5）measurement
（6）bust
（7）blouse
（8）shirt
（9）knitted-wear
（10）mass production
（11）shopping
（12）occupation
（13）counter
（14）chest
（15）inch
（16）scarf
（17）cotton
（18）design
（19）chain store
（20）ready-wear
（21）apparel
（22）tight fitted
（23）T-shirt
（24）summer suit
（25）material
（26）bulk production
（27）clothing
（28）pressing
（29）solid color
（30）shell fabric
（31）knit rib collar
（32）loose-fitted blouse
（33）classic look
（34）commodity
（35）bath towel
（36）workmanship
（37）discount
（38）dressing room
（39）polyester
（40）delivery date
（41）wrinkle
（42）edge stitching
（43）garment fault
（44）culottes
（45）selvedge
（46）care label
（47）tailored jacket
（48）eyelet

2. Fill dialogue “B” with your suggestion.

（1）A: Good afternoon, may I help you?
B:______________________________

（2）A: What’s your bust measurement?
B:______________________________

（3）A: What dose such a smashing tie cost?
B:______________________________

（4）A: What are the prices?
B:______________________________

（5）A: Can I take a look at that of shirt?
B:______________________________

（6）A: Do you know your size in inches?

B: ______________________________

（7）A: How much is this ready-to-wear one?

B: ______________________________

## 6.2 Telephone Conversation 电话交谈

### 6.2.1 Situation Ⅰ 情景Ⅰ

A：Hello.

B：Is this the Rykiel Fashion Chain Store？

A：Yes，it is.

B：I'd like to speak to Mr. Wu.

A：Hold the line please. I'll call him.

B：Yes，thanks.

A：I'm sorry，he is out.

B：When do you expect him back？

A：I'm not sure，but he'll probably not come back before four. Shall I ask him to call you back？①

①我没有把握，但他也许4点前不会回来。要不要我叫他给你回电话？

B：Yes，thanks. Tell him to call Mr. Chen. My telephone number is 9988467.

A：O.K. I'll tell him.

B：Thank you very much. Bye-bye.

### 6.2.2 Situation Ⅱ 情景Ⅱ

A：I'm Mr. Ray. Is Mr. Wu here？

B：Sorry，Mr. Wu is not here. Can I take a message？

A：Thanks，but I'd rather have him call later.

B：Does Mr. Wu know your telephone number？

A：Yes，he does.

B：And your name is Ray，did you say？

A：Yes，I am Mr. Ray. Spelling is R-A-Y.

B：I've got it. Mr. Ray. I'll tell him to call you back later.

A：Thank you.

### 6.2.3 Situation Ⅲ 情景Ⅲ

A：Hello，Is that Mr. Chen？

B：Yes，speaking，what can I do for you？

A：Did you get our catalogues？

B：Yes，You know that your products interest us very much.

A：I am very pleased to hear that. I'll bring some of our samples for your reference and inspection. What time would you like me to come？②

②听你这样说我非常高兴。我将拿些样品给你参照和检查。你希望我几点来？

B：Well，you may drop in this afternoon，is that suitable？

A：That's just perfect. I'll see you then. Goodbye.

B：Goodbye.

### 6.2.4 Situation Ⅵ 情景Ⅳ

A：Hello. Is that 9988768？

B：Yes，it is.

A：May I speak to Mr. Wang，please？

B：Yes，I am. Who's speaking？

A：Good morning. Mr. Wang. This is Chen speaking.

B：Good morning Mr. Chen，what's up？

A：Yes，we want to buy some pure silk shirts. Do you have any in stock？

B：Yes，we always have that in stock.

A：What are the prices？

B：The prices run from about fifty to one hundred dollars each.

A：Will you come over here with catalogues？

B：With pleasure，shall I call on you at once？

A：Well，I am waiting for you. Goodbye.

### 6.2.5 Situation Ⅴ 情景Ⅴ

A：Can I speak to the manager，please？

B：May I know who's calling？

A：This is Mr. Peter of the Polo Fashion Co.

B：I beg your pardon！ I can't hear you well.

A：I am Mr. Peter of the Polo Fashion Co. Can I speak to your manager，Miss？

B：Yes，hold the line please.（Covers phone）Mr. Wu，there's a Mr. Peter of

③ 好的，请不要挂线。（盖住电话）吴先生，Polo时装公司的Peter先生找你。

the Polo Fashion Co. on the line.[③]

Mr. Wu：Oh，yes，say I'm busy，you deal with him，and he's only trying to convince me to buy their goods.

B：( uncovers phone ) I'm sorry，Mr. Peter. Our manager is busy at the moment. Can I do anything for you ?

A：Yes I'd like to show some samples of our products. May I come and see you now ?

B：I think I'll be able to find some time for you. You can come and see me here. But I'm not sure if I can be of any help to you.

A：That's very kind of you. Let me visit your Chain Store next time.

## *Words and Expressions*

catalogue ['kætəlɔg] 目录
sample ['sæmpl] 样板
reference ['refərəns] 参考
product ['prɔdəkt] 产品
inspection [in'spekʃən] 检查
perfect ['pə:fikt,] 完美
suitable ['sju:təbl] 适当的
pure silk [pjuə silk] 真丝
stock [stɔk] 库存
pleasure ['pleʒə] 愉快，快乐
manager ['mænidʒə] 经理
fashion ['fæʃən] 时装
convince [kən'vins] 说服，使信服
chain store [tʃein stɔ:] 连锁店

## *Exercises*

1. Translate the following terms into Chinese.

（1）sample
（2）catalogue
（3）inspection
（4）quality
（5）stock
（6）fashion
（7）chain store
（8）pure silk
（9）shirt
（10）manager
（11）product
（12）fabric
（13）material
（14）skirt
（15）collar
（16）sleeve
（17）front
（18）pants
（19）measurement
（20）goods
（21）evening dressing

2. Fill dialogue "B" with your suggestion.

（1）A: Have you got any 100% cotton trousers like that?
B:________________________________________

（2）A: Can you recommend some other color for me?
B:________________________________________

（3）A: Is this color in trend?
B:________________________________________

（4）A: What is the price?

B:______________________________________________

（5）A: Could you please tell me what your best-selling products are?

B:______________________________________________

（6）A: What trends do you expect for the coming year?

B:______________________________________________

## 6.3 Business Demonstration 商务洽谈示范

### 6.3.1 Before the Meeting 洽谈前

（1）How do you do？ I'm Kevin.（Nice to meet you.）

How do you do，I'm Alex.（Nice to meet you，too.）

（2）This is my card.

Thank you.

（3）This is our manager，Mr. Chen.

Nice to meet you.

（4）Do you know our production manager，Mr. Tom？

Yes，I do. We met 6 months ago.

（5）Where are you from？

I'm from India.

（6）Welcome to our company.

Thank you.

（7）Hello，Mr. Chen.

Good morning，Miss Wang，nice to see you again.

（8）How are you？

Very well，thank you. And you？

（9）How about your business？

Not so bad. How about you？

（10）When did you arrive？

Yesterday afternoon.

（11）Did you have a good flight？

Yes，it was fine. Thank you.

（12）Where are you staying？

At the West-Lake Hotel.

（13）How long will you stay？

Maybe two weeks.

### 6.3.2 Beginning the Meeting 开始洽谈

（1）Good morning，nice to see you all again.

Thanks for your seeing us.

（2）Could you tell me a bit about your business？

Of course. I'd like to tell you more about our business.

（3）What does your enterprise manufacture？

We manufacture a wide range of goods involving woven and knitted. ①

①我们公司的产品范围涉及机织与针织服装。

（4）What is your company concerned with？

Our company is concerned with trading，manufacturing，marketing，etc.

（5）Where does your company export to？

We export to USA，Australia，Japan and Europe，etc.

（6）How many employees have you got？

We have 250 employees.

（7）When was your enterprise established？

Our enterprise was established in 1992.

（8）What does your enterprise mainly deal with？

We deal mainly with under-wears，knit fashion and woven product.

（9）How about your annual turnover？

It is $150 million.

（10）Who are the market leaders in your field？

We are one of the market leaders in our field.

（11）How competitive are your prices？

We are one of the cheapest.

（12）Have you got any other product lines？

We also produce knitted-wears.

（13）Would you mind telling me what your best selling products are？

Well，here is our product catalogue. The women's dress is the best-selling products in this quarter.

### 6.3.3 About the Enterprise 关于企业

（1）Could you tell me who your main customer is？

Our main customer is Polo Enterprise Co., Rykiel Trading Co., etc.

(2) What are your major products?

Our major products are woven fashion about women's wears. ②

②我们的主要产品是机织女装。

(3) Where is your head office?

Our head office is in Hong Kong.

(4) Where are your main branches?

Our major branches are in China and Japan.

(5) How many factories do you have?

We have five factories.

(6) What are you in charge of?

I am in charge of the marketing department.

(7) Who do you deal with mainly?

I deal mainly with overseas buyers.

### 6.3.4 Discussing the Products 讨论产品

(1) We are looking for someone to produce knitwear, would you tell me which are the most fashionable styles and the best selling, please? ③

③我们在寻找能生产针织服装的企业。你能告诉哪些是流行款式与畅销产品吗?

Yes, here you are. This is our latest catalogue.

(2) Have you got a catalogue to show me?

Yes, certainly. Here is our latest sample.

(3) Which are your most popular goods in knitwear?

These are very fashionable.

(4) I'm looking for a factory to produce these. Could you make something similar?

I'm sure we could. Here are samples from past orders.

(5) Sorry, but we will have to make some changes here if possible.

We would be able to modify the product according to your requirements.

(6) What are the special features of this sample?

Well, the fabric is modern and up to date.

(7) What kind of market are you aiming at?

We are aiming at the young range of this market.

(8) How do you compare with the rest of the market on this product?

I would say our quality is better.

(9) Could you produce something like this?

Yes, I'm sure we could. We can make some samples for you. Then we can talk about the details.

### 6.3.5 Further information about the Products 关于产品的更多资料

（1）Could you produce something like this？ Here are the design and specifications.

I'm sure we could. We will make up a sample for you.

（2）What size do you need？

It is four sizes — small, medium, large and extra large.

（3）How many can you produce per month？

We can produce 10 thousand pieces a month.

（4）How many do you need at most？

We will need at least 10 thousand pieces monthly.

（5）What makes you think that your products are better than the others？

We have the good quality product and a prompt delivery.

（6）What would you expect the cost to be？

I expect the cost to be about $80 per unit.

（7）Our Minimum order would be 10 thousand, and for the large orders we will expect the price to be reduced proportionally. [④] The prices decrease according to the unit of the order. It works out to be very cost-effective.

④我们要求最小订单数量必须是1万件，对于大的订单我们将给一定比例的降价优惠。

（8）Good. I look forward to doing business with you.

Me too. Thanks.

### 6.3.6 Talking about Prices 议价

（1）I would like to place an order. What is the unit price of the trousers, and style No. is MT-1523？

Well, it depends on the quantity of your order. For the first orders over $1000 we offer a 5% discount.

（2）Can you give me a special price on these？

I'm sorry, I can't. That is the lowest I can handle.

（3）Does the price include delivery？

Yes, it does. / No, it doesn't.

（4）Does the price include insurance？

Yes, it covers all risks including loss and damages.

(5) And what about payment?

We usually prefer a L/C or a cheque.

(6) How would you like payment?

Whatever is most convenient and quickest.

(7) When will you invoice me?

I'll invoice you when you place your order before delivery.

(8) When do you want payment?

I expect payment within a month.

(9) What are the terms?

It's cash on delivery.

(10) When do I need to confirm my order?

It's better to do it as soon as possible because we are very busy at the end of the year. Can you decide now?

(11) I'll have to speak to my boss before I give you an answer about the order quantity and payment.

That's fine. I look forward to hearing from you.

### 6.3.7 Talking about Delivery 交货谈论

(1) May I ask what your delivery time is on these products?

The delivery time is four weeks.

(2) How long does it take to deliver?

It takes four weeks to deliver.

(3) When would you need a confirmed order?

I need it by mid-May.

(4) I hope there won't be any delay in shipment.

The delay is impossible if the letter of credit is on time.

(5) Can you guarantee that there will be no delivery problems?

Yes, I can promise you there won't be any problems.

(6) How will the consignment be sent to me?

We'll send it by air or overland.

(7) How long will it take?

It'll take four weeks.

(8) When will the shipment arrive?

The shipment will arrive on May10th.

（9）What does the price include？

It includes port charges, import duties, insurance and handling.

### 6.3.8 Developing Trends 发展趋势

（1）Could you give me a few details on the enterprise？

With pleasure. As you can see some details from this graphs…

（2）How much have your production costs fallen since 2007？

Our production costs have fallen by 15% since 2007.

（3）What has demand been like this year？

The prices of raw materials have reduced this year, and the expenses have decreased too. ⑤

⑤今年原材料的价格已经下降了，并且各项开支也下降了。

（4）What level of production have you achieved throughout the year？

There has been a high level of production throughout the year.

（5）What was the production output this year？

The output increased to $50 million this year.

（6）What has the inflation level been like？

There has been a sharp increase in inflation since 2007.

（7）What do you expect for the coming year？

We expect the labor cost will increase because of lacking human resources.

### 6.3.9 Negotiating with Price and Delivery 商讨价格与交货

（1）What do you think about it？

If you don't mind my saying, I think we should alter the contract.

（2）What's your opinion on it？

In my opinion, we have to raise the price.

（3）What is your feeling about it？

I would prefer to discuss the final price later.

（4）Could you settle your account earlier？

Of course, we could. /Well, actually, that may be a bit difficult.

（5）We were rather surprised of the price increasing.

I'm afraid I can't agree with you entirely.

（6）It is difficult to accept your putting up the price.

I'm very sorry, it won't happen again.

(7) With all due respect, you don't seem to understand. But I really must point out…

I'm sorry but we have done our best.

(8) How come the delivery is three weeks behind schedule?

We've had a slight problem with quality control.

(9) Is there any other way we could do it?

Don't worry. I'm sure it won't be difficult to arrange it.

(10) What do you think about my suggestion?

I agree with you entirely.

(11) I don't quite agree with you.

In that case I have no other alternative but to cancel the order. [6]

(12) Would you be prepared to make some changes?

We would like to try.

⑥那样的话，我别无选择，只能取消订单。

### 6.3.10 Ending the Meeting 结束洽谈

(1) Perhaps we should meet again next month?

Great, that'll be fine.

(2) How about meeting at your office next time?

Perfect.

(3) What do you hope to be able to do?

I hope we'll be able to work out a proposal.

(4) What are your plans?

We plan to do the final research in the coming week.

(5) When shall we have another meeting?

I think we should make it next month.

(6) Shall we meet for lunch tomorrow?

I think that should be okay.

(7) Where would you like to have lunch? How about at the West-Lake Hotel?

Great. I'll reserve a table at 2.00 p.m.

(8) Shall we discuss it again tomorrow?

Yes, let's…

(9) It was very nice to meet you.

Me too. I'll see you next time.

## Words and Expressions

concerned with [kən'sə:nd wið] 相关的
manufacture [mænju'fæktʃə] 制造
marketing ['mɑ:kitiŋ] 市场
export to [eks'pɔ:t, tu:] 出口
employee [emplɔi'i:] 雇员
established [is'tæbliʃt] 建立
under-wears ['ʌndə - wεə] 内衣
annual turnover ['ænjuəl 'tə:n,əuvə] 年营业额
market leaders ['mɑ:kit 'li:dəz] 市场领导者
catalogue ['kætəlɔg] 目录
branched offices [brɑ:ntʃt'ɔfisiz] 分公司
best-selling [best-'seliŋ] 最好销的
customer ['kʌstəmə] 顾客
overseas ['əuvə'si:z] 海外
enterprise ['entəprɑiz] 企业
fashionable ['fæʃənəbl] 流行的，时髦的
guarantee [gærən'ti:] 担保
design [di'zɑin] 设计
specification [ˌspesifi'keiʃən] 规格，要求
delivery [di'livəri] 交货期
quality ['kwɔliti] 质量
cost-effective [kɔst -i'fektiv] 有效成本
insurance [in'ʃuərəns] 保险
payment ['peimənt] 付款方式
confirmed order [kən'fə:md 'ɔ:də] 确认订单
shipment ['ʃipmənt] 装船，出货
handling ['hændliŋ] 处理，搬运
raw materials [rɔ: mə'tiəriəlz] 原材料
inflation [in'fleiʃən] 通货膨胀
human resource ['hju:mən ri'sɔ:siz] 人力资源
negotiating with [ni'gəuʃieit wið] 商讨，谈判
contract ['kɔntrækt] 合同
alternative [ɔ:l'tə:nətiv] 可选择的
proposal [prə'pəuzəl] 提议，建议
research [ri'sə:tʃ] 研究

## Exercises

1. Translate the following terms into Chinese.

（1）market leader
（2）branched office
（3）best-selling
（4）marketing dept.
（5）enterprise
（6）sample
（7）fashionable
（8）trading office
（9）guarantee
（10）fashion design
（11）supervising
（12）specification
（13）delivery
（14）insurance
（15）payment
（16）shipment
（17）letter of credit
（18）raw material
（19）manufacturing
（20）marketing
（21）export
（22）employee
（23）under-wear
（24）annual turnover
（25）human resource
（26）pants
（27）slacks
（28）T-shirt
（29）shell fabric
（30）contract
（31）proposal
（32）research
（33）R & D
（34）interlining
（35）CIF
（36）FOB
（37）retailer
（38）wholesaler
（39）agent
（40）distributor
（41）manufacturer
（42）consumer
（43）end user
（44）L/C
（45）customer

(46) vendor
(47) client
(48) buyer
(49) contractor
(50) sub-contractor

2. Fill dialogue "B" with your suggestion.

(1) A: Nice to meet you.
B: ______

(2) A: Could you tell me a bit about your business?
B: ______

(3) A: How long does it take to deliver?
B: ______

(4) A: What are you in charge of?
B: ______

(5) A: What would you expect the cost to be?
B: ______

(6) A: What are your best-selling products?
B: ______

(7) A: Can you give me a special price on these?
B: ______

(8) A: May I ask you who your competitors are?
B: ______

(9) A: How would you like payment?
B: ______

# 译文

# 第六章　商务英语

## 6.1　连锁店实用英语

### 6.1.1　陪客巡看

（提示：S 代表销售员，C 代表顾客）

S：欢迎光临本店。你看似很焦急的样子。怎么了？

C：这是我第一次来广东。我想为我母亲买些东西。但我去机场之前仅有一小时时间购物。我可以请你帮忙吗？

S：好的，有什么需要我帮忙的？

C：我想买一件女式罩衣和一条围巾，但我又怕错过班机。

S：不用担心，请你跟我来，我保证你可以买到你想要的东西并且不会误了航班。

C：你真是好人。我感激不尽呀。

S：我很乐意为你效劳。服装和针织类商品在三楼。这边请。（他们上楼了。）顺便问一下，你是来自哪里的？

C：我来自美国。

S：你的工作是什么？

C：我是一名教师。

S：噢，是个很好的职业哦。你母亲喜欢什么颜色呢？

C：除了蓝色其他都可以。

S：她的胸围尺码多大？

C：大概 38 英寸。

S：这款罩衫是纯棉的，质量好且款式漂亮。你觉得怎样呢？

C：很不错，我喜欢。现在几点了？

S：我们才花了五分钟。不用担心，我们去那边看看。

C：我想买一条黄色的和一条红色的围巾。

S：还有十多分钟的时间。请联系楼下的服务部叫出租车带你到机场。

C：十分感谢，再见。

S：祝旅途愉快。

### 6.1.2 在“JW”连锁店

（提示：A 代表销售员，B 代表顾客）

A：先生，早上好。请问有什么可以帮到你呢？

B：是的，我在找些东西作为礼物送人。

A：这些漂亮的领带怎样？我们这有来自法国、意大利、美国和瑞士的高品质领带。它们设计非凡，每条都很特别，颜色和品位都很不错。你可以把它们配成对比色。每条领带都是艺术品。

B：像这样一条漂亮的领带要多少钱？

A：要 95 元人民币，先生。

B：价格太贵啦。

A：但是先生，看一下它的质量、设计和颜色，每条领带都花费了大量的心血和努力。它是一件艺术品，而不是批量生产的普通货。

B：是的，它们真的很好看，很特别。好的，给我拿四条吧。我要那条红色带有蓝黄圆点的，以及粉红色带有蓝绿圈的那条。

A：先生，你的品位真高。这两条都很特别。

B：可以帮我包装好吗？我要作为礼物送人的。

A：当然可以，先生。

### 6.1.3 购买服装

（提示：A 代表销售员，B 和 C 代表顾客）

A：下午好，有人招呼你吗？

B：没有。我在找一些轻便的夏装套装。这里有成衣产品适合我吗？

A：让我帮你量一下尺寸，先生。我想我们的成衣尺码对你太小了。我们可以给你挑选合身的裤子，但衬衫肩宽有点窄。

B：好的，我可以试一下吗？或许可以改一下。

A：当然，先生。这边请。这件衬衫是最大码的。请你穿上试试。

B：我明白。这还是太小。我喜欢这件的面料。

A：好的，先生，我们让裁剪部给你做一件怎样？成衣和量身定制之间的价格没有多大的差异。

B：这件成衣要多少钱？

A：250 元。我想你花 200 元就可以量身定做一件了。

B：好的，这不错。那让我去跟你们的裁缝师说说。

A：好的，先生。这边请，我来给你介绍我们的裁缝师。

B：谢谢。

A：不客气。

C：请问，今年流行这个颜色吗？

A：是的，这个颜色现在很流行。

C：谢谢你的意见，我就要这件。

A：还需要其他的吗？

C：我想买一些浴巾。

A：你想要什么颜色的呢？

C：两条白色的和三条红色的。

A：它们的质地都很好。

C：价格是多少呢？

A：白色的每条 7 元；红色的每条 8.5 元。

C：还有没有便宜些的呢？

A：这里有白色的 4.5 元一条，红色 5 元一条。

C：好的，那我就买便宜的吧。

A：谢谢。还要不要再看看别的东西？

C：不用了，谢谢。一共多少钱？

A：一共 124 元。

C：给你 125 元。

A：谢谢，这是你的 1 元零钱。

### 6.1.4 议价

（提示：C 代表顾客，S 代表售货员）

C：这里有像这条一样的纯棉裤子吗？

S：这是最后一条了。

C：但这条裤子有小疵点。价格还可以便宜点吗？

S：请等一下。我去问问老板。

C：好的。

S：不好意思让你久等了。我老板同意给你 5% 的折扣。

C：5% 的折扣？还是有点贵。不能再便宜点吗？

S：对不起，不可以了。

C：顺便说一句，这裤子不是流行的。

S：但这裤子面料很好，并且很合适你。

C：但我对这条裤子的做工不太放心。

S：对于这种情况，你可以另外享受 5% 折扣。我们百货公司规定，顾客发现疵品可得到 5% 的折扣。

C：好的，我就买下吧。

### 6.1.5 认同顾客

（提示：S 代表销售员，C 代表顾客 / 吴女士）

S：吴女士，你是想要纯毛的还是毛涤混纺的套装呢?

C：我喜欢毛涤套装。它们的保形性好，不像纯毛产品易起皱。我想买工作套装。我想毛涤面料更适合日常穿着。

S：关于毛涤这点，我想你说的太对了。你看上的这件套装真的很不错。这套海军式是经典款式，不是吗?

C：是的，真的很好看。

S：你是说要一件经典款式的套装吗?

C：是的，我想那是最好的职业装。

S：你说你穿 L 码的吗?

C：好的，那要看看那件套装的裁剪。但 L 码比较合适。

S：这里有两件 L 码的海军蓝经典款式套装。你喜欢哪件?

C：我想右边那件会好些。

S：你不喜欢领边缉面线的那件吗?

C：那正是吸引我的地方。

S：怎么不去试一下呢？不用花多少时间的，更衣室就在右边。

## 6.2 电话交谈

### 6.2.1 情景 I

A：您好。

B：请问是 Rykiel 服装连锁店吗?

A：是的。

B：我想找吴先生。

A：请不要挂线。我去帮你叫他。

B：好的，谢谢。

A：对不起，他出去了。

B：他什么时候回来呢?

A：我没把握，但他也许 4 点前不会回来。要不要我叫他给你回电话?

B：好的，谢谢。麻烦你叫他打给陈先生。我的电话号码是 9988467。

A：好的，我会告诉他的。

B：谢谢你，再见。

### 6.2.2 情景Ⅱ

A：我是 Ray 先生，请问吴先生在吗？

B：抱歉，吴先生不在。可以为你留言吗？

A：谢谢，我还是稍后等他打电话给我吧。

B：吴先生知道你的电话号码吗？

A：是的，他有我电话。

B：你刚说你叫 Ray 先生是吗？

A：是的，我是 Ray 先生。拼作 R-A-Y。

B：我知道了，Ray 先生。我会告诉吴先生稍后给你打电话。

A：谢谢。

### 6.2.3 情景Ⅲ

A：您好，请问你是陈先生吗？

B：是的，请问有什么可以帮到你呢？

A：请问你拿到我们的目录表吗？

B：是的，我们对贵公司的产品很感兴趣。

A：听你这样说我非常高兴。我将拿些样品给你参照和检查。你希望我几点来？

B：嗯，你可以今天下午就拿过来的，方便吗？

A：那太好了。到时见，再见。

B：再见。

### 6.2.4 情景Ⅳ

A：您好，请问这个号码是 9988768 吗？

B：是的。

A：请问王先生在吗？

B：我就是了。请问你是谁？

A：早上好，王先生。我是陈先生。

B：早上好，陈先生。请问有什么事呢？

A：有，我想买一些真丝衬衫。你那有存货吗？

B：有的，我们一直有一些库存。

A：请问价格多少呢？

B：每款价格有 50 美元到 100 美元的。

A：你可以带上产品目录过来给我看看吗？

B：十分乐意，我可以现在过去吗？

A：嗯，我随时恭候。再见。

### 6.2.5 情景Ⅴ

A：请问你们的经理在吗？

B：请问是谁打来的呢？

A：我是 Polo 时装公司的 Peter 先生。

B：可以麻烦你再说一遍吗！我听不清楚。

A：我是 Polo 时装公司的 Peter 先生。我想要找你们的经理，女士，可以吗？

B：好的，请不要挂线。（盖住电话）吴先生，Polo 时装公司的 Peter 先生找你。

吴先生：哦，你帮我打发他一下吧，跟他说我现在很忙，他只是想叫我买他们的产品而已。

B：（放开话筒）不好意思，Peter 先生。我们的经理现在很忙。请问还有什么事可以帮到你的。

A：是的，我想给你们看一下我们的样品。请问我现在可以过去拿给你看吗？

B：我想我会给你另找时间的。你可以过来拿给我看。但是我没把握能否帮到你。

A：你真是好人，那我下次拜访你们的连锁店吧。

## 6.3 商务洽谈示范

### 6.3.1 洽谈前

（1）你好？我是 Kevin（很高兴见到你。）

你好，我是 Alex。（也很高兴见到你。）

（2）这是我的名片。

谢谢。

（3）这位是我们的经理陈先生。

很高兴见到你。

（4）Tom先生，你认识我们的生产部经理？

是的，我们在六个月前见过面。

（5）你来自哪里呢？

我来自印度。

（6）欢迎来到我们公司。

谢谢。

（7）你好，陈先生。

早上好，王小姐，很高兴再次见到你。

（8）你好吗？

很好，谢谢。你呢？

（9）你的业务做得还好吗？

不差。你的呢？

（10）你什么时候到的呢？

昨天下午。

（11）你还满意这次的航班吧？

是的，很好。谢谢。

（12）你现在住哪里？

在西湖酒店。

（13）你会待多久呢？

大概两个星期。

### 6.3.2 开始洽谈

（1）早上好，很高兴再次见到你。

谢谢光临。

（2）跟我说一下你们的业务吧？

好的。我很乐意告诉你更多有关我们的业务。

（3）你们公司的产品范围有哪些？

我们公司的产品范围涉及机织与针织服装。

（4）你们公司业务涉及哪些方面？

我们公司涉及外贸、生产制作和销售等。

（5）你公司的产品出口到哪里的？

我们的产品出口到美国、澳洲、日本、欧洲等。

（6）你们公司有多少员工？

有 250 名员工。

（7）你们公司是什么时候成立的？

我们公司是在 1992 年的时候成立的。

（8）你们公司主要是做什么产品？

我们主要做内衣类、针织类和机织类。

（9）你们公司每年的营业额是多少？

有 1.5 亿元。

（10）哪个企业是你们这行的领军企业？

我们是这行的领军企业之一。

（11）你们的价格有竞争力吗？

我们是最低价格的企业之一。

（12）你们有没有其他的产品？

我们还生产针织服装。

（13）你介意告诉我你们最畅销的产品吗？

好的，这是我们的产品目录。女士裙装是我们这季最畅销的产品。

### 6.3.3 关于企业

（1）可以告诉我你们主要顾客吗？

我们的主要顾客有 POLO 公司、Rykiel 贸易公司等。

（2）你们主要生产什么产品呢？

我们的主要产品是机织女装。

（3）你们的总部在哪里？

我们的总部在中国香港。

（4）贵公司的主要分公司是在哪里？

主要分布在中国大陆和日本。

（5）贵公司有多少工厂呢？

我们有五间工厂。

（6）你负责哪方面工作？

我负责市场销售这块。

（7）你一般接待什么样的顾客呢？

我一般都接待外商。

### 6.3.4 讨论产品

（1）我们在寻找能生产针织服装的企业。你能告诉我哪些是流行款式与畅销产品吗？

好的，给你。这是我们最新的产品目录。

（2）可以给我看一下你们的产品目录吗？

好的，当然可以。这是我们最新的样板。

（3）哪款是你们最畅销的针织产品呢？

这些都是最流行的款式。

（4）我在找一间工厂加工这些产品。你们可以帮我找到类似的工厂吗？

我想可以的。这是先前订单的样板。

（5）不好意思，如果可以的话。我想在这些地方做些改变。

我们可以根据你的要求去修改产品。

（6）这件样板有什么特征？

嗯，它的面料很新颖并很时尚。

（7）你们的目标市场是？

我们的目标市场是年轻一代。

（8）这种产品跟同类市场的产品比较有什么优势？

我觉得我们的质量会好些。

（9）你们可以生产类似的产品吗？

是的，我们可以。我们可以先给你制版。然后再跟你谈一些细节的问题。

### 6.3.5 关于产品的更多资料

（1）你们可以生产类似的产品吗？这是设计图和规格表。

我想我们可以。我们会先制作一个样板给你。

（2）你需要什么尺码？

要四个尺码，小码、中码、大码和加大码。

（3）你们每个月的产量是多少？

我们一个月可以生产 1 万件。

（4）你们至少要多少件？

我们每个月至少要 1 万件。

（5）有什么可以让我觉得你们生产的服装比别人好？

我们生产的产品质量好，供货快速。

（6）你期待的价格是怎样的？

我觉得单价为每件 80 美元。

（7）我们要求最小订单数量必须是 1 万件，对于大的订单我们将给一定比例的降价优惠。

总价会根据订单数量而有所下降。所以很实惠。

（8）好的，我很期待跟你们合作。

我也是，谢谢。

### 6.3.6 议价

（1）我想要下一个订单，款号为 MT-1523，每件的单价是多少钱？

嗯，那要看你的订单数量了。若一次下定超过 1000 美金的话可以享受 5% 的折扣。

（2）这还能给个特价么？

不好意思，不可以了。这是我们可以给的最低的价格了。

（3）这个价格包括运输费用在内吗？

是的，包括了。/ 不，没有。

（4）这个价格包括保险费用吗？

是的，这是包括了丢失和损坏赔偿的保险费用。

（5）那付款方式是？

我们通常是用 L/C 或支票方式付款。

（6）你想要怎样的付款方式呢？

采用最方便、快捷的方式。

（7）你什么时候给我开发票？

在出货前下好订单，我就给你开发票。

（8）你想什么时候付款？

我会在一个月内付款。

（9）采用什么付款方式？

现金支付交货。

（10）什么时候我需要确认订单呢？

你最好尽快，因为我们年尾会很忙。你现在可以做决定吗？

（11）我要先跟我的老板谈一下再给你回复我们订单的数量和支付方式。

好的，我期待你的回复。

### 6.3.7 交货谈论

（1）我想问一下这批货的交货时间有多长？

要四个星期吧。

（2）这批货的船期是多长时间？
大概要花四个星期运送。
（3）你什么时候需要落实订单？
我想在五月中旬的时候。
（4）我希望船期不会延误。
如果信用证准时的话就不会延误的。
（5）你能保证交货期不会有问题吗？
是的，我可以向你保证不会有任何问题。
（6）你们用怎样的方式把货物送到我那？
我们会采用空运或是陆运的方式。
（7）货物要多久才到达？
要花四个星期吧。
（8）那批货的交付期是什么时候呀？
交付期是 5 月 10 日。
（9）交付费用包括了什么？
包括了港口停泊费用、进口税、保险费用和手续费用。

### 6.3.8 发展趋势

（1）你可以给我介绍一下你们公司的具体情况吗？
非常的荣幸。你可以从这个表看出……
（2）从 2007 年以来你们生产的费用减少了多少？
自从 2007 年来我们生产费用减少了 15%。
（3）今年需求怎样？
今年原材料的价格已经下降了，并且各项开支也下降了。
（4）今年全年你们的产量水平达到什么程度？
今年全年的产量维持在高水平。
（5）今年的产量情况如何？
今年的产量上升到了 5 亿美元 。
（6）现今的通货膨胀水平有什么状况？
从 2007 年来通胀就升的很厉害。
（7）你认为来年会有什么情况？
因为缺乏人力资源，我估计劳工成本会增加。

### 6.3.9 商讨价格与交货

（1）你认为如何？

如果你不介意的话，我想我们应该改一下合同。

（2）你有什么建议吗？

在我看来，我们必须提高价格。

（3）你觉得怎样？

我想先讨论一下最后定价问题。

（4）请问你们可不可以快点把货款过账？

当然可以。/ 嗯，说实话可能有点难了。

（5）我们对价格的提升感到很惊讶。

我恐怕不能完全同意你的想法。

（6）我很难接受你提升的价格。

非常的抱歉，不会有下次的了。

（7）出于尊重，你好像还没明白。

但是我真的有必要指出……对不起，我们已经尽力了。

（8）怎么这批货的交付期会比原计划迟了 3 个星期？

我们在质量控制上出了些问题。

（9）还有没有别的办法了呢？

不用担心。我保证不难安排。

（10）你觉得我的建议怎样？

我完全同意你。

（11）我不是十分同意。

那样的话，我别无选择，只能取消订单了。

（12）你准备做一些改变吗？

我们想试一下。

### 6.3.10 结束洽谈

（1）或许下个月我们还会再见面？

嗯，那样会很不错。

（2）下次在你的办公室见面怎样？

太好了。

（3）你觉得我们能干什么？

我希望可以完成这个计划书。

（4）你们打算怎样？

我们打算在下星期做最后一次调查。

（5）下次会议是什么时候？

我想还是下个月开吧。

（6）明天我们一起吃中餐？

好的，没问题。

（7）你喜欢去哪里吃中餐？西湖酒店怎样？

好的，那我预定下午2点的台吧。

（8）我们明天再讨论怎样？

好的，让我们……

（9）很高兴认识你。

我也是，下次见。

## 本章小结

- 系统介绍了服装连锁店内不同情况的英语沟通。
- 在不同的情况下，如何通过电话与客人沟通。
- 综合地介绍了服装企业业务洽谈技巧与方式。
- 此章节重点训练与提高学生在服装连锁店工作、电话交谈方式以及业务洽谈方面的英语沟通与口语表达能力。

**CHAPTER 7**

# BUSINESS LETTER IN CLOTHING INDUSTRY 服装行业商业信函

**课题名称：** BUSINESS LETTER IN CLOTHING INDUSTRY
服装行业商业信函

课题内容：Introduction 简介
Envelope 信封
The Parts of Business Letters 商业信函的构成
Model Letters 示范信函
E-mail 电子邮件

**课题时间：** 6课时

**教学目的：** 让学生了解规范的商业信函的基本知识，并掌握有关信封表达、信函内容的基本构成。掌握各种情形的商业信函写法与应用，重点掌握服装行业电子邮件的编写与应用。

**教学方式：** 结合PPT多媒体课件，以教师课堂讲述为引导，学生进行商业信函的撰写训练为主。

**教学要求：** 1. 熟悉规范的商业信函的基本要求与信封格式。
2. 熟悉各种商业信函内容的基本构成与编写。
3. 熟悉规范的商业电子邮件的基本要素与撰写。

**课前（后）准备：** 结合专业知识，课前预习课文内容。课后熟读课文主要部分，掌握各种典型的服装行业信函与电子邮件的撰写与应用。

# CHAPTER 7

# BUSINESS LETTER IN CLOTHING INDUSTRY 服装行业商业信函

## 7.1 Introduction 简介

①规范的商业信函有助于扩展业务和增加利润，但不规范的商业信函会令公司失去发展机会。

②要明了、简洁、有条理性。

It is very important for the attractive business letter in English. Good business letters help increase business and profit, but bad business letter will lose business opportunities. ① The following general guideline will be useful when writing the good business letter in clothing industry.

（1）Be clear, brief and businesslike. ②

（2）Don't write confused, overlong or pointless letter.

（3）Be polite, friendly and informal.

（4）Write to communicate.

（5）Write concise and purposeful letter.

## 7.2 Envelope 信封

### 7.2.1 Envelope Pattern 信封样式

The envelope should match the letterhead in color, quality and other tastes. The first line of the address should be centered slightly below the center of the envelope itself. And the inside address should be typed in the same style on the envelope（Table 7-1）.

**Table 7-1 Envelope Pattern**

| | |
|---|---|
| *Return Address:*<br>Polo Co. Ltd.<br>36, Oxford St.<br>London, W, 5<br>England | Stamp |
| | VIA AIR MAIL<br>*To*:<br>Merchandise Group<br>26 California Avenue<br>Seattle, WA 6288<br>USA |
| Attention of Mr. Smith | |

### 7.2.2 Envelope Remark 信封备注

The remark words should be typed in the lower left corner of the envelope, such as the following words:

（1）Air Mail, Printed Matter, Express, Registered, etc.

（2）By Courtesy of Mr. …or By Favors of Miss … etc.

（3）Attention of Mr. … or Recommending Mr. … etc.

## *Words and Expressions*

attractive [ə'træktiv] 吸引的
brief [bri:f] 摘要，大纲
businesslike ['biznislaik] 有条理的，有效率的
overlong [əuvəlɔŋ] 太长的
pointless ['pɔintlis] 无意义的
concise [kən'sais] 简明的，简练的
letterhead ['letəhed] 信头
book rate / printed matter [buk reit / 'printed 'mætə] 印刷品
personal ['pə:sənl] 亲启，个人的
capital ['kæpitl] 大写字母
registered ['redʒistəd] 挂号
Air Mail [eə meil] 航空
express [iks'pres] 特快，速递
by courtesy of [bɑi 'kə:tisi ɔv,] 托某人面交
recommending [rekə'mend] 推荐，托付
envelope pattern ['envələup 'pætən] 信封样式

## *Exercises*

1. Translate the following terms into Chinese.

（1）domestic
（2）by courtesy of
（3）recommending
（4）apparel
（5）department
（6）sales contract
（7）license
（8）letter of credit
（9）shipping marks
（10）fabric content
（11）invoice
（12）account

2. Write out some special notes of envelope.

3. Give an example of illustrated envelope.

## 7.3 The Parts of Business Letters 商业信函的构成

### 7.3.1 Letter Pattern 信文样式

Generally，international business letters should follow a definite sequence that is convenient for the reader，and it follows a standard pattern（Table7-2）.

Table 7-2 Full Block Style

| LETTERHEAD |
|---|
| Reference Number                                    Date |
| Inside Name & Address |
| ..................... |
| ......... |
| Attention Line |
| ......... |
| Salutation |
| Subject Heading |
| Body of Letter ........................................................................................ |
| ........................................................................................................ |
| ........................................................................................................ |
| ............... |
| ............... |
| Complimentary Close |
| Company's Name |
| Writer's Signature |
| Writer's Typed Name |
| Title |
| Postscripts/ Enclosure |

### 7.3.2 Explaining of Letter Pattern 样式说明

（1）Letterhead：General，the letterhead contents should include Name & Address，Cable Address，Trade Mark，etc. [①] And the letterhead design must be in good taste，and the writing size well situated.

①信头的内容包括名字和地址、电报地址、贸易商标等。

（2）Dateline：The date should be written in one straight line. There are two forms in dateline writing.

USA：May 1，2020 or 1 May 2020

UK：1st May 2020

（3）Reference Number：refers to Document Number.

（4）Inside Name & Address：The contents should contain title，name and address of the recipient，it is usually placed at the left margin above the body of the letter. Below details are shown by the written sequence：

Name

Title

Company Name

Street Address ( P.O. Box No. )

City State ( Province )

Country

( 5 ) Attention Line: the letter receiver.

( 6 ) Salutation: The salutation should be written flush with the first line of the inside address. The punctuation needed after salutation is the comma " , ". And the following is the formal & informal salutation:

Formal Salutation: Dear Sir, My Dear Madam, etc.

Informal Salutation: Hi Joe, Hello May , etc.

( 7 ) Subject Heading: a brief about the contents of the message allows the receiver to read quickly.

( 8 ) Body of Letter: It is the main part of the message that consists of only one paragraph or few paragraphs, as you need to convey your message clearly and pleasantly.

( 9 ) Complimentary Close: the position of the complimentary close depends on the length of the signature. It should begin to the right of the vertical center and not extend beyond the right margin. ② It should be placed at least two spaces below the last line of the letter body. The types of general complimentary close should be: Respectfully yours, Sincerely yours, Truly yours, Yours faithfully, etc.

②结束语的位置根据签名的长度来确定。它应该从垂直中心的右边开始且不能越过右边界。

( 10 ) Writer's Information: the information of the person, who wrote or dictated the letter, and signature should always be written by hand. It should contain the following: Company's Name, Writer's Signature, Writer's Typed Name and Title, etc.

( 11 ) Postscripts / Enclosure: such as price list, goods catalogue, or other special wording, etc.

## *Words and Expressions*

definite ['definit] 明确的，确定的

convenient [kən'vi:njənt] 方便的，合适的

inside address [in'said ə'dres] 信内地址

attention [ə'tenʃən] 注意（指定受信人）

salutation [sælju(:)'teiʃən] 称呼

body of the letter ['bɔdi ɔv, ðə 'letə] 信文，正文

complimentary close [;kɔmpliment(ə)ri

klәuz,] 结束语，结尾敬语
signature ['signitʃә] 署名
postscript / P.S. ['pәustskript] 附启
cable address ['keibl ә'dres] 电报挂号
trade mark [treid mɑ:k] 贸易商标
situated ['sitjueitid] 位于，境地
contain [kәn'tein] 包含
abbreviation [әbri:vi'eiʃәn] 缩写，略语
recipient [ri'sipiәnt] 领受的，接受者
margin ['mɑ:dʒin] 边缘，栏外
flush [flʌʃ] 齐头的
punctuation [pʌŋktju'eiʃen] 标点
comma ['kɔmә] 逗号
Madam ['mædәm] 夫人
subject matter ['sʌbdʒikt 'mætә] 提出问题
consist [kәn'sist] 组成
paragraph ['pærәgrɑ:f] 段落
convey [kәn'vei] 传递，通知
vertical ['vә:tikәl] 垂直位置
respectfully yours [ris'pektfuli jɔ:z] 敬上（用于政府高官或长辈）
truly yours / yours very truly ['tru:li jɔ:z / jɔ:z әveri 'tru:li] 敬上（用于正式公文函件）
sincerely yours / yours faithfully [sin'siәli jɔ:z / jɔ:z 'feiθfuli] 敬上（用于一般商业书信）
cordially yours ['kɔ:djәli jɔ:z] 敬上（用于知己好友）
dictated [dik'teitid] 口述，听写
informal [in'fɔ:mәl] 不正式的
good taste [gud teist] 好的品位

## *Exercises*

1. Translate the following terms into Chinese.

(1) inside address
(2) respectfully yours
(3) salutation
(4) complimentary
(5) signature
(6) postscript
(7) yours faithfully
(8) cable address
(9) trade mark
(10) cordially yours
(11) abbreviation
(12) schedule
(13) full block style
(14) knit wear
(15) supplier
(16) document
(17) fashion
(18) business letter

2. Write out the major points of a business letter.

3. Write a letter with full block style by your suggestion (any topic).

## 7.4 Model Letters 示范信函

### 7.4.1 Enquiry and Order 询问与订购

There is have the simplicity & clearing as a rule in the general enquiry and order letter.① The content must be shown with category list, price list, unit price and sample swatch, etc. And the enquiry and order letters should be brief and to the point, telling the receiver exactly what the enquiry is about.②

①普通的询问与订购信应以简洁和清晰为原则。

②询问与订购信函要简明扼要，让收信人明确查询内容。

### Case Ⅰ: First Enquiry 第一次询问

18 May 2020

Garment Products Inc.
312 Western Highway
Austin
USA

Dear Sirs,

We should be glad to know whether you supply lots of cotton men's shirts, blouses, shorts, overalls and jeans, etc.③

We also require lots reversible wool blankets, please send us the category of any goods that you recommend or supply from stock.

We look forward to your early reply.

Yours faithfully,

R& k Trading Office
***Ivy Wang***
Merchandiser

③我们想了解你们是否可提供大量的纯棉男装衬衫、罩衣、短裤、工装裤和牛仔裤等。

### Case Ⅱ: Reply to First Enquiry 答复第一次询问

28 May 2020

Mrs. Ivy Wang
R& k Trading Office
8 Whit Street
Manchester England

Dear Mrs. Wang,

Thank you for your enquiry of 18 May. We are pleased to hear that you are interested in our products.

We are sending you a copy of our latest catalogues under separate cover, together with samples of some of our products.

We also manufacture a wide range of leather garments in which you may be interested. They are fully illustrated in our catalogue and are of the same quality as our woven products. Mrs. Liu will be able to show you samples when she visits.④

We look forward to receiving an order from you.

Yours Sincerely,

Garment Products Inc.
***Kevin Zhang***
Manager

④我们还生产各种皮衣服装，贵公司可能会感兴趣。这些产品的插图都刊登在我们的产品目录中，并且与我们的机织产品同样具有高质量。刘女士拜访时将向您展示样品。

### Case Ⅲ: Confirming an Order（Ⅰ） 确认订购（Ⅰ）

18 July 2020

Jennet Garment Inc.

Shenzhen

China

Dear Sirs,

We are pleased to say that the mohair rugs, which you dispatched on 15th July were delivered in good condition on 17th July.⑤

The care and promptness with which you have attended to our order are very much appreciated.⑥

Yours faithfully,

R& k Trading Office

***Ivy Wang***

Merchandiser

⑤贵公司在7月15日发出的马海毛地毯，我们已于7月17日收到，完整无缺。

⑥对于你们对我们订单的快捷处理与关照，深表感激。

### Case Ⅳ: Confirming an Order（Ⅱ） 确认订购（Ⅱ）

5 June 2020

Mr. CK

Merchandiser

Wool Garment Inc.

Bradford & Yorkshire

UK

Dear Sir,

We want to say how pleased we were to receive your order of 18th May for cotton prints and welcome you as one of our customers.

We confirm supply of the prints fabric at the prices stated in your letter and are arranging for dispatch next week by ship. When the products reach you, we fell confident you will be completely satisfied with them, in fact, at these prices offered they represent added value.⑦

As you may not be aware of the wide range of products we deal in, we are enclosing a copy of our catalogue and hope that our handling of your first order with us will lead to further business between us and mark the beginning of a happy working relationship.⑧

Yours sincerely,

Menton Garment Inc.

***Gary Chen***

Senior Merchandiser

⑦贵公司收到产品时，我们有信心保证你们会感到完全满意，事实上，以这个价格，实在是物有所值。

⑧您也许不清楚我们公司经营的产品范围，现附上一份目录表，并希望彼此间合作的第一张订单能增进彼此之间的业务来往，展开愉快的合作关系。

## Words and Expressions

enquiry [in'kwaiəri] 调查，询问
simplicity [sim'plisiti] 简单
information [infə'meiʃən] 资料
category ['kætigəri] 种类，目录表
price list [prɑis list] 价目表
unit price ['ju:nit prɑis] 单价
reversible [ri'və:səbl] 正反面
blanket ['blæŋkit] 毛毯
prints [prints] 印花布
customer ['kʌstəmə] 顾客
confirm [kən'fə:m] 证实，批准
confident ['kɔnfidənt] 有自信的
cotton prints ['kɔtn prints] 印花棉布
sample swatch ['sæmpl swɔtʃ] 样板
overalls ['əuvərɔ:l] 工装裤
men's shirt [men's ʃə:t] 男装衬衫
blouse [blɑuz] 女装罩衫
shorts [ʃɔ:ts] 短裤
trousers ['trauzəz] 长裤，西裤
bath towel [bɑ:θ 'tauəl] 毛巾，浴巾
supply [sə'plai] 供应
describe [dis'krɑib] 叙述，描述
recommend [rekə'mend] 推荐
stock [stɔk] 库存
straw hats [strɔ: hæts] 草帽
enclosed [in'kləuzd] 附寄，附件
appreciated [ə'pri:ʃieitid] 感激
polyester ['pɔliestə] 聚脂纤维，涤纶
blends [blendz] 混纺
mohair ['məuheə] 马海毛
rugs [rʌgs] 围毯，小地毯
dispatched [dis'pætʃt] 派遣，调度
represent [repri'zent] 表示，说明
exceptional [ik'sepʃənl] 例外的
value ['vælju:] 价值
be aware of [bi: ə'weə ɔv,] 注意到的
relationship [ri'leiʃənʃip] 联系，关系
handling ['hændliŋ] 处理，管理
offering ['ɔfəriŋ] 提供，报价
unit price ['ju:nit prɑis ] 单价
delivery [di'livəri] 交付
regular way ['regjulə wei] 常规方法
satisfaction [ˌsætis'fækʃən] 满意
coat [kəut] 上衣，外套
satisfactory [ˌsætis'fæktəri] 满意的

## Exercises

1. Translate the following terms into Chinese.

(1) information
(2) shirt
(3) blouse
(4) straw hats
(5) cotton
(6) trousers
(7) shorts
(8) slacks
(9) jeans
(10) bath towel
(11) garment
(12) unit price
(13) sample swatch
(14) price list
(15) total price
(16) category
(17) stock
(18) goods
(19) schedule
(20) trade
(21) quota
(22) contract
(23) specification
(24) FOB

2. List the contents of an enquiry letter.

3. Write an enquiry letter by your suggestion (the topic relative to clothing industry).

### 7.4.2 Offers 报价

The offering letter contains all the necessary information in a concise form, such as Style, Order No., Size & Color, Quantity, Unit price & Discount, Total price, Payment, and Delivery, etc.

**Case Ⅰ: Making a Firm Offer 报实价**

1 May 2020

Mr. Ray
Purchasing Manager
Star Garages
Hill St.
England

Dear Sir,

Your kind letter of 15th March came duly to hand. According to your request, we have pleasure in sending you a price list of the goods we handle, and any orders that you may kindly place with us will be promptly delivered in the regular way.⑨

We shall be very glad to have you call in and look over our stock, because we are constantly improving the goods in every possible way, and will offer them at the lowest prices we can give you.⑩

Yours very truly,

***Alice Luo***
Export Manager

***Enc. Price List***

⑨根据您的需求，我们非常乐于寄一份我们经营的产品价目表给你们，若你们需要订购，我们将依照常规迅速供应。

⑩非常欢迎您前来参观，因为我们一直不断地尽可能提高产品质量，同时，我们可以给您提供最低价格。

**Case Ⅱ: Making a Counter Offer 还价**

16 August 2020

Mrs. Elsa
Export Manager
Brisbane
USA

**Continued**

Dear Mrs. Elsa,

We are in receipt of your letter of 21 July offering us 1000 dozens of 100% silk women's blouse at $240 per dozen on the usual terms.⑪

We regret to inform you that our buyers find your price much too high. We are informed that some lots of India silk products have been sold here at a level about 20% lower than yours. To facilitate the transaction, we counter offer as follows.⑫

*1000 dozens of 100% silk women's blouse, at FOB $200 per dozen, other terms as per your letter of 21 July.*

As the market price is falling, we recommend your immediate acceptance.

Yours very truly,

***Jack Luo***

Merchandising Manager

⑪承蒙7月21日来函报盘1000打100%真丝女装罩衣，以每打240美元按一般条款交易，谨此致谢。

⑫我们得悉本地现正出售的印度丝绸产品，价格较上述报盘约低20%。为达成交易，现谨还价如下。

## *Words and Expression*

commodity [kə'mɔditi] 商品名称 *

shipment ['sipmənt] 装船方式

payment ['peimənt] 付款方式

insurance [in'ʃuərəns] 保险

rayon shirt ['reiɔn ʃə:t] 黏胶丝衬衫

pants [pænts] 裤子

discount ['diskaunt] 折扣

promptness [prɔmptnis] 敏捷，迅速

facilitate [fə'siliteit] 推动，促进

transaction [træn'zækʃən] 交易

counter offer ['kauntə 'ɔfə] 还盘，还价

## *Exercises*

1. Translate the following terms into Chinese.

(1) Commodity
(2) shipment
(3) insurance
(4) brand
(5) wool-polyester
(6) discount
(7) mohair
(8) rugs
(9) blanket
(10) order sheet
(11) confirmation
(12) CIF
(13) L/C
(14) international
(15) cashmere
(16) unit price
(17) delivery
(18) discount
(19) payment
(20) order date
(21) item
(22) total price
(23) stock
(24) shell fabric
(25) blouse
(26) waistcoat
(27) baby wear
(28) striped fabric
(29) handling
(30) yardage

（31）approved sample （32）specification （33）package

2. Write an order letter about shell fabric by your suggestion.
3. Write an offering letter about jeans by your suggestion.

### 7.4.3 Payment and Shipment 付款与装运

The letter must be brief and to the point, and tabulation of the essential details relating to the order clarified terms agreed to by both the vendor and the purchaser.[13] When writes a payment and shipment letter, you must consider the following points.

（1）Identify the order details.

（2）Tabulate all the relevant details about product.

（3）End with a friendly message.

⑬这类信函必须简明扼要，并列出订单的主要细节，以阐明买卖双方商定的条款。

**Case Ⅰ: Confirming a Purchase 确认购货**

7 March 2020

Mr. Michael Chan
82 Changi Road
Singapore

Dear Mr. Chan,

As a result of our recent exchange of e-mail, we wish to confirm our merchandise from you of 15000 pieces pants.[14] The following terms and conditions will apply.

1. Unit Price: FOB US$ 20 per piece.
2. Total Amount: FOB US$ 300000.
3. Packing: Individual Poly-bag for each pants.
4. Payment: By confirmed and irrevocable letter of credit in you favor, payable by draft at sight.[15]

We are pleased that this first transaction with your company has come to a successful conclusion. We look forward to a continuing and mutually beneficial trade between our companies.

Yours Sincerely,

Star Garment Ltd.
*Mary Roberts*
Merchandiser

⑭依据近来双方电子邮件往来结果，现欲确认向贵公司订购15000条裤子。

⑮开立不可撤销的信用证，以贵公司为受益人。

**Case Ⅱ: Adjusting a Delivery Date 调整船期**

8 September 2020

Mrs. Julia
Ant-Kids Inc.
8 High Road
Clapham
UK

Dear Mrs. Julia,

We refer to your order on Ant-686 for 1000 pieces 100% cotton men's shirt.

Owing to problems at the port, we will not be able to meet the agreed delivery date of 20 October. We are doing everything possible to ship your order but the contracted date has now become unrealistic.⑯

We believe, however, that we will be able to meet a 25 October delivery deadline. We apologize for the inconvenience, but the delay is due to circumstances beyond our control.⑰

Yours faithfully,

Rykiel Fashion Ltd.
***TK Kong***
Export Manager

⑯ 我们已尽力安排，但始终无法如期装运货物。

⑰ 然而，如贵公司能把限期延迟至10月25日，我们深信必可按时交货。此次事件纯属意外，对贵公司所造成的不便，恳请原谅。

## Words and Expressions

tabulation [tæbju:læʃən] 表格
relevant ['relivənt] 有关的，相应的
merchandise ['mə:tʃəndaiz] 商品，货物
irrevocable letter of credit [irrevocable 'letə ɔv 'kredit] 不可撤销信用证
payable ['peiəbl] 可付的，应付的
mutually ['mju:tʃuəli] 互相地，互助
beneficial [beni'fiʃəl] 有益的，受益的
port [pɔ:t] 港口
unrealistic ['ʌnriə'listik] 不切实际的，不实在的
deadline ['dedlain] 最终期限
circumstance ['sə:kəmstəns] 环境，详情，境况

## Exercises

1. Translate the following terms into Chinese.

(1) stock
(2) straw hats
(3) polyester
(4) blends
(5) mohair
(6) rugs
(7) category
(8) price list
(9) unit price
(10) blanket
(11) FOB
(12) prints
(13) overalls
(14) men's shirt
(15) blouse
(16) shorts
(17) trousers
(18) bath towel

2. Write a payment letter by your suggestion.
3. Write a shipment letter by your suggestion.

### 7.4.4 Complaint Letters 投诉信件

After selling goods, enterprises may receive various complaint letters, it is very important to handle these complaint letters carefully.

**Case Ⅰ: Delaying Complaint 延期投诉**

16 August 2020

Miss Helga Bond
Jack Garment Ltd.
18 Dorado Drive
Dallas
Texas
USA

Dear Miss,

We are surprised that our order of the July 18 for men's shirts has not yet come to hand.

Our customers are writing to us every day, asking for an explanation, because they are urgently requiring the shirts.[18] If you value your trade with us, please send men's shirts to us immediately, and notify us by phone or e-mail.

Yours faithfully,

Rykiel Fashion Ltd.
***Roy Che***
Senior Merchandiser

⑱ 由于我们的顾客急需这些衬衫，所以他们每天都写信给我们，让我们解释有关缺货情况。

**Case Ⅱ: Complaint about Quality 有关质量问题的投诉**

16 May 2020

Bandy Fashion Inc.
112 Bukit Timah Street
Sigapore

**Continued**

Dear Sirs,

We duly received 180 dozens of 100% cotton pants you sent us, but regret finding on examination that 6 dozens of them were poor quality, being material defects or manufacturing problems.⑲

We would be glad if you will look into the matter, and let us know what you propose to do.

Yours Sincerely,

Polo Fashion Ltd.

***May Li***

Merchandiser

⑲我们按时收到贵公司发出的180打100%全棉裤子，但是经过检测后遗憾地发现，其中6打裤子存在品质差、面料瑕疵或工艺疵点等问题。

**Case Ⅲ: Answering a Complaint　回复投诉**

16 May 2020

Polo Fashion Ltd.

87 Jalan Arnap

Kuala Lumpur

Malaysia

Dear Sirs,

A dealer in Hong Kong sold $250 worth of Polo shirts in three days. This ought to answer your complaint that our shirts don't sell well.⑳

We are enclosing a copy of that dealer's letter to show you, too. It can really make money through selling Rykiel brand shirts.

Yours very truly,

Rykiel Fashion Ltd.

***Elsa Li***

Merchandiser

⑳中国香港的一家经销商在短短三天内共卖出价值250美元的樽领衬衫。这可以回答你关于我们的衬衫不好销的抱怨了吧。

## *Words and Expressions*

sales letter [seils 'letə] 推销函件
characteristic [kæriktə'ristik] 特征，特性
arouse interest [ə'rauz 'intrist] 引起兴趣
create desire [kri(:)'eit di'zaiə] 引起需求
carry conviction ['kæri kən'vikʃən] 使人信服
induce action [in'dju:s 'ækʃən] 导致行动
dealer ['di:lə] 销售公司，商人
complaint [kəm'pleint] 控诉
dozen ['dʌzn] 打（数量）
pants [pænts] 裤子
regret [ri'gret] 遗憾，失望
examination [igzæmi'neiʃən] 调查，检查
quality ['kwɔliti] 质量
defect [di'fekt] 疵点
manufacturing [ˌmænju'fæktʃəriŋ] 制造
propose [prə'pəuz] 申请，计划
explanation [eksplə'neiʃən] 说明，解析
urgently ['ə:dʒəntli] 紧急地

## *Exercises*

1. Translate the following terms into Chinese.

（1）sales letter　（5）quality　（9）rayon
（2）dealer　（6）dozen　（10）pants
（3）brand　（7）size label　（11）material
（4）main label　（8）cotton　（12）retailer

2. Write a sales letter of garment products by your suggestion.

## 7.5 E-mail 电子邮件

Recently, the rapid expansion of the Internet has the effect of bringing a large number of people together under one network. Sending e-mail is relatively cheap and quick. For example, a short message can be sent around the world in less than a minute, and the message can be sent to many people at the same time. In clothing industries, many members of large companies will have already been familiar with sending electronic messages via their company computer network. ①

①在服装界内，很多大机构的员工已经熟悉通过公司的计算机网络传递电子邮件。

### 7.5.1 E-mail Element 邮件要素

The standard e-mail elements need to contain following information.

（1）To: the e-mail address of the person that you want to send the message to. E-mail addresses are made up of two parts: the 'username' which is found to the left of the @ character, and domain which is to the right of the @ character. Such as ray@pub.huizhou.gd.cn.

(2) Subject: a brief statement about the contents of the message allows the receiver to read quickly through his list of incoming mail and identify messages of special importance.② It also allows the receiver to save certain messages and delete others.

②在收件人快速浏览所收到的信件时，一个简明扼要的标题会让他立刻注意到关键信息。

(3) Cc: addresses of those people who will receive a copy of the message. As normal business communication, "cc" stands for "carbon copy".

(4) Body Text: This part comprises the full text of the message you are sending. Start this section with an appropriate greeting, followed by the message itself, and finish with an appropriate close.

(5) Attached File: Most e-mail programs allow you to attach word documents, program files or pictures to your message. This serves the same purpose as the "Enc." section often found at the end of a business letter.

### 7.5.2 Case Study 案例分析

In summary, remember to follow these main guidelines when writing e-mail program:

(1) Keep one subject per message, and don't cover various issues in a single e-mail.

(2) Use a descriptive subject heading.

(3) Keep messages short and to the point, and write short sentences.

(4) Use bulleted lists to break up complicated texts.

(5) Quote from the original e-mail if required and target dates.

(6) Conclude your message with actions required and target dates..

**Case Ⅰ: Participation for Sales Presentation 参加展销**

Date: 10th December, 2020

To: ray@pub.huizhou.gn.cn

From: james@granford.co.uk

Subject: Participation for Sales Presentation

---

Dear Sir,

Thanks for your impressive sales presentation about leisure wear on 15th May. It was very informative and we obtained lots of new buyers.

**Continued**

In the meeting of management board, we have discussed your proposals very thoroughly. Unfortunately, after considering our financial situation and market situation, we decided not to participate in this sales presentation.

Over the next few months, we will be slimming down our operations to compensate for the downturn in trading and market. When the economy picks up, we shall certainly be interested in talking to your company again.③

③在接下来几个月内，我们需紧缩业务开销以补偿贸易与市场销售额的下降。当经济变好时，我们一定有兴趣与你们公司再谈相关事宜。

In any event, please contact with us during six months, when we might be interested in talking about further matters.

Thanks for your time and efforts.

Yours faithfully,

***Bandy Liu***

Merchandising Manager

Rykiel Fashion company

**Case Ⅱ: Announcing a Price Increase　通知客户调整价格**

Date: Tuesday, August 20, 2020

To: ivan@antkid.com.cn

Cc: james@163.com

From: Alex@winpap.co.uk

Subject: Announcing a Price Increase

---

Dear Sir,

We enclose our new catalogue and price list. The revised prices will apply from 1 September 2020.

You will be aware that inflation is affecting the whole clothing industry. We have been affected like everyone else and some price about woven collection increases have been unavoidable. We have not, however, increased our prices across the board. In many cases, there is a small price increase, but in others, none at all.④

④由于通货膨胀影响整个服装行业，连带机织系列产品价格上涨，我们和每家公司一样都受到此冲击。虽然如此，我们并未全面提升价格，调整幅度也不大。

We can assure you that the quality of our product has been maintained at a high standard.

We look forward to receiving your orders.

Yours faithfully,

***Alex Wang***

Merchandising Manager

Polo Trading Ltd.

*Enc.: Product Catalogue and Price List*

## *Words and Expressions*

brief [bri:f] 大纲，摘要
identify [ɑi'dentifɑi] 识别，确定
carbon copy / Cc ['kɑ:bən 'kɔpi] 副本
attachment [ə'tætʃmənt] 附件
guideline [gaidlain] 方针，导向
sales presentation
  [seils prezen'teiʃən] 展销
financial [fɑi'nænʃəl,] 财政的
proposal [prə'pəuzəl] 提议，计划
slimming ['slimiŋ] 减轻
compensate ['kɔmpenseit] 补偿，偿还
downturn ['daʊntɜ:n] 低迷时期
economy [i(:)'kɔnəmi] 经济
opportunity [ɔpə'tju:niti] 机会
revise [ri'vaiz] 修正，修改
inflation [in'fleiʃən] 通货膨胀

## *Exercises*

1. Translate the following terms into Chinese.

(1) internet
(2) economy
(3) financial
(4) cotton fabric
(5) electronic mail
(6) communication
(7) network
(8) sales presentation
(9) brief
(10) national costume
(11) uniform
(12) business wear
(13) student wear
(14) leather garment
(15) fur garment
(16) denims
(17) apparel
(18) silhouette

2. List the contents of e-mail.

3. Write a piece of e-mail in the clothing industry by your suggestion.

# 译文

# 第七章　服装行业商业信函

## 7.1　简介

具有吸引力的英文商业信函是非常重要的。规范的商业信函有助于扩展业务和增加利润，但不规范的商业信函会令公司失去发展机会。下面几条方针有助于写好服装行业的商业信函。

（1）要明了、简洁、有条理性。

（2）不要混淆、过长或无意义的信件。

（3）要有礼貌、友善并正式。

（4）明确沟通。

（5）精炼且目的明确。

## 7.2　信封

### 7.2.1　信封样式

信封应与信头颜色、质量，以及其他品位保持一致。地址的第一行应该写在信封中间偏离少许的位置。并且信内地址应与信封的格式一样（表 7-1）。

表 7-1　信封样式

| | |
|---|---|
| *回信地址：* | 邮票 |
| Polo Co. Ltd.<br>36, Oxford St.<br>London, W, 5<br>England | |
| | 航空信件<br>*去信地址：*<br>Merchandise Group<br>26 California Avenue<br>Seattle, WA 6288<br>USA |
| Smith 先生收启 | |

### 7.2.2 信封备注

备注词语应打印在信封左下角位置，如下面词语：

（1）航空邮件、印刷品、快递、挂号等。

（2）由……先生托交或……小姐托交等。

（3）由……先生收启或推荐……先生等。

## 7.3 商业信函的构成

### 7.3.1 信文样式

为了方便读者阅读，一般国际商务函件应按一定的顺序撰写，下面是一个标准样式（表 7-2）。

**表 7-2 齐头样式**

<u>信头</u>

参考编号　　　　　　　　　　　　　　　　　　　　日期

信内地址

…………………

………

指定收件人

………

称呼

主题

正文 ……………………………………………………………………

……………………………………………………………………………

……………………………………………………………………………

……………

……………

结束语

公司名称

笔者署名（签名）

笔者姓名（打字）

头衔

附件

### 7.3.2 样式说明

（1）信头：信头的内容包括名字和地址、电报地址、贸易商标等。并且信头的设计必须品位高，书写大小要适中。

（2）日期：日期填写成一直线。一般有两种日期写法。

美式写法：May 1, 2020 or 1 May 2020

英式写法：1st May 2020

（3）参考编号：指文件的编号。

（4）信内地址：包括了头衔、人名、收信人的地址，这些都通常写在信文上面左边位置。具体细节应按以下顺序写：

人名

头衔

公司名

街道地址（邮政编号）

城市（州/省）

国家

（5）指定收件人：收信人。

（6）称呼：称呼应该与信内地址第一行齐头编写。称呼后的标点为逗号“,”。下面是一些正式与不正式的称呼：

正式的称呼：Dear Sir, My Dear Madam 等。

不正式的称呼：Hi Joe, Hello May 等。

（7）主题：主题是正文的概括，方便收信人快速了解信的主要内容。

（8）正文：信件的主要组成部分，可包括一个或几个自然段落，目的是清晰愉快地表达你的信息。

（9）结束语：结束语的位置根据签名的长度来确定。它应该从垂直中心的右边开始且不能越过右边界。结束语应该写在信文最后一行低两行以上的位置。一般信函的结束语包括：Respectfully yours、Sincerely yours、Truly yours、Yours faithfully 等。

（10）寄信人信息：指写信人或口述人的信息，署名必须用手写。信息包括公司名、笔者的手写签名、打印名和头衔等。

（11）附件：如价格表、产品目录或其他特殊备注等。

## 7.4 示范信函

### 7.4.1 询问与订购

普通的询问与订购信应以简洁和清晰为原则。内容必须包括了产品目录表、价格表、单价和样板等。询问与订购信函要简明扼要，让收信人明确查询内容。

**案例Ⅰ：第一次询问**

2020 年 5 月 18 日

Garment Products Inc.
312 Western Highway
Austin
USA

先生：

您好，我们想了解你们是否可提供大量的纯棉男装衬衫、罩衣、短裤、工装裤和牛仔裤等。

我们还需要大量的双面羊毛地毯，请将仓库有的或你推荐的所有产品目录寄送给我们。

期待你尽早回复。

敬上，

R& k Trading Office
***Ivy Wang***
跟单员

**案例Ⅱ：答复第一次询问**

2020 年 5 月 28 日

Mrs. Ivy Wang
R& k Trading Office
8 Whit Street
Manchester England

王女士：

您好，谢谢你 5 月 18 日的询问信件。很高兴你对我们的产品感兴趣。

我们正将一份最新在独立封面下的目录表复印件连同产品样板一起寄送给你。

我们还生产各种皮衣服装，贵公司可能会感兴趣。这些产品的插图都刊登在我们的产品目录中，并且与我们的机织产品同样具有高质量。刘女士拜访时将向您展示样品。

我们希望尽早接到您的订单。

敬上，

Garment Products Inc.
***Kevin Zhang***
经理

**案例Ⅲ：确认订购（Ⅰ）**

2020年7月18日

Jennet Garment Inc.
Shenzhen
China

先生：

您好，贵公司在7月15日发出的马海毛地毯，我们已于7月17日收到，完整无缺。

对于你们对我们订单的快捷处理与关照，深表感激。

敬上，

R& k Trading Office
***Ivy Wang***
跟单员

**案例Ⅳ：确认订购（Ⅱ）**

2020年6月5日

Mr. CK
Merchandiser
Wool Garment Inc.
Bradford & Yorkshire
UK

先生：

您好，我们非常高兴收到您在5月18日寄来的有关印花棉布的订单，同时也欢迎您成为我们的顾客。

我们确认以信中规定的价格提供印花棉布，并安排在下周装船发货。贵公司收到产品时，我们有信心保证你们会感到完全满意，事实上，以这个价格，实在是物有所值。

您也许不清楚我们公司经营的产品范围，现附上一份目录表，并希望彼此间合作的第一张订单能增进彼此之间的业务来往，展开愉快的合作关系。

敬上，

Menton Garment Inc.
***Gary Chen***
高级跟单员

### 7.4.2 报价

报价信函以精简的表格包含所有必要的信息，如款式、订单号、尺码和颜色、数量、单价和折扣、总价、付款方式和交付期等。

**案例Ⅰ：报实价**

2020年5月1日

Mr. Ray
Purchasing Manager
Star Garages
Hill St.
England

先生：

您好，您于3月15日寄出的信件已经收到。根据您的需求，我们非常乐于寄一份我们经营的产品价目表给你们，若你们需要订购，我们将依照常规迅速供应。

非常欢迎您前来参观，因为我们一直不断地尽可能提高产品质量，同时，我们可以给您提供最低价格。

敬上，

***Alice Luo***
出口部经理

*附件：价目表*

**案例Ⅱ：还价**

2020年8月16日

Mrs. Elsa
Export Manager
Brisbane
USA

Elsa女士：

您好，承蒙7月21日来函报盘1000打100%真丝女装罩衣，以每打240美元按一般条款交易，谨此致谢。

我们得悉本地现正出售的印度丝绸产品，价格较上述报盘约低20%。为达成交易，现谨还价如下。

*1000打100%真丝女装罩衣，FOB价为每打200美元，其他条款与7月21日信件中所提一致。*

因为市场的价格一直再下调，我们建议您尽快接受。

敬上，

***Jack Luo***
跟单部经理

### 7.4.3 付款与装运

这类信函必须简明扼要，并列出订单的主要细节，以阐明买卖双方商定的条款。在写付款与装运信函时，必须考虑下面几点。

（1）明确订单细节。

（2）表格形式列出产品相关细节。

（3）用礼貌用语结尾。

**案例Ⅰ：确认购货**

2020 年 3 月 7 日

Mr. Michael Chan
82 Changi Road
Singapore

陈先生：

您好，依据近来双方电子邮件往来结果，现欲确认向贵公司订购 15000 条裤子。以下是交易方式与条款。

1. 单价：FOB 价每条 20 美元。
2. 总价：FOB 价 $ 300000 美元。
3. 包装：每条裤子用一个独立透明包装袋包装。
4. 付款：开立不可撤销的信用证，以贵公司为受益人。

我们很高兴与你们公司第一次交易成功。希望我们双方彼此之间继续保持互惠的合作贸易。

敬上，

Star Garment Ltd.
***Mary Roberts***
跟单员

**案例Ⅱ：调整船期**

2020 年 9 月 8 日

Mrs. Julia
Ant-Kids Inc.
8 High Road
Clapham
UK

**续表**

Julia 女士：

您好，我们来谈谈关于 Ant-686 订单的 1000 件 100% 棉男装。

由于港口的问题，我们无法根据合约在 10 月 20 日交货。我们已尽力安排，但始终无法如期装运货物。

然而，如贵公司能把限期延迟至 10 月 25 日，我们深信必可按时交货。此次事件纯属意外，对贵公司所造成的不便，恳求原谅。

敬上，

Rykiel Fashion Ltd.
***TK Kong***
出口部经理

### 7.4.4 投诉信件

产品出售后，公司可能会收到各种各样的投诉信。认真处理这些投诉信是非常重要的。

**案例Ⅰ：延期投诉**

2020 年 8 月 16 日

Miss Helga Bond
Jack Garment Ltd.
18 Dorado Drive
Dallas
Texas
USA

小姐：

您好，我们对 7 月 18 日有关男装衬衫的订单直到现在还没有到达感到惊讶。

由于我们的顾客急需这些衬衫，所以他们每天都写信给我们，让我们解释有关缺货情况。如果您还看重与我们的交易，请马上将这批男装衬衫寄送给我们，并用电话或电邮通知我们。

敬上，

Rykiel Fashion Ltd.
***Roy Che***
高级跟单员

**案例Ⅱ：有关质量问题的投诉**

2020年5月16日

Bandy Fashion Inc.
112 Bukit Timah Street
Sigapore

先生：

您好，我们按时收到贵公司发出的180打全棉裤子，但是经过检测后遗憾地发现，其中6打裤子存在品质差、面料瑕疵或工艺疵点等问题。

如果你们能来看看究竟，我们会很高兴，也可以让我们知道你们打算如何解决。

敬上，

Polo Fashion Ltd.
***May Li***
跟单员

**案例Ⅲ：回复投诉**

2020年5月16日

Polo Fashion Ltd.
87 Jalan Arnap
Kuala Lumpur
Malaysia

先生：

您好，中国香港的一家经销商在短短三天内共卖出价值250美元的樽领衬衫。这可以回答你关于我们的衬衫不好销的抱怨了吧。

我们也可以附上该经销商的信件给你看看，销售我们Rykiel品牌的衬衫是真的可以赚钱的。

敬上，

Rykiel Fashion Ltd.
***Elsa Li***
跟单员

## 7.5 电子邮件

目前，互联网迅速发展，使许多人使用互联网。发邮件既经济实惠又快捷。例如，一通简短的信息可在一分钟内发送到世界各地并让许多人在同时收到。在服装界内，很多大机构的员工已经熟悉通过公司的计算机网络传递电子邮件。

### 7.5.1 邮件要素

标准的邮件需要涵盖以下信息：

（1）发送给：收件人的电邮地址。邮件地址由两部分组成：在符号@左边为用户名，符号@右边为网络运营商的域名。例如，ray@pub.huizhou.gd.cn。

（2）主题：在收件人快速浏览所收到的信件时，一个简明扼要的标题会让他立刻注意到关键信息。同时可以让收件人存储有关信息并删除其他。

（3）副本转送：这些人的电邮地址可收到一份信件副本。用于一般商务交流，“cc”为“carbon copy”的简写。

（4）正文：这部分涵盖了你想要发送的所有内容，正文开始必须有问候语，接着是正文内容，最后有结束语。

（5）附件：许多邮件容许你附带文字文档、程序文件或是图片。其作用与信件末尾中常见的“附件”一样。

### 7.5.2 案例分析

总之，写邮件时切记如下指导方针：

（1）一封邮件只说明一个内容，一个单独邮件不要涵盖多个内容。

（2）运用一个描述性的正文标题。

（3）信息内容简短到位，尽量用简短语句。

（4）用明确的标题列表代替复杂的正文内容。

（5）必要时可引用原邮件发送。

（6）用必要举措和目标日期作为信文的总结。

**案例Ⅰ：参加展销**

Date: 10th December, 2020
To: ray@pub.huizhou.gn.cn
From: james@granford.co.uk
Subject: Participation for Sales Presentation

先生：

您好，谢谢你们在 5 月 15 日举行的让人印象深刻的休闲服展销会。该展销会信息丰富，使我们获得了很多新客户。

在管理层会议上，我们非常彻底地讨论了你们的计划。可惜，考虑我们财务状况与市场形式后，我们决定不参加这次的商品展销会。

在接下来几个月内，我们需紧缩业务开销以补偿贸易与市场销售的下降。当经济状况变好时，我们一定有兴趣与你们公司再谈相关事宜。

不管怎样，在六个月内请与我们保持联系，那时我们可能有兴趣再进一步商讨有关事宜。

承蒙关照，不胜感激。

敬上，

***Bandy Liu***
跟单部经理
Rykiel Fashion company

**案例Ⅱ：通知客户调整价格**

Date: Tuesday, August 20, 2020
To: ivan@antkid.com.cn
Cc: james@163.com
From: Alex@winpap.co.uk
Subject: Announcing a Price Increase

先生：

您好，现附上我们的最新产品目录与价目表。修订价格从 2020 年 9 月 1 日起适用。

由于通货膨胀影响整个服装行业，连带机织系列产品价格上涨，我们和每家公司一样都受到此冲击。虽然如此，我们并未全面提升价格，调整幅度也不大。

我们向您保证我们的产品是高质量的。

我们希望接到您的订单。

敬上，

***Alex Wang***
跟单部经理
Polo Trading Ltd.

*附件：产品目录与价目表*

## 本章小结

- 简单介绍了的商业信函的基本知识、信封表达以及信函内容的基本构成。
- 通过不同的案例，介绍了各种规范的商业信函撰写与应用。
- 简单介绍了商业电子邮件的基本要素与要求，并以案例形式明确了商业电子邮件撰写与应用。
- 此章节重点培养与提高学生对各种规范的商业信函与电子邮件的撰写与应用能力。

# REFERENCE
# 参考文献

[ 1 ] 王传铭．英汉服装服饰词汇［M］．北京：中国纺织出版社，2007.
[ 2 ] 中华人民共和国国家标准国家技术监督局．服装工业术语［M］．北京：中国轻工业出版社，1985.
[ 3 ] 张宏仁，梁娟，张小良．服装英语实用教材［M］．3 版．北京：中国纺织出版社，2015.
[ 4 ] 陈霞，张小良，等．服装生产工艺与流程［M］．2 版．北京：中国纺织出版社，2014.
[ 5 ] 霍尔特，格里戈尔，桑普森．国际商业书信大全［M］．北京：外语教学与研究出版社，2009.
[ 6 ] 周叔安．英汉 / 汉英服装分类词汇［M］．北京：中国纺织出版社，2009.
[ 7 ] 冯麟．服装跟单实务［M］．2 版．北京：中国纺织出版社，2015.
[ 8 ] CLAIRE SHAEFFER. SEWING FOR THE APPAREL INDUSTRY［M］．UK：PRENTICE HALL INC.，2001.

* APPENDIX Ⅰ~Ⅲ

## APPENDCES 附录

## STYLE DESCRIPTION 款式描述

**课题名称：** STYLE DESCRIPTION 款式描述

**课题内容：** Fashion Style 服装款式
Fashion Details 服装细节
Garment Parts 服装部件

**课题时间：** 4 课时

**教学目的：** 让学生了解并掌握服装部件、时装细节以及服装部件等相关专业词汇。进一步掌握不同服装款式的英文描述。

**教学方式：** 结合 PPT 与音频多媒体课件，以教师课堂讲述为引导，学生进行款式描述训练为主。

**教学要求：** 1. 熟悉服装部件、时装细节与部件等专业词汇。
2. 熟悉不同服装款式的英文描述。

**课前（后）准备：** 结合专业知识，课前预习相关内容。课后熟记相关专业词汇，掌握不同服装款式的英文描述。

# APPENDICES
# 附录

## Appendix Ⅰ: Fashion Style 附录Ⅰ: 服装款式

### Dress 连衣裙，套装裙

1. ballet ['bælei,] dress 芭蕾舞裙装
2. bare-Back [beə-bæk] dress 露背裙装
3. bare midriff [beə 'midrif] dress 露腰套装
4. beach [bi:tʃ] dress 海滩装
5. camisole ['kæmisəul] dress 露背吊带裙
6. cape [keip] dress 披肩礼服
7. coat [kəut] dress 外套式连衣裙
8. culottes [kju(:)'lɔts] dress 套装裙裤
9. evening ['i:vniŋ] dress 晚礼服
10. halter ['hɔ:ltə] dress 吊带领装
11. peplum ['pepləm] dress 腰褶裙
12. petticoat ['petikəut] dress 衬裙，内裙

13. trumpet ['trʌmpit] dress 紧身喇叭连衣裙
14. safari [sə'fɑːri] dress 狩猎装
15. shirtwaist ['ʃɜːtweist] dress 衬衫裙
16. sweater ['swetə] dress 编织连裙装
17. slip [slip] dress 衬裙式连衣裙
18. torso ['tɔːsəu] dress 低腰紧身连衣裙
19. trapeze [trə'piːz] dress 梯形裙装
20. princess [prin'ses] dress 公主装

## Jacket 夹克

## Sweater 羊毛衣，毛线衫

1. anorak ['ɑːnərɑːk] jacket 滑雪溜冰衫
2. battle ['bætl] jacket 军服夹克
3. blazer [bleizə] jacket 宽松外衣
4. bolero [bə'lɛərəu] jacket 前胸撇开短上衣
5. cardigan ['kɑːdigən] jacket 开胸夹克衫
6. reefer ['riːfə(r)] jacket 女式紧身双排扣上衣
7. saddle ['sædl] jacket 骑马外套
8. shearling ['ʃiəliŋ] jacket 羊皮外套
9. smocking ['smɔkiŋ] jacket 半正式晚礼服
10. spencer ['spensə] jacket 短外套夹克
11. Aran ['ær ən] sweater 阿兰毛衣
12. bolero [bə'lɛərəu] sweater 波蕾若开襟毛衫
13. cardigan ['kɑːdigən] sweater 开襟式毛衫
14. coat [kəut] sweater 外套式毛衫
15. crew [kruː] sweater 船员毛衣

16. cowl-neck [kaul-nek] sweater 垂褶领毛衣
17. dolman ['dɔlmən] sweater 德尔曼毛衣
18. pull-on [pul-ɔn] sweater 套头毛衣
19. sweater ['swetə] 羊毛衫
20. twin [twin] sweater 成对式毛衣

## Underwear 内衣，里衬　Swimwear/ Swimsuit 游泳衣

## Coat 大衣，外套

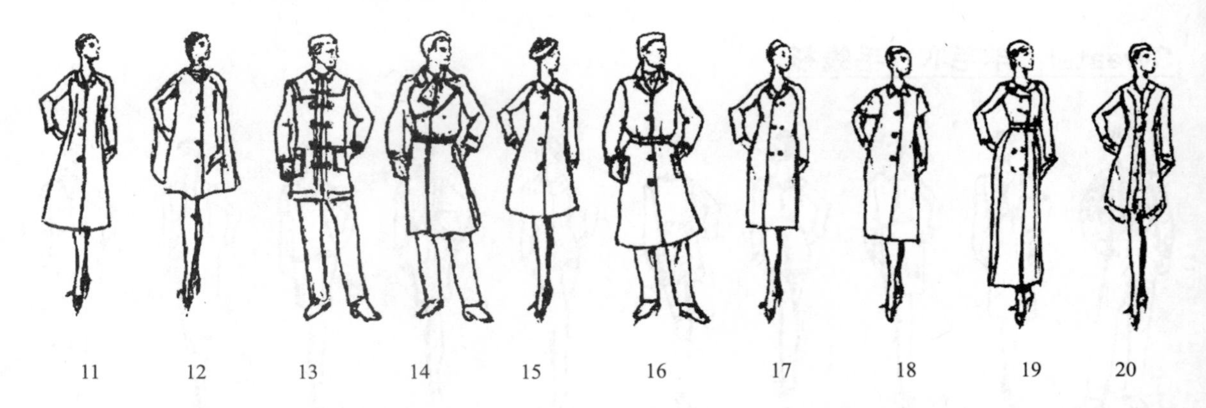


1. body hugger ['bɔdi hʌg] 亵衣
2. briefs [bri:fs] 男弹力短内裤
3. camisole ['kæmisəul] 花边胸衣
4. chemise [ʃi'mi:z] 女睡衣
5. corset ['kɔ:sit] 紧腰衣
6. bikini [bi'ki:ni] 比基尼泳衣
7. blouse [blɑuz] 宽松式连胸罩泳衣
8. cabana sets [kə'ba:nə sets] 卡巴拿沙滩装
9. string bikini [striŋ bi'ki:ni] 捆带系比基尼泳衣
10. tank [tæŋk] 传统女式泳衣
11. A-Line [ei-lɑin] coat A 字形外套
12. cape [keip] coat 披风大衣
13. duffel ['dʌfəl] coat 粗呢大衣
14. duster ['dʌstə] coat 风衣
15. flared [flɛəd] coat 宽摆式大衣
16. chesterfield [tʃestəfi:ld] coat 软领长大衣
17. classic ['klæsik] coat 传统外套
18. inverness [ˌinvə'nes] coat 披肩外套
19. maxi ['mæksi] coat 特长式大衣
20. midi ['midi] coat 迷第外套

## Vest/ Waistcoat 背心, Suit 套装

## Shirt 衬衫

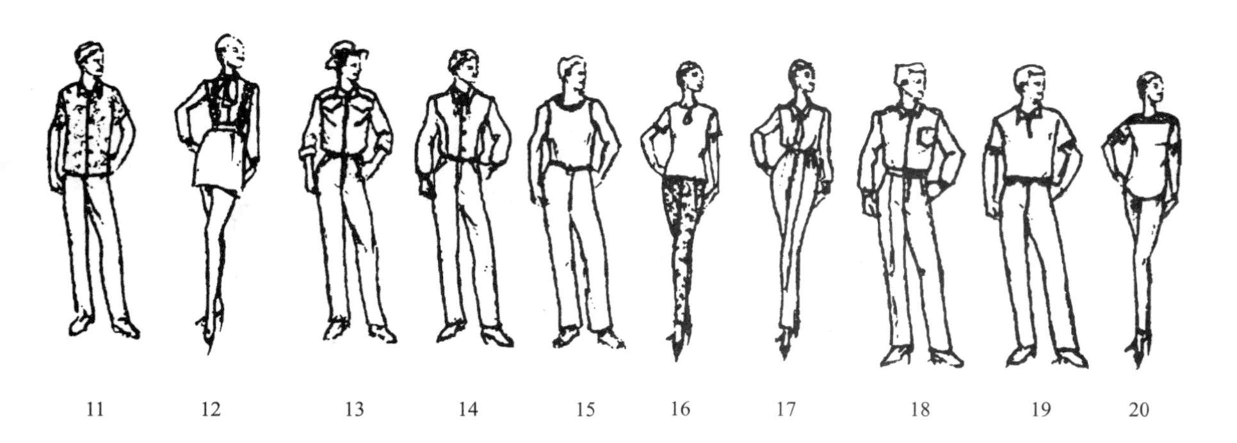


1. down [dɑun] vest 羽绒背心
2. formal ['fɔːməl] vest 礼服背心
3. hunting ['hʌntiŋ] vest 猎装背心
4. jerkin ['dʒəːkin] vest 紧身皮袄
5. sweater ['swetə] vest 背心式毛衣
6. sleeveless ['sliːvlis] suit 无袖上衣套装
7. jumper ['dʒʌmpə] suit 跳伞装
8. tailored ['teiləd] suit 西式套装
9. tunic ['tjuːnik] suit 束腰套装
10. vested ['vestid] suit 有马夹套装
11. Hawaiian [hɑː'waiiən] shirt 夏威夷式衬衫
12. ascoted [ə'æskət] shirt 阿司阔衬衫
13. cowboy ['kaʊbɔi] shirt 牛仔衬衫
14. body ['bɔdi] shirt 紧身衬衫
15. athletic [æθ'letik] shirt 运动背心
16. Henley ['henli] shirt 亨利汗衫
17. Ivy ['aivi] shirt 藤纹衬衫
18. Pilot ['pɑilət] shirt 飞行员衬衫
19. Polo T-Shirt ['pəuləu 'tiː,-ʃəːt] 马球衫
20. Tee-Shirt ['tiː,-ʃəːt] T 恤衫

## Pyjamas 睡衣

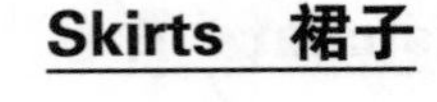

## Skirts 裙子

## Shorts 短裤

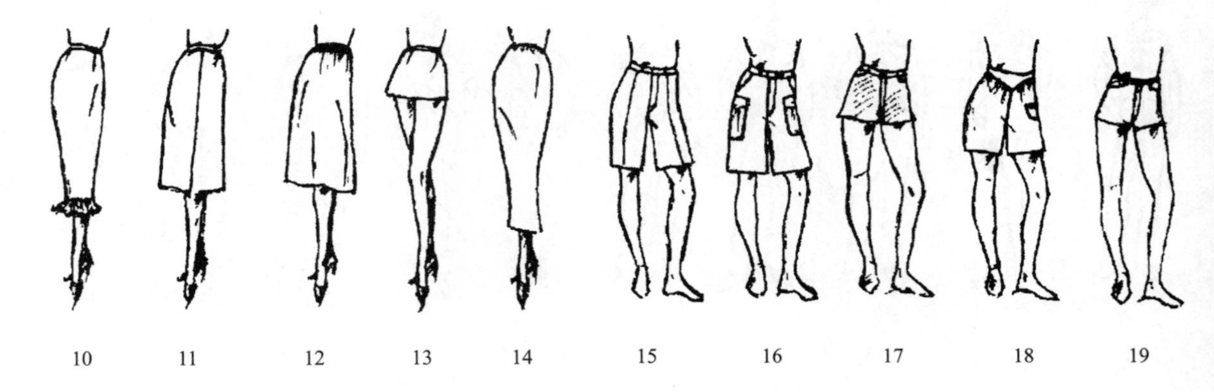


1. Chinese [ˈtʃaiˈni:z] pyjama 中式睡衣
2. coat-style [kəut-stail] pyjama 外套形睡衣
3. karate [kəˈrɑ:ti] pyjama 空手道服式睡衣
4. pyjamas set [pəˈdʒɑ:məz set] 睡衣套装
5. A-Line [ə-ˈlain] skirt A 字裙
6. balloon [bəˈlu:n] skirt 气球形裙
7. bell [bel] skirt 钟形裙
8. circular [ˈsə:kjulə] skirt 圆台裙
9. draped [dreip]skirt 垂褶裙
10. flounced [flauns] skirt 裙脚绉边裙
11. four gored [fɔ: gɔ:] skirt 四片裙
12. gathered [ˈgæðəd] skirt 缩褶裙
13. mini [ˈmini] skirt 迷你裙
14. tulip [ˈtju:lip] skirt 郁金香裙
15. Bermuda [bə(:)ˈmju:də] shorts 百慕大短裤
16. cargo [ˈkɑ:gəu] shorts 大贴袋短裤
17. cut-offs [kʌt-ɔ:fs] shorts 毛边牛仔短裤
18. pleated [pli:t] shorts 打褶裥短裤
19. short [ʃɔ:t] shorts 超短裤

## Pants 裤子

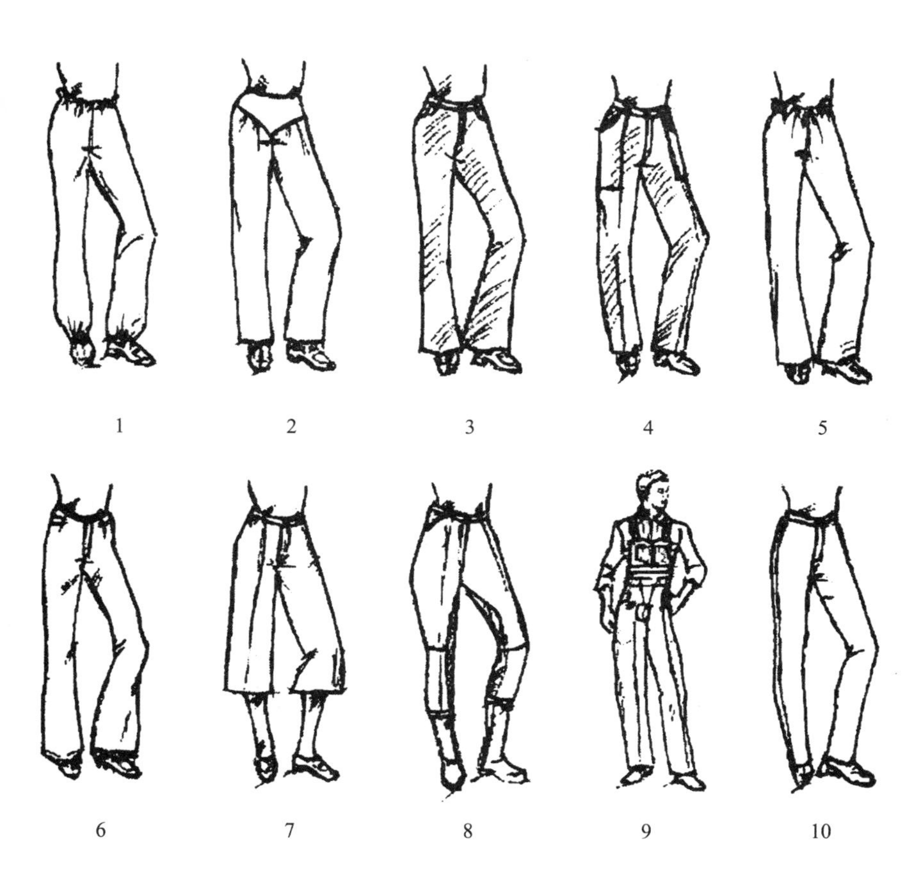


1. ankle-tied ['æŋkl-taid] pants 束脚裤
2. baggy ['bægi] pants 袋状裤
3. bell-bottom [bel-'bɔtəm] pants 喇叭裤
4. cargo-pocket ['kɑːgəu-'pɔkit] pants 大贴袋牛仔裤
5. drawstring ['drɔːstriŋ] pants 束带紧腰裤
6. flared [flɛəd] pants 喇叭裤
7. gaucho ['gautʃəu] pants 短长裤
8. jodhpurs ['dʒɔdpəz] pants 马裤
9. overalls ['əuvərɔːlz] pants 工装裤
10. stovepipe [stəuv paip] pants 烟筒形长裤

# Appendix Ⅱ: Fashion Details 附录Ⅱ：服装细节

## Front Fastening / Front Opening 前开口

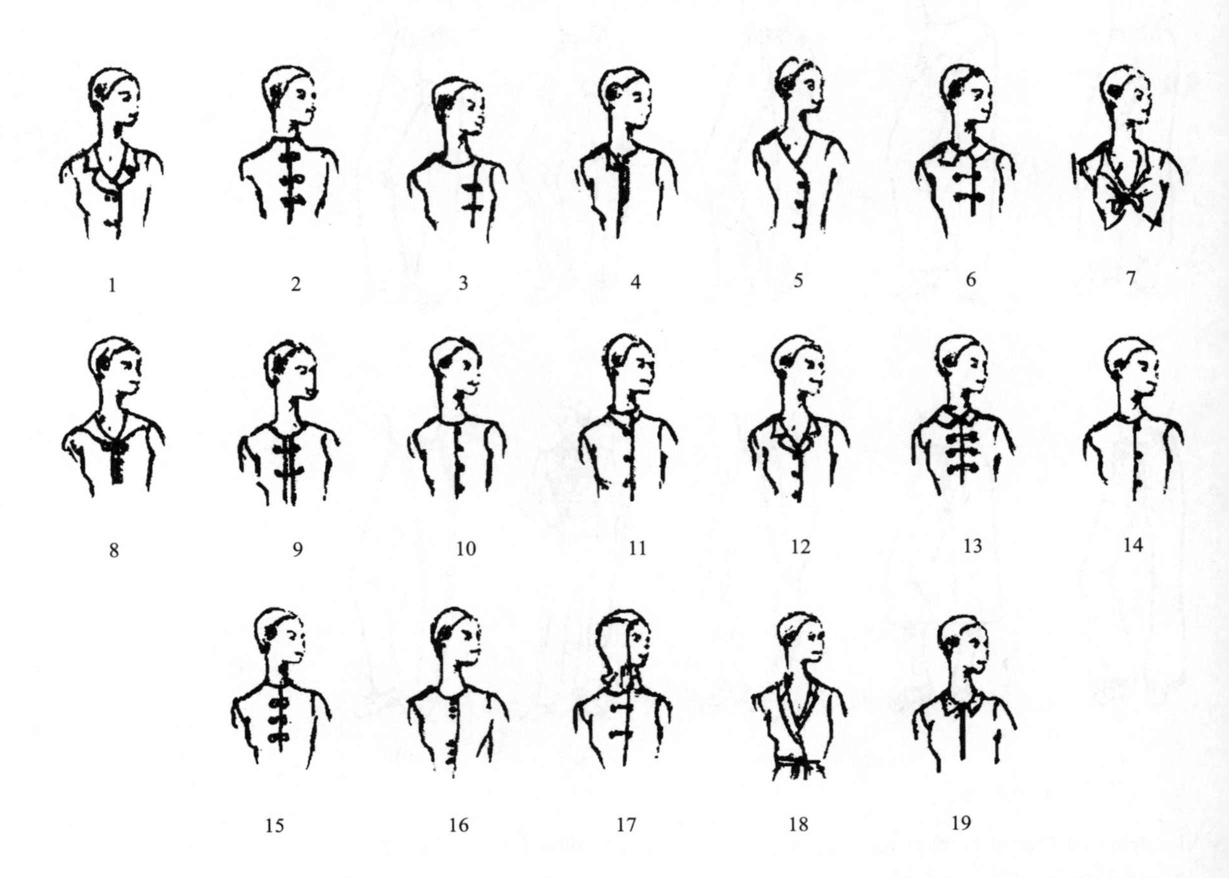


1. double-breasted ['dʌbl-brestid] fastenings 双纽门襟
2. braid [breid] button 编结扣
3. buckled ['bʌkld] 带扣
4. fly [flɑi] front opening 暗开襟
5. crossed-over [krɔst-'əuvə] front closing 叠门襟
6. frogging ['frɔgiŋ] button 纺锤形纽扣
7. knotting ['nɔtidtai] 编结物
8. laced-up [leist-ʌp] front opening 系带式门襟
9. linked [liŋkt] button 连接饰物
10. looped [lu:pt] fastening 绳圈式衣扣
11. side-buttoned [said-bʌtnd] 侧边扣孔
12. single button ['siŋgl bʌtnd] 单边纽扣
13. strap [stræp] button 条扣
14. tab [tæb] button 扣襻
15. bow [bɑu] button 蝴蝶结
16. three-button grouping [θri:-'bʌtn gru:piŋ] 三粒组纽扣
17. toggled ['tɔgld] button 索结绳纽
18. wrapped & tied [əræpt & taid] button 叠门襟系带
19. zipper ['zipə] 拉链

## Neckline 领圈

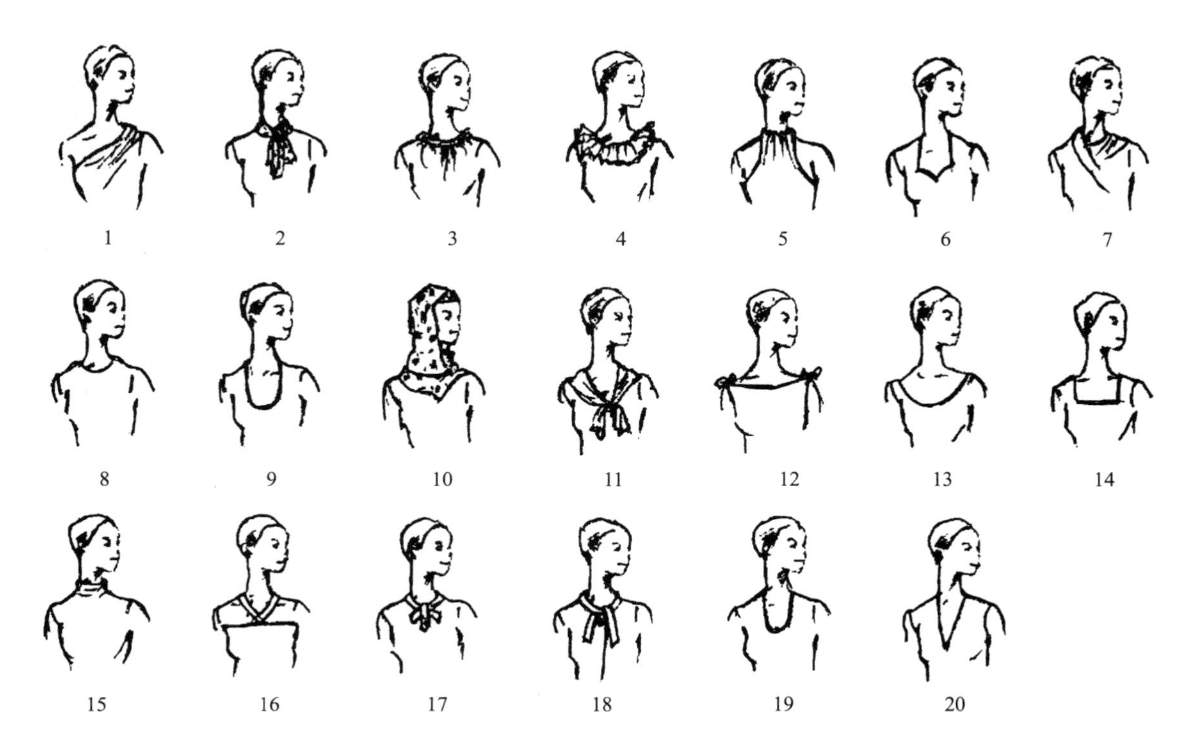


1. asymmetric [æsi'metrik] neckline 不均齐形领口
2. scarf [skɑːf] neckline 围巾形领口
3. drawstring ['drɔːstriŋ] neckline 伸缩型领口
4. frilled [frild] neckline 绉边领
5. halter ['hɔːltə] neckline 套索系领
6. sweat-heart [swet-hɑːt] neckline 爱心形领口
7. high & low cowl [hɑi & ləu kaul] neckline 高低垂领
8. round [raund] neckline 圆领
9. horseshoe ['hɔːʃʃuː] neckline 马蹄形领口
10. hooded ['hʊdid] neckline 头巾形领口
11. draped [dreip] neckline 打褶形领口
12. bateam or boat ['bættiːm ɔː bəut] neckline 舟形领口
13. scoop [skuːp] neckline 椭圆形领口
14. square [skwεə] neckline 四边形领口
15. stand-away [stænd ə'wei] neckline 离颈领口
16. strap [stræp] neckline 狭条布条形领口
17. tab [tæb] neckline 扣襻形领口
18. tie [tai] neckline 打结形领口
19. U-shape [juː-ʃeip] neckline U 形领口
20. V-shape [viː-ʃeip] neckline V 形领口

## Collar 领子

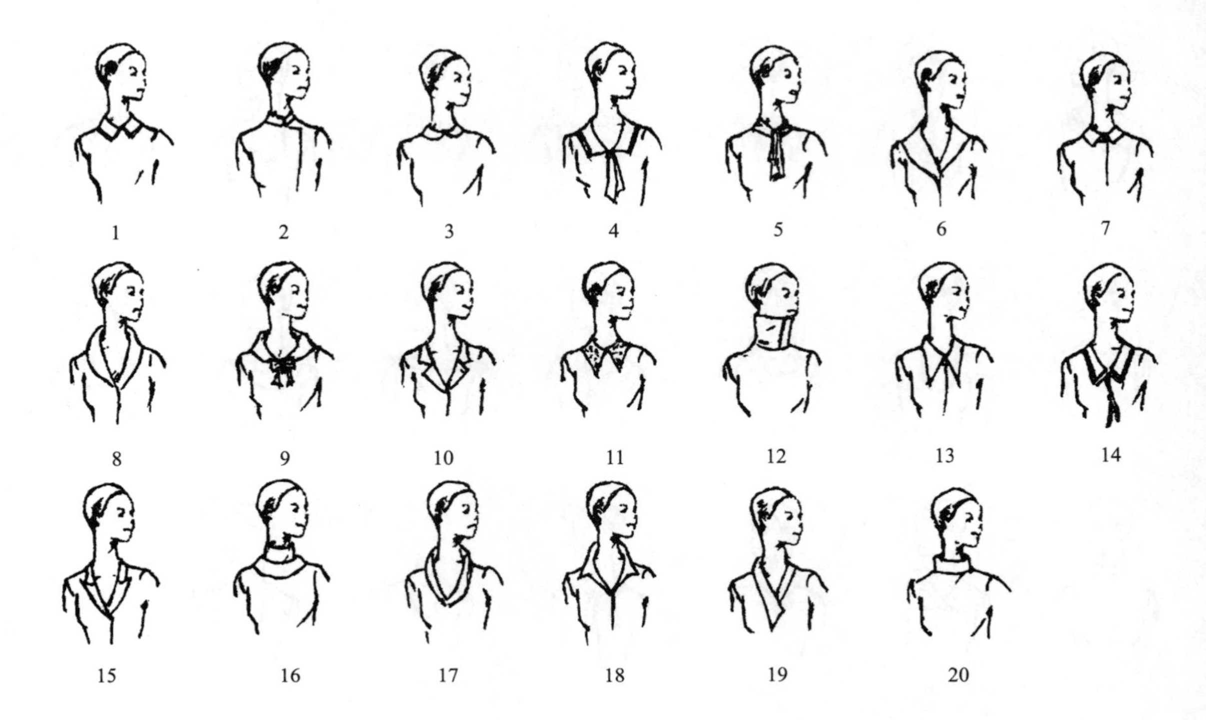


1. Eton ['iːtn] collar 伊顿领
2. mandarin ['mandarin] collar 旗袍领，中式领
3. Peter Pan ['piːtə pæn] collar 彼得·潘领，铜盘领，小圆领
4. sailor ['seilə] collar 领巾领
5. scarf [skɑːf] collar 围巾领
6. shawl [ʃɔːl] collar 肩巾领、青果领
7. collar with stand ['kɔlə wið stænd] collar 分上下领形
8. roll [rəul] collar 翻边领
9. picture ['piktʃə] collar 象形领
10. tailored ['teiləd] collar 西装领
11. button-down ['bʌtn-dɑun] collar 领下角有扣
12. funnel ['fʌnəl] collar 漏斗形领
13. peaked [pikt] collar 尖形领
14. swallow ['swɔləu] collar 燕子领
15. coat [kəut] collar 大衣领
16. polo ['pəuləu] collar 马球形领
17. stand-up [stænd-ʌp] collar 直立领
18. Milan [mi'læn] collar 米兰式领
19. cross-over [krɔs-'əuvə] collar 叠领
20. turtle ['təːtl] collar 龟形领

## Sleeve 袖子

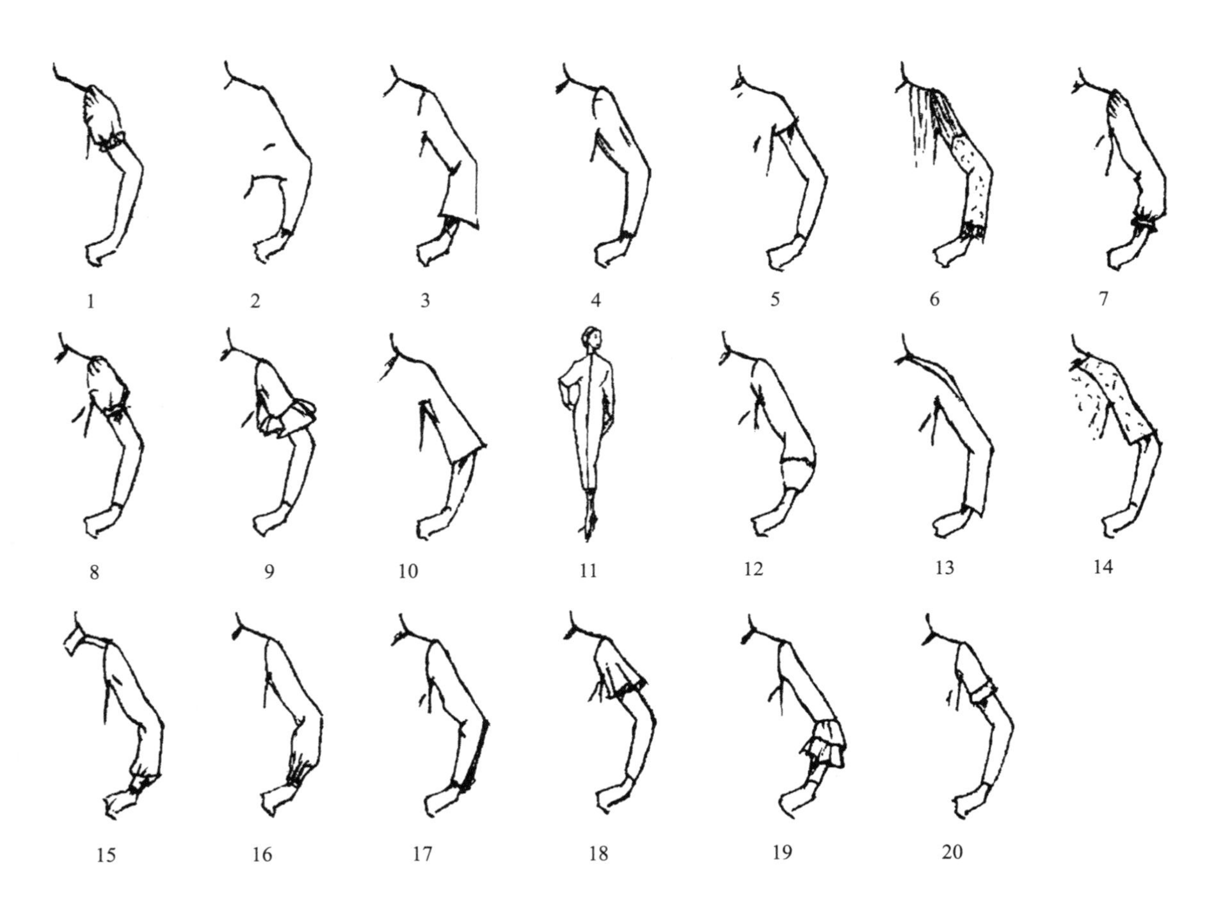


1. balloon [bə'luːn] sleeve 气球袖
2. batwing ['bætwiŋ] sleeve 蝙蝠袖
3. bell [bel] sleeve 钟形袖
4. bishop ['biʃəp] sleeve 主教式袖
5. cap [kæp] sleeve 顶盖式
6. cuffed ['kʌft] sleeve 翻边袖
7. drawstring ['drɔːstriŋ] sleeve 伸缩袖
8. puff [pʌf] sleeve 泡泡袖
9. flared [flɛəd] sleeve 喇叭袖
10. kimono [ki'məunəu] sleeve 和服式袖
11. kite [kɑit] sleeve 风筝形袖
12. lantern ['læntən] sleeve 灯笼袖
13. magyar ['mægjɑː] sleeve 连身出袖
14. raglan ['ræglən] sleeve 连肩袖，插肩袖
15. shirt [ʃəːt] sleeve 衬衫袖
16. shirred [ʃəː] sleeve 平行绉缝式袖
17. two-piece [tuː-piːs] sleeve 两片袖
18. cape [keip] sleeve 披肩式袖
19. tiered [tiəd] sleeve 层列式袖
20. set-in [set-in] sleeve 装袖 / 有袖窿袖

## Pockets 口袋

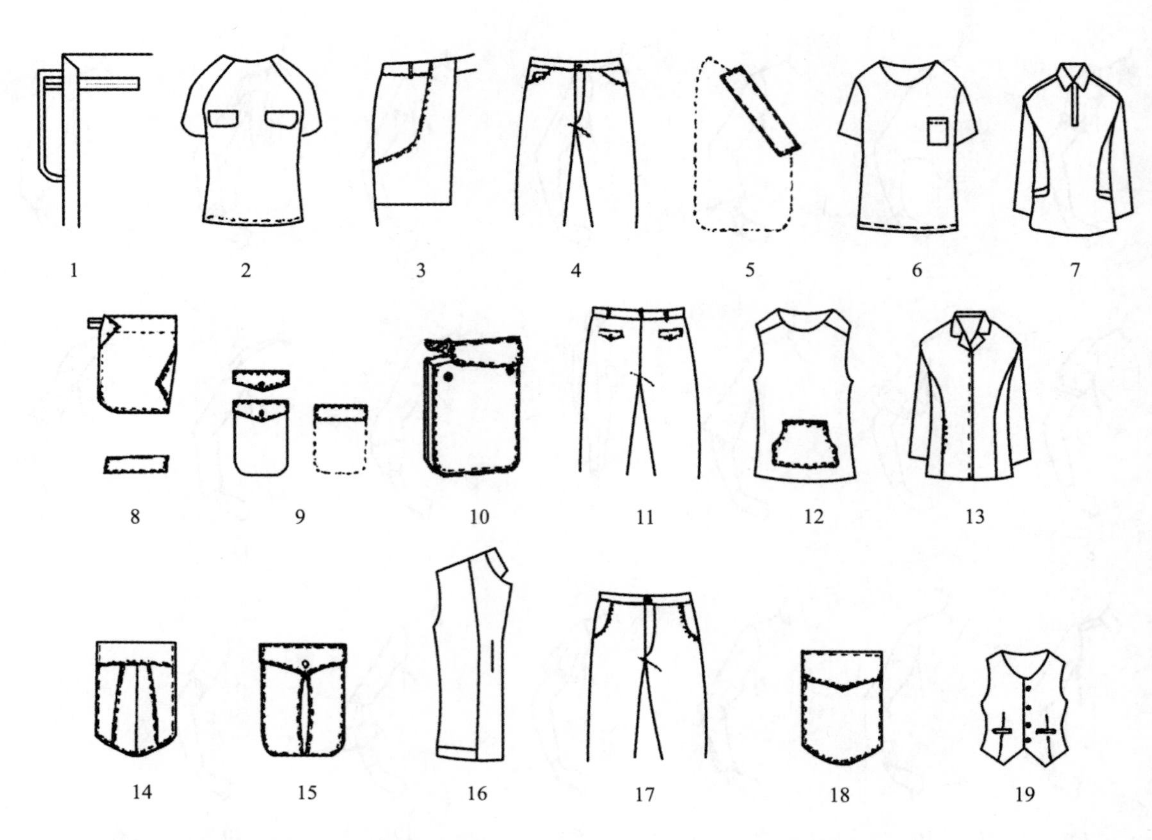


1. bound [bɑund] pocket 嵌线口袋
2. breast [brest] pocket 胸袋
3. cargo ['kɑːgəu] pocket 大贴袋
4. change [tʃeindʒ] pocket 零钱袋
5. slash [slæʃ] pocket 切缝口袋
6. patch [pætʃ] pocket 明贴袋
7. cross [krɔs] pocket 横开口袋
8. welt [velt] pocket 单贴边袋，西装袋
9. flap [flæp] pocket 盖式口袋
10. gusset ['gʌsit] pocket 接裆口袋
11. hip [hip] pocket 裤后袋
12. kangaroo [ˌkæŋgə'ruː] pocket 袋鼠式口袋
13. seam [siːm] pocket 摆缝口袋
14. Pin Tucks [pin-tʌks] pocket 针纹褶饰口袋
15. pleated [pliːtd] pocket 褶饰口袋
16. ticket ['tikit] pocket 内袋
17. western ['westən] pocket 西部型口袋
18. western-flap ['westən flæp] pocket 牛仔型盖口袋
19. watch [wɔtʃ] pocket 表袋

## Pleats 褶裥

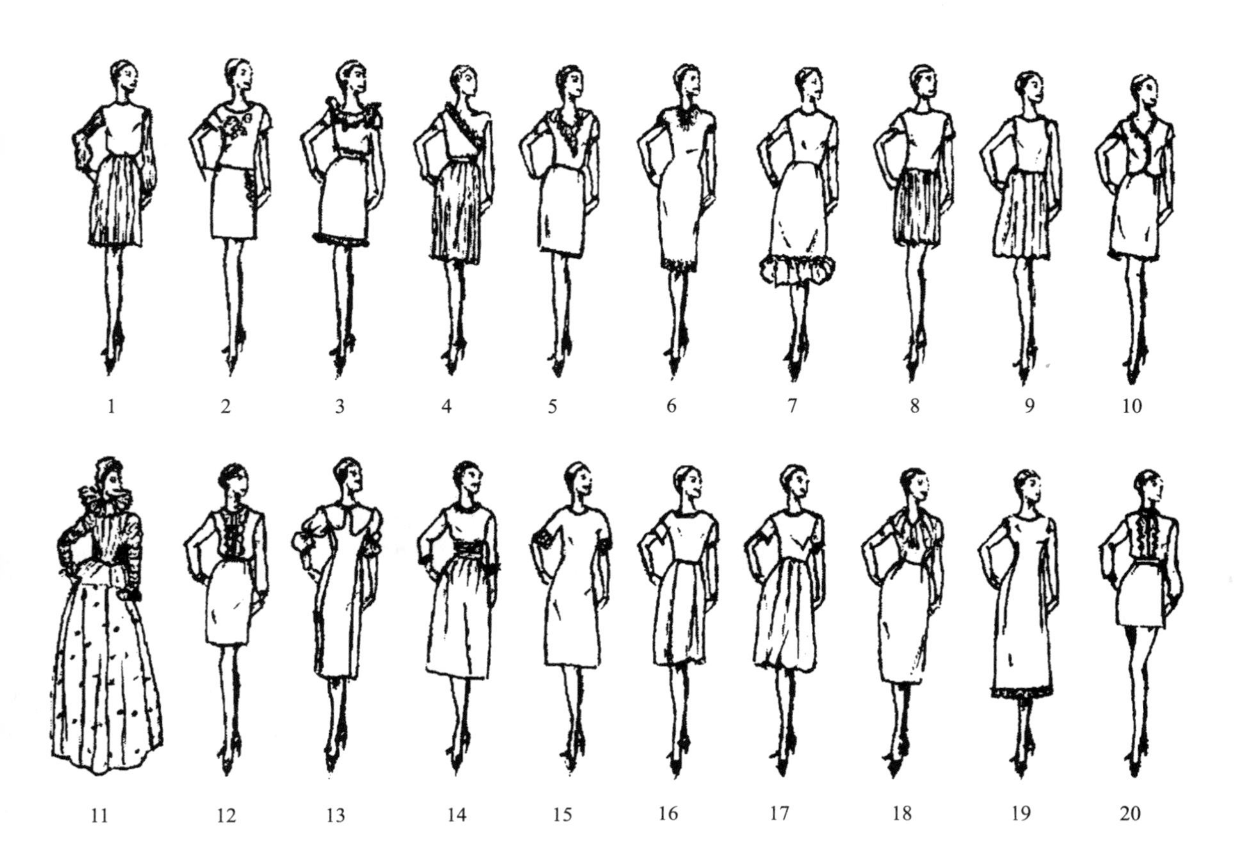


1. accordion [ə'kɔːdjən] pleats 风琴褶
2. beading [biːdiŋ] pleats 珠绣褶
3. bound-edge frill [baund-edʒ fril] pleats 镶边绉褶
4. box [bɔks] pleats 箱形褶裥
5. cascade [kæs'keid] pleats 小瀑布式褶
6. embroidery [im'brɔidəri] pleats 刺绣
7. flounce [flauns] pleats 裙脚皱褶
8. knife [naif] pleats 剑褶裥
9. sunray ['sʌnrei] pleats 太阳褶
10. pleated frill [pliːtd fril] pleats 折叠花边
11. ruff [rʌf] pleats 绉领褶
12. ruffle ['rʌfl] pleats 蓬松皱边
13. shirring ['ʃɜːriŋ] pleats 平行皱缝
14. smocking ['smɔkiŋ] pleats 缀线衣裥
15. cording ['kɔːdiŋ] pleats 布边花纹
16. double ['dʌbl] pleats 叠褶
17. tuck [tʌk] pleats 褶缝裥
18. waterfall ['wɔːtəfɔːl] pleats 大瀑布式褶
19. tie frill [pai fril] pleats 混杂绉边
20. jabot [ʒæ'bəu] pleats 衬衫绉边

# Appendix Ⅲ: Garment Parts　附录Ⅲ：服装部件

## Men's Shirt　男衬衫

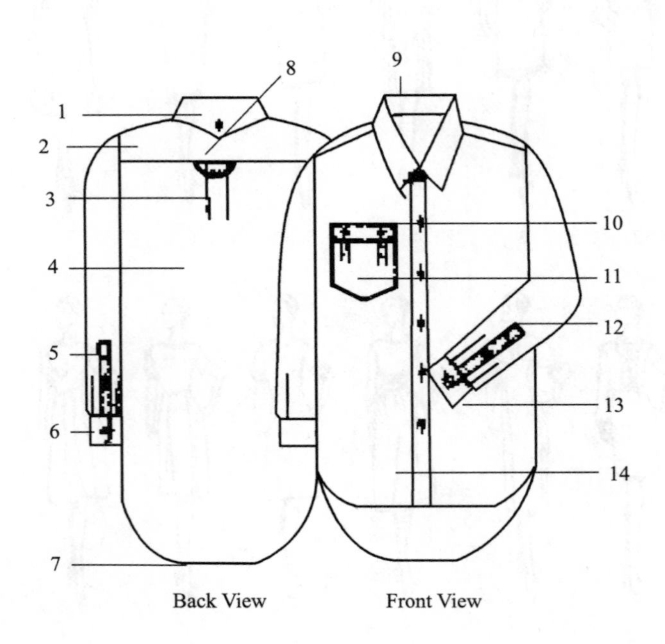


1. update trend collar [ʌp'deit trend 'kɔlə] 新潮领子
2. yoke [jəuk] 过肩，育克（广东话：担干）
3. box pleat [bɔks pliːt] 箱形褶裥（广东话：工字褶）
4. back panel [bæk 'pænl] 后幅，后片
5. sleeve placket [sliːv 'plækit] 袖衩（广东话：三尖袖衩）
6. single cuff ['siŋgl 'kʌf] 单层克夫（广东话：单层介英）
7. round bottom [raund 'bɔtəm] 圆形下摆
8. loop tab [luːp tæb] 襻带（广东话：耳圈）
9. collar stand ['kɔlə stænd] 领座
10. pocket flap ['pɔkit flæp] 袋盖
11. patch pocket [pætʃ 'pɔkit] 明贴袋
12. one piece sleeve [wʌn piːs sliːv] 一片袖
13. front panel [frʌnt 'pænl] 前衣片
14. front placket [əfrʌnt 'plækit] 前明门襟

## Jacket 夹克

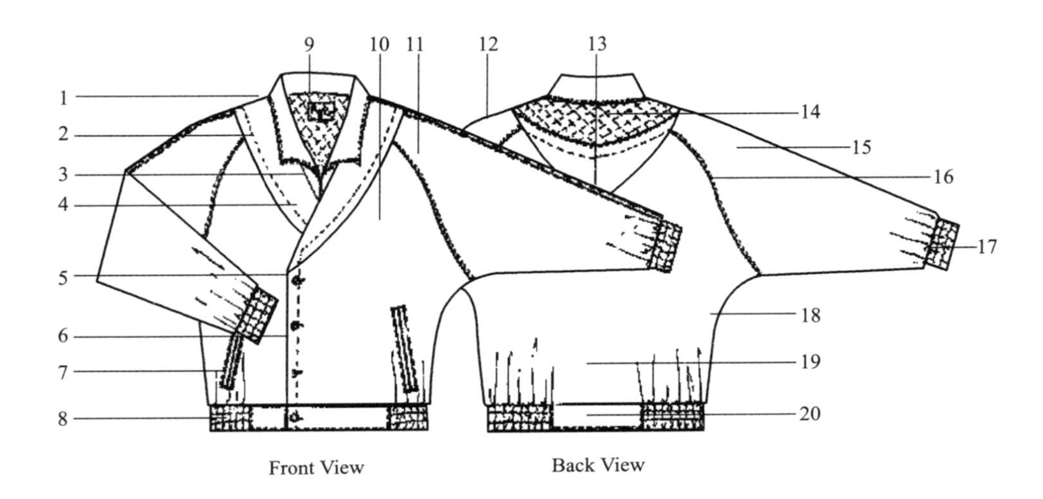


1. one piece collar [wʌn piːs 'kɔlə] 一片领
2. lapel [lə'pel] 襟贴，挂面
3. center front with zipper ['sentə frʌnt wið 'zipə] 拉链式前中
4. break line [breik lɑin] 翻折线
5. break point [breik pɔint] 翻折点
6. front edge [frʌnt edʒ] 门襟子口
7. double jetted pocket ['dʌbl 'dʒetid 'pɔkit] 双嵌线袋（广东话：双唇袋）
8. elastic bottom [i'læstik 'bɔtəm] 松紧带下摆
9. main label [mein 'leibl] 主商标
10. front [frʌnt] 前幅，前衣片
11. front sleeve / raglan sleeve [frʌnt sliːv / 'ræglən sliːv] 前袖 / 插肩袖
12. shoulder seam ['ʃəuldə siːm] 肩缝
13. hood [hud] 帽子
14. lining ['lɑiniŋ] 里料
15. back sleeve [bæk sliːv] 后袖
16. armscye seam [ɑːmsai siːm] 袖窿弧线
17. elastic cuff [i'læstik 'kʌf] 松紧带袖口
18. underarm seam ['ʌndər ɑːm] 袖底线
19. back [bæk] 后幅，后衣片
20. back bottom [bæk 'bɔtəm] 后幅下摆

## Knitted Jacket 针织夹克

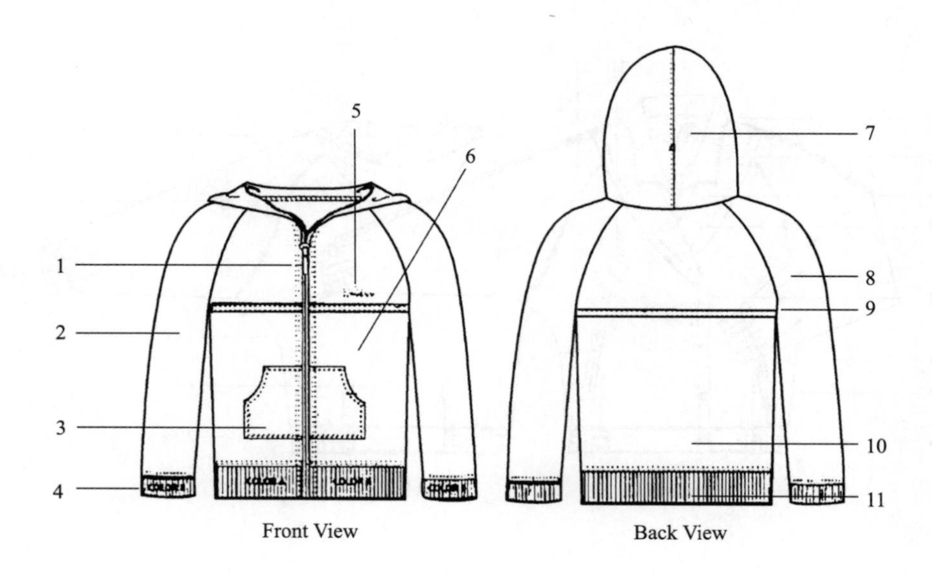


1. center front zipper ['sentə frʌnt 'zipə] 前中拉链
2. front raglan sleeve [frʌnt 'ræglən sli:v] 前插肩袖
3. extra size pocket ['ekstrə saiz 'pɔkit] 超大型口袋
4. ribbing cuff ['ribiŋ 'kʌf] 罗纹袖口
5. embroidery logo [im'brɔidəri lɔgəu] 刺绣商标
6. front [frʌnt] 前幅，前衣片
7. hood [hud] 帽子
8. back raglan sleeve [bæk 'ræglən sli:v] 后插肩袖
9. color tape ['kʌlə teip] 色带条
10. back [bæk] 后幅，后衣片
11. ribbing bottom ['ribiŋ 'bɔtəm] 罗纹下摆

## Suit Jacket 西装夹克

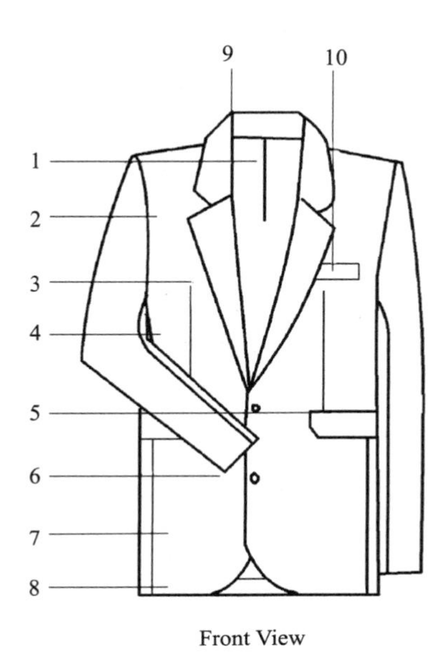


Front View

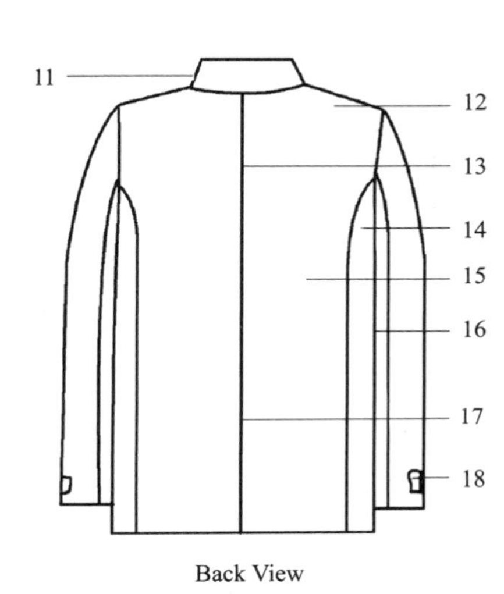


Back View

1. lining ['lainiŋ] 里料
2. collar notch ['kɔlə nɔtʃ] 领子串口
3. lapel [lə'pel] 襟贴，挂面，驳头
4. front dart [frʌnt dɑ:t] 前腰省
5. pocket flap ['pɔkit flæp] 袋盖
6. eyelet & button ['ailit & bʌtn] 扣眼与纽扣
7. front part [frʌnt pɑ:t] 前衣片
8. bottom ['bɔtəm] 下摆
9. collar stand ['kɔlə stænd] 领座
10. breasted welt pocket ['brestid velt 'pɔkit] 胸袋
11. collar fall ['kɔlə fɔ:l] 领面
12. shoulder seam ['ʃəuldə si:m] 肩缝
13. back part [bæk pɑ:t] 后幅
14. under sleeve ['ʌndə sli:v] 小袖，底袖
15. side panel [said 'pænl] 侧面嵌片（广东话：小身，侧片）
16. top sleeve [tɔp sli:v] 大袖，面袖
17. center back ['sentə bæk] 后中
18. sleeve opening [sli:v 'əupəniŋ] 袖开口

## Men's Vest 男装背心

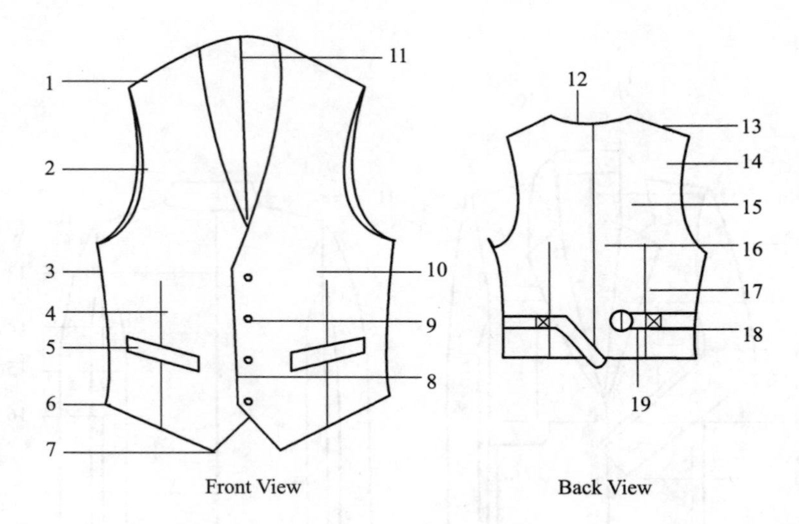


1. shoulder point [ˈʃəuldə pɔint] 肩点
2. front armhole [frʌnt ˈɑːmhəul] 前幅袖窿
3. side seam [said siːm] 侧缝
4. front waist dart [frʌnt weist daːt] 前腰省
5. welt pocket [velt ˈpɔkit] 挖袋，腰袋
6. side slit [said slit] 侧缝衩位
7. bottom [ˈbɔtəm] 下摆
8. center front edge [ˈsentə frʌnt edʒ] 前中边位
9. button & buttonhole [ˈbʌtn & ˈbʌtnhəul] 纽扣与扣眼
10. front [frʌnt] 前幅，前衣片
11. lining [ˈlɑiniŋ] 里料
12. back neck [bæk nek] 后领圈
13. shoulder line [ˈʃəuldə lɑin] 肩线
14. back armhole [bæk ˈɑːmhəul] 后夹圈
15. back [bæk] 后幅，后衣片
16. center back [ˈsentə bæk] 后中
17. back waist dart [bæk weist daːt] 后腰省
18. waist belt [weist belt] 腰带
19. buckle [ˈbʌkl] 带扣

## Mandarin Dress 旗袍

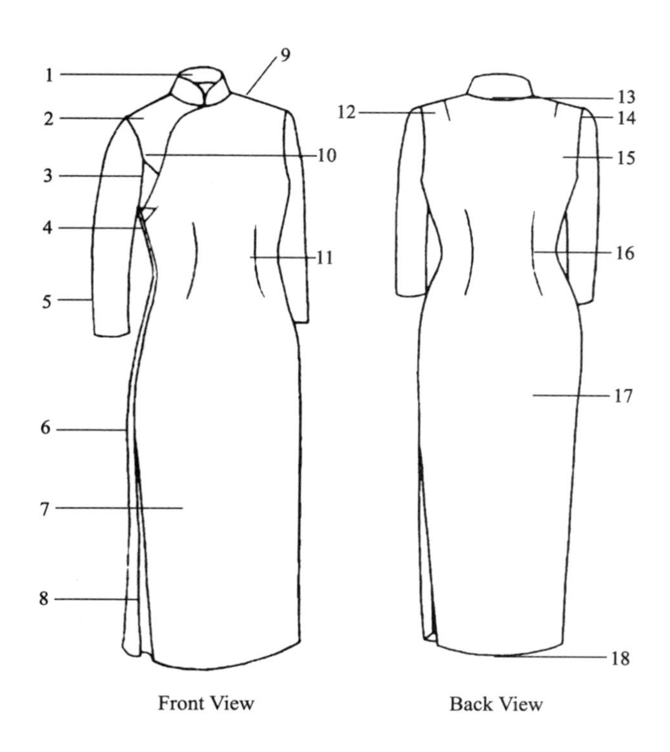


1. under collar ['ʌndə 'kɔlə] 领里
2. under facing ['ʌndə 'feisiŋ] 底襟
3. cross bust dart [krɔs bʌst da:t] 横胸省
4. zipper opening ['zipə 'əupəniŋ] 拉链开口
5. sleeve [sli:v] 袖子
6. side seam [said si:m] 侧缝
7. front bodice [frʌnt 'bɔdis] 前衣身
8. side slit [said slit] 摆缝衩位
9. shoulder line ['ʃəuldə lain] 肩线
10. top facing edge [tɔp 'feisiŋ edʒ] 大襟边
11. front waist dart [frʌnt weist da:t] 前腰省
12. back shoulder dart [bæk 'ʃəuldəda:t] 后肩省
13. top collar [tɔp 'kɔlə] 面领
14. sleeve head [sli:v hed] 袖头
15. armhole ['ɑ:mhəʊl] 袖窿
16. back waist dart [bæk weist da:t] 后腰省
17. back bodice [bæk 'bɔdis] 后衣身
18. bottom ['bɔtəm] 下摆

## Overalls 吊带工装裤

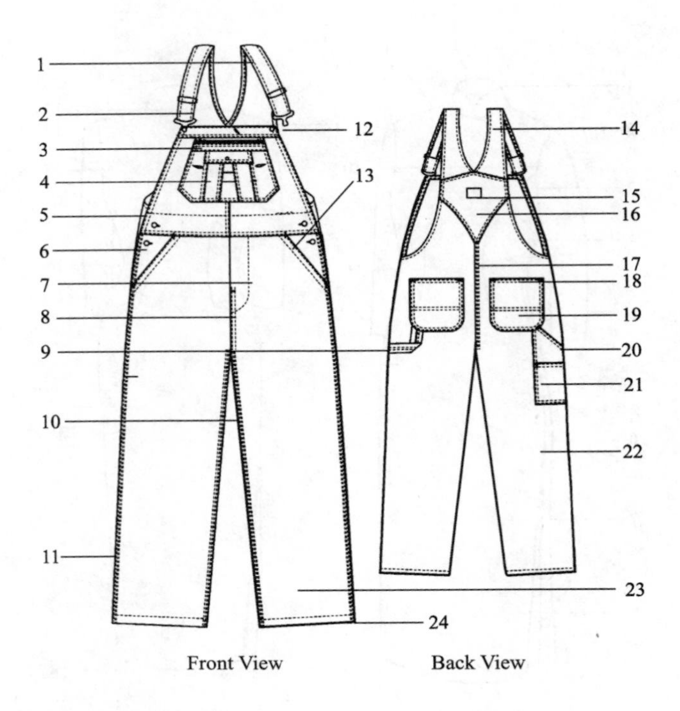


1. under suspender ['ʌndə sə'spendə(r)] 吊带底
2. buckle ['bʌkl] 带扣
3. zipper pocket ['zipə 'pɔkit] 拉链袋
4. chest patch pocket [tʃest pætʃ 'pɔkit] 胸贴袋
5. side seam opening [said siːm 'əupəniŋ] 摆缝开口
6. pocket facing ['pɔkit 'feisiŋ] 袋贴
7. front fly [frʌnt flai] 前门襟
8. crotch [krɔtʃ] 裤裆（广东话：小浪）
9. decorative strap ['dekərətiv stræp] 装饰带条
10. in seam [in siːm] 内接缝（广东话：内浪，内长）
11. out seam [aut siːm] 外接缝（广东话：侧缝，外长）
12. press stud [pres stʌd] 揿扣（广东话：工字扣）
13. slant pocket [slɑːnt 'pɔkit] 斜袋
14. top suspender [tɔp sə'spendə(r)] 吊带面
15. brand label [brænd 'leibl] 主商标
16. back tops [bæk tɔps] 后上衣身
17. seat seam [said siːm] 后裤裆缝（广东话：后浪骨）
18. back patch pocket [bæk pætʃ 'pɔkit] 后贴袋
19. decorative stitching ['dekərətiv 'stitʃiŋ] 装饰面缝线迹
20. pocket mouth ['pɔkit mauθ] 袋口
21. cargo pocket ['kɑːgəu 'pɔkit] 大贴袋
22. back [bæk] 后幅，后衣片
23. front [frʌnt] 前幅，前衣片
24. bottom ['bɔtəm] 下摆（广东话：裤脚）

## 5-Pkt. Jeans 牛仔裤

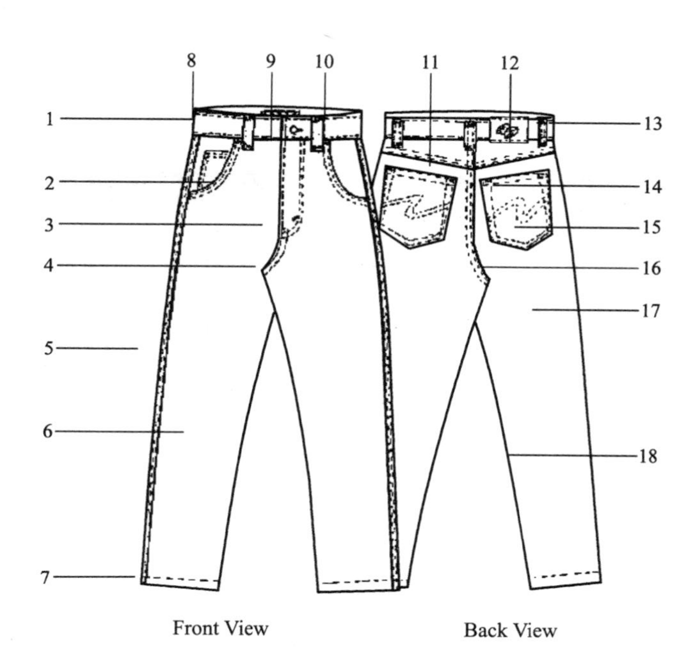


1. waistband ['weistbænd] 腰头
2. front curve pocket [frʌnt kə:v 'pɔkit] 前弯袋
3. fly [flɑi] 门襟（广东话：纽牌）
4. crotch [krɔtʃ] 裤裆
5. side seam [said si:m] 侧缝
6. front [frʌnt] 前幅，前衣片
7. bottom ['bɔtəm] 下摆（广东话：裤脚）
8. change pocket/ coin pocket [tʃeindʒ 'pɔkit kɔin 'pɔkit] 表袋，零钱袋
9. eyelet & press stud ['ailit & pres stʌd] 扣眼与工字扣
10. pocket facing ['pɔkit 'feisiŋ] 袋贴
11. yoke [jəuk] 育克（广东话：机头）
12. leather label ['leðə 'leibl] 皮牌
13. belt-loop [belt-lu:p] 襻带，裤襻（广东话：裤耳）
14. rivet ['rivit] 铆钉，包头钉（广东话：撞钉）
15. hip pocket [hip 'pɔkit] 后袋
16. seat seam [si:t si:m] 后裤裆缝(广东话：后浪骨）
17. back [bæk] 后幅，后衣片
18. in seam [in si:m] 内接缝（广东话：内浪，内长）

## Shorts 短裤

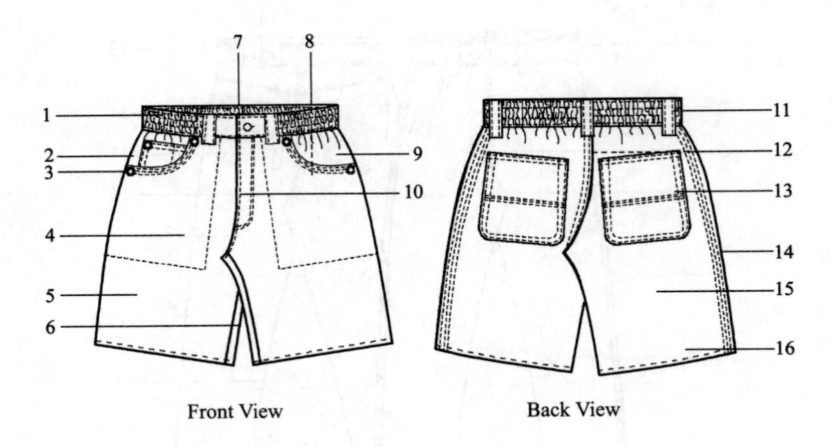


1. elastic waistband [i'læstik 'weistbænd] 松紧带裤腰
2. change pocket [tʃeindʒ 'pɔkit] 表袋，零钱袋
3. rivet ['rivit] 铆钉，包头钉（广东话：撞钉）
4. pocket bag ['pɔkit bæg] 袋布
5. front part [frʌnt pɑ:t] 前幅，前衣片
6. in seam [in si:m] 内接缝（广东话：内浪，内长）
7. button & button hole ['bʌtn & 'bʌtnhəul] 纽扣与扣孔
8. front curve pocket with double needle [frʌnt kə:v 'pɔkit wið 'dʌbl 'ni:dl] 双针前弯袋
9. pocket facing ['pɔkit 'feisiŋ] 袋贴
10. fly [flɑi] 门襟
11. belt loop [belt lu:p] 襻带，裤襻（广东话：裤耳）
12. seat seam with double needle [si:t si:m wið 'dʌbl 'ni:dl] 后裆双针线迹
13. round hip patch pocket [raund hip pætʃ 'pɔkit] 圆角后贴袋
14. side seam with three needles [said si:m wið 'θri: ni:dlz] 侧缝三针线迹
15. back part [bæk pɑ:t] 后幅，后衣片
16. bottom ['bɔtəm] 裤脚

## Skirt 西装裙

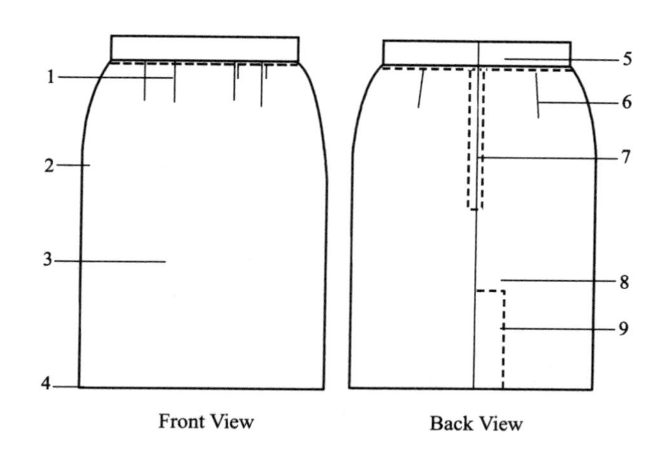


1. front dart [frʌnt daːt] 前腰省
2. side seam [said siːm] 侧缝
3. front panel [frʌnt 'pænl] 前幅，前衣片
4. bottom ['bɔtəm] 下摆（广东话：裙脚）
5. waistband ['weistbænd] 腰头
6. back dart [bæk daːt] 后腰省
7. zipper opening ['zipə 'əupəniŋ] 拉链开口
8. back panel [bæk 'pænl] 后幅，后衣片
9. vent [vent] 裙衩

## Jacket & Trousers　夹克套装

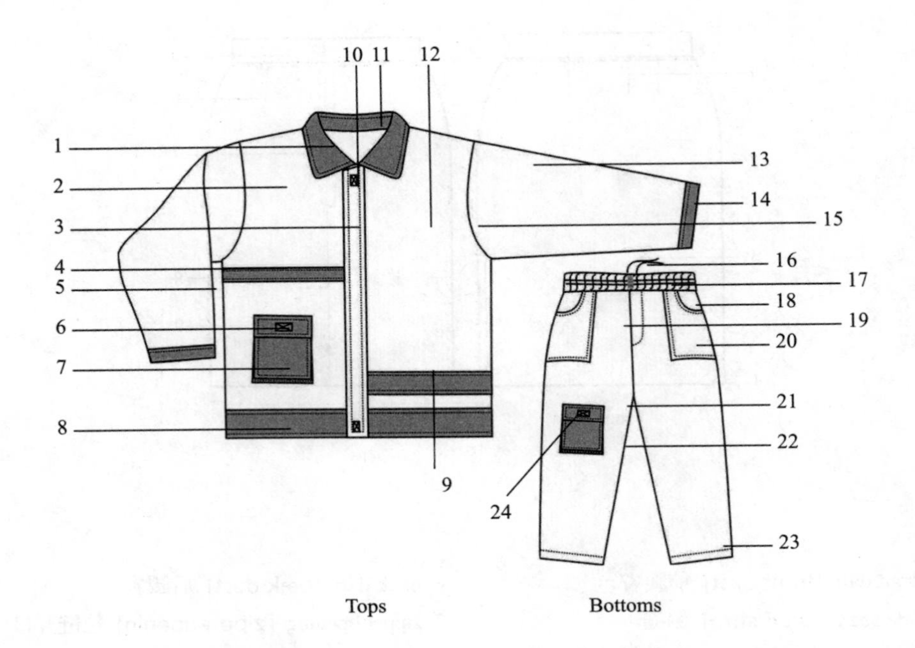


1. top collar [tɔp 'kɔlə] 领面
2. front [frʌnt] 前幅，前衣片
3. fly opening [flɑi 'əupəniŋ] 暗门襟
4. contrast strip ['kɔntræst, strip] 撞色带条
5. side seam [said siːm] 侧缝
6. pocket flap ['pɔkit flæp] 袋盖
7. bellows pocket ['beləuz 'pɔkit] 风箱式袋口
8. bottom ['bɔtəm] 下摆
9. color strip ['kʌlə strip] 色带
10. center front ['sentə frʌnt] 前中
11. under collar ['ʌndə 'kɔlə] 底领
12. left front [left frʌnt] 左前幅
13. sleeve [sliːv] 袖子
14. sleeve opening [sliːv 'əupəniŋ] 袖口
15. armhole ['ɑːmhəul] 袖窿（广东话：夹圈）
16. cotton strings ['kɔtn striŋz] 棉绳
17. elastic waistband [i'læstik 'weistbænd] 松紧带裤腰
18. pocket facing ['pɔkit feisiŋ] 袋口贴边
19. fly [flɑi] 门襟
20. side patch pocket [said pætʃ 'pɔkit] 侧贴袋
21. crotch point [krɔtʃ pɔint] 裤裆点
22. in seam [in siːm] 内接缝（广东话：内浪，内长）
23. bottom ['bɔtəm] 下摆（广东话：裤脚）
24. velcro ['velkrəu] 魔术贴，尼龙搭扣

## Sports Wear 运动服

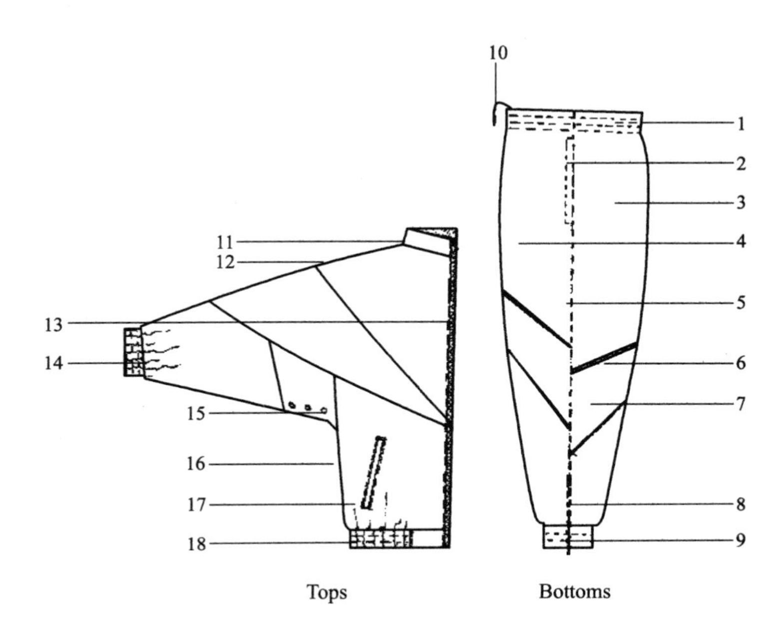


1. self-turned waistband [self-tə:nd 'weistbænd] 原身裤腰
2. straight pocket [streit 'pɔkit] 直袋
3. back part [bæk pɑ:t] 后幅，后衣片
4. front part [frʌnt pɑ:t] 前幅，前衣片
5. side seam [said si:m] 侧缝
6. seam with top-stitching [si:m wið 'tɔp-stitʃ] 面缝线迹（广东话：缉面线）
7. cutting portion ['pɔ:ʃən 'pɔ:ʃən] 切片
8. bottom opening ['bɔtəm 'əupəniŋ] 下摆开口（广东话：裤脚开口）
9. bottom ['bɔtəm] 下摆（广东话：裤脚）
10. cotton string ['kɔtn striŋ] 棉绳
11. stand collar [stænd 'kɔlə] 立领
12. shoulder seam ['ʃəuldə si:m] 肩缝
13. center front with zipper ['sentə frʌnt wið 'zipə] 前中拉链
14. elastic cuff [i'læstik 'kʌf] 松紧带袖口
15. eye-let [ɑi-let] 孔眼
16. under arm seam ['ʌndə ɑ:m si:m] 袖底缝
17. side pocket [said 'pɔkit] 侧口袋
18. elastic bottom [i'læstik 'bɔtəm] 松紧下摆

*APPENDIX Ⅳ~Ⅷ

## APPLICATION FORM AND OTHERS　求职应聘申请表及其他

**课题名称：** APPLICATION FORM AND OTHERS　求职应聘申请表及其他

课题内容：Organization Chart　组织结构

Position or Title　职位与头衔

Application Form　求职应聘申请表

Trading Terms　贸易术语

Abbreviation of Term　缩略语

**课题时间：** 2 课时

**教学目的：** 让学生了解并熟记企业组织架构与职位名称，掌握企业求职应聘申请表的设计与填写。

**教学方式：** 结合 PPT 与音频多媒体课件，以教师课堂讲述为引导，学生课堂练习为主。

**教学要求：** 1. 熟悉组织架构图、职位名称等专业词汇。

2. 熟悉求职应聘申请表的格式与应用。

**课前（后）准备：** 结合专业知识，课前预习相关内容。课后熟记相关专业词汇，掌握服装企业应聘申请表的填写与应用。

## Appendix IV: Organization Chart 附录IV：组织结构

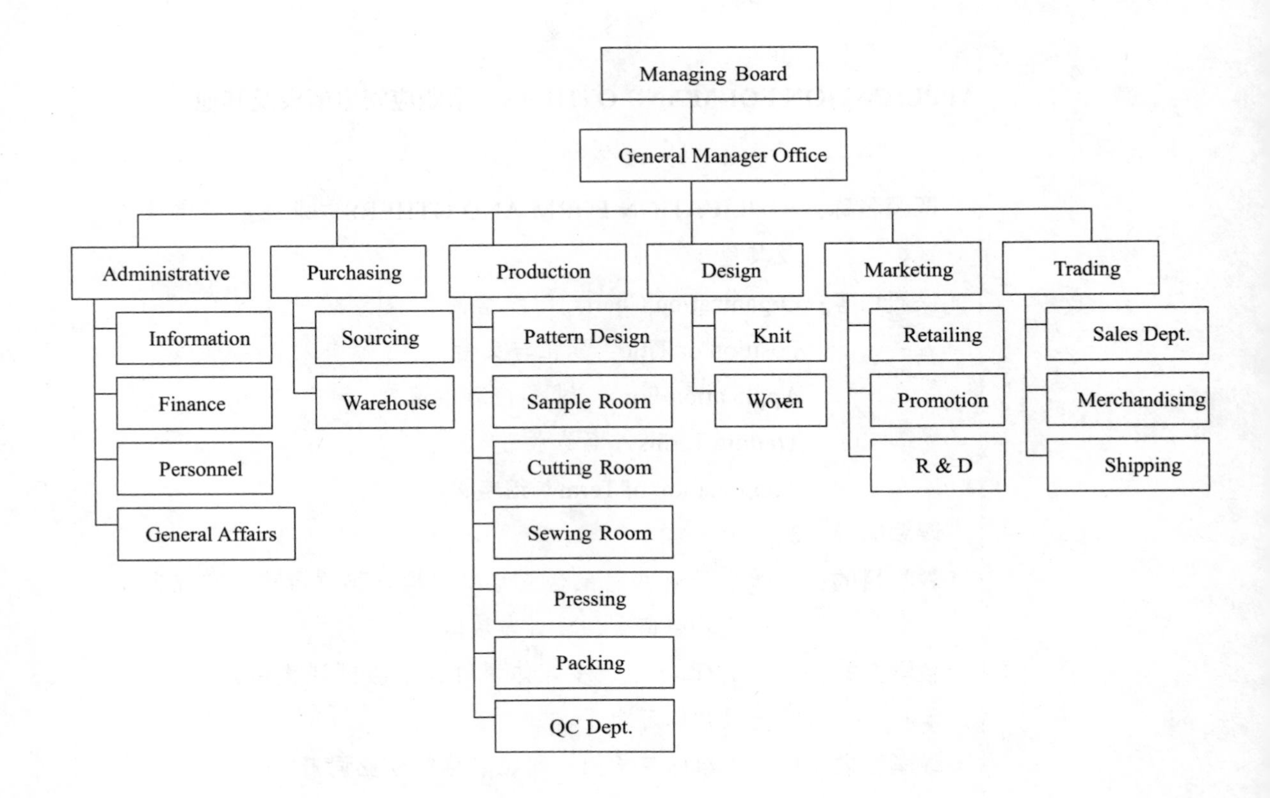


## Appendix Ⅴ: Position or Title 附录Ⅴ：职位与头衔

1. president ['prezidənt] 总裁
2. vice president [vais 'prezidənt] 副总裁
3. CEO/ chief executive officer [tʃi:f ig'zekjutiv 'ɔfisə] 首席执行官
4. CFO/ chief finance officer [tʃi:f fai'næns 'ɔfisə] 财务总监
5. COO/ chief operation officer [tʃi:f ɔpə'reiʃən 'ɔfisə] 首席营运官
6. CKO/ chief knowledge officer [tʃi:f 'nɔlidʒ 'ɔfisə] 首席知识管理执行官
7. chairman ['tʃeəmən] 主席，董事长
8. vice chairman [vais 'tʃeəmən] 副主席，副董事长
9. individual director [indi'vidjuəl di'rektə] 独立董事
10. managing director ['mænidʒiŋ di'rektə] 董事总经理
11. financial general manager [fɑi'nænʃəl, 'dʒenərəl 'mænidʒə] 财务总经理
12. administrative general manager [əd'ministrətiv 'dʒenərəl 'mænidʒə] 行政总监
13. senior manager ['si:njə mænidʒə] 高级经理
14. production manager [prə'dʌkʃən 'mænidʒə] 生产经理

15. marketing manager ['mɑ:kitiŋ 'mænidʒə] 市场部经理
16. purchasing manager ['pə:tʃəs 'mænidʒə] 采购部经理
17. administrative manager [əd'ministrətiv 'mænidʒə] 行政部经理
18. shipping manager ['ʃipiŋ 'mænidʒə] 船务部经理
19. sales manager [seils 'mænidʒə] 销售经理
20. account manager [ə'kaunt 'mænidʒə] 会计部经理
21. QC manager ['mænidʒə] 质检部经理
22. merchandising manager ['mə:tʃəndaiziŋ 'mænidʒə] 采购部经理
23. vice manager [vais 'mænidʒə] 副经理
24. chief designer [tʃi:f di'zainə] 首席设计师
25. designer [di'zainə] 设计师
26. assistant designer [ə'sistənt di'zainə] 助理设计师
27. senior supervisor ['si:njə 'sju:pəvaizə] 高级主任
28. production supervisor [prə'dʌkʃən 'sju:pəvaizə] 生产主任
29. marketing supervisor ['mɑ:kitiŋ 'sju:pəvaizə] 市场部主任
30. purchasing supervisor ['pə:tʃəs 'sju:pəvaizə] 采购部主任
31. administrative supervisor [əd'ministrətiv 'sju:pəvaizə] 行政部主任
32. shipping supervisor ['ʃipiŋ 'sju:pəvaizə] 船务部主任
33. sales supervisor [seilz 'sju:pəvaizə] 销售主任
34. account supervisor [ə'kaunt 'sju:pəvaizə] 会计部主任
35. QC supervisor [QC 'sju:pəvaizə] 验货部主任
36. senior merchandiser ['si:njə 'mə:tʃəndaizə] 高级跟单员，高级采购员
37. merchandiser ['mə:tʃəndaizə] 跟单员，采购员
38. assistant merchandiser [ə'sistənt 'mə:tʃəndaizə] 助理跟单，助理采购员
39. clerks [klɑ:k,z] 职员
40. offices ['ɔfisiz] 文员
41. QC / quality controller ['kwɔliti kən'trəulə] 品质控制员
42. quality inspector ['kwɔliti in'spektə] 质检员
43. pattern maker ['pætən meikə] 制板工
44. sample maker ['sæmpl meikə] 样衣工
45. marker maker ['mɑ:kə meikə] 排料员
46. machinist [mə'ʃi:nist] 机修师
47. mechanic [mi'kænik] 技工
48. operator ['ɔpəreitə] 操作工
49. spreader ['spredə] 拉布工
50. sewing operator ['səuiŋ 'ɔpəreitə] 车工
51. pressing operator/ presser ['presiŋ 'ɔpəreitə / presə] 熨烫工
52. packing operator/ packer ['pækiŋ 'ɔpəreitə/ 'pækə] 包装工
53. mender ['mendə] 修补工
54. cutter ['kʌtə] 裁剪工
55. bundling worker ['bʌndliŋ 'wə:kə] 捆扎工人
56. warehouse keeper ['wεəhaus 'ki:pə] 仓库保管员
57. helper ['helpə] 帮工

# Appendix Ⅵ : Application Form　求职应聘申请表

## Application Form（1） 求职应聘申请表（1）

**Application for Full-Time Employment**

Please write down the true message in English

| | | | | |
|---|---|---|---|---|
| Name | | | | Recent Photo |
| Date of Birth | | | | |
| Sex | | | | |
| Age | | Height | CM | |
| Census Register | | | | |
| Nationality | | | | |

| | | | | |
|---|---|---|---|---|
| ID Card No. | | Marital Status | | |
| Current Location | | E-Mail | | |
| Contact No. | | Mobile No. | | |
| Address | | | Postal Code | |
| Apply Position | | Date Available | | |
| Expected Salary | | Specialty | | |
| Expected Working Location | | Language | | |

**Education**

| Start—End | Name of School | Major & Certificate |
|---|---|---|
| | | |
| | | |
| | | |

**Working Experience**

| Name of Company | Phone | Start—End | Last Position | Reason for Leaving |
|---|---|---|---|---|
| | | | | |
| | | | | |
| | | | | |

**Introduce Yourself in 50 Words**

**Remark**

Applicant:

Date:

## Application Form（2） 求职应聘申请表（2）

**EMPLOYEE REGISTRATION AND APPLICATION FORM**

**BRANCH COMPANY**

OUR REF. ______________________

POSITION APPLIED FOR ______________

Photos

**A. PERSONAL DATA**

NAME:（in English）__________ （in Chinese）__________

ADDRESS: ______________________________

TELEPHONE NO.: __________ DATE OF BIRTH: ________ SEX: ________

PLACE OF BIRTH: __________ I.D. CARD/PASSPORT NO.: __________

MARITAL STATUS: SINGLE____ MARRIED____ DIVORCED____ OTHERS____

NAME OF KIN: ______________ RELATIONSHIP: __________

ADDRESS: ______________ TEL. NO.: ______________

**B. ACADEMIC QUALIFICATIONS**

| | NAME OF SCHOOL | MONTH/YEAR | CERTIFICATE OBTAINED |
|---|---|---|---|
| a. | | | |
| b. | | | |
| c. | | | |
| d. | | | |

**C. EMPLOYMENT HISTORY**

| | COMPANY | POSITION | EMP. PERIOD | SALARY | REASON FOR LEAVING |
|---|---|---|---|---|---|
| a. | | | | | |
| b. | | | | | |

**D. KNOWLEDGE（STATE EXCELLENT, GOOD, FAIR OR POOR）**

ENGLISH WRITTEN __________ SPOKEN __________

CHINESE WRITTEN __________ CANTONESE __________

MANDARIN __________ OTHERS __________

**E. REFERENCES**

a. NAME __________ OCCUPATION __________

ADDRESS __________ TEL. NO. __________

b. NAME __________ OCCUPATION __________

ADDRESS __________ TEL. NO. __________

**Continued**

**F. OTHER**

EXPECTED SALARY ________________ DATE AVAILABLE ________________

BANK NO. __________________________

* HEREBY CONFIRM THAT ALL INFORMATIONS ARE CORRECT AND UNDERSTAND THAT ANY MISREPRESENTATION OF THE ABOVE MAY BE LOST TO INSTANT DISMISSAL*

DATE : ____________________ SIGNATURE OF APPLICANT: ________________

FOR OFFICE USE ONLY

| DATE | INTERVIEWER | COMMENT |
| --- | --- | --- |
| | | |

POSITION RECOMMENDED ____________________ DEPT. ____________________

SUGGESTED SALARY ________________________ STARTING DATE ____________

REMARKS ________________________________________________

APPROVED BY:

| ________________ | ________________ | ________________ |
| --- | --- | --- |
| DEPT. MANAGER | PERSONNEL MANAGER | DIRECTOR |

# Appendix Ⅶ: Trading Terms　附录Ⅶ：贸易术语

1. agent [ˈeidʒənt] 代理商
2. agreement [əˈgriːmənt] 书面协议，同意书
3. amendment [əˈmendmənt] 修正书
4. application for conversion [æpliˈkeiʃən fɔː ˌkənˈvəːʃən] 折换申请书
5. application for negotiation of draft under L/C [æpliˈkeiʃən fɔː, nigəuʃiˈeiʃən ɔv, drɑːft ˈʌndə el/siː] 出口押汇申请书
6. appropriation [əˌprəupriˈeiʃən] 拨款
7. award of bid [əˈwɔːd ɔv, bid] 决算，中标，定标
8. buyer [ˈbaiə] 买家
9. bid bond [bid bɔnd] 押标金
10. bid [bid] 标单，报价
11. bilateral trade [baiˈlætərəl treid] 双边贸易
12. bill of import exchange [bil ɔv, imˈpɔːt iksˈtʃeindʒ] 进口结汇单
13. bill of purchasing [bil ɔv, ˈpəːtʃəs] 出口结汇单
14. bill to purchase [bil tuː ˈpəːtʃəs] 进口结汇
15. black market rate [blæk ˈmɑːkit reit] 黑市汇率
16. buying rate [ˈbaiiŋ reit] 买入汇率
17. consignee [kənsaiˈniː] 收货人
18. capture market [ˈkæptʃə ˈmɑːkit] 争取市场
19. certificate of advance surrender for export exchange [səˈtifikit ɔv, ədˈvɑːns səˈrendə fɔː ˌeksˈpɔːt, iksˈtʃeindʒ] 预缴外汇证明书
20. clean bill bought [kliːn bil bɔːt] 购光票
21. client [ˈklɑiənt] 客人
22. collections [kəˈlekʃənz] 进出口托收
23. commercial procurement [kəˈməːʃəl prəˈkjuəmənt] 商业采购
24. cross rate [krɔs reit] 套汇率
25. contract [ˈkɔntrækt,] 合约
26. conversion rate [kənˈvəːʃən reit] 折合率，转换率
27. copy document [ˈkɔpi ˈdɔkjumənt] 副本单据
28. correspondence biding [kɔrisˈpɔndəns bidiŋ] 通信投标
29. correspondent [kɔrisˈpɔndənt] 代理银行
30. counter offer [ˈkauntə ˈɔfə] 还价(还盘)
31. counterpart fund [ˈkauntəpɑːt fʌnd] 相对基金
32. drummer [ˈdrʌmə] 驻外代表
33. delivery order [diˈlivəri ˈɔːdə] 发货单
34. developed country [diˈveləpt ˈkʌntri] 发达国家
35. developing country [diˈveləpiŋ ˈkʌntri] 发展中国家
36. direct trade [diˈrekt treid] 直接贸易
37. discount rate [ˈdiskaunt reit] 贴现率
38. discrepancy [disˈkrepənsi] 差异
39. distributor [disˈtribjutə] 分销商
40. documentary bill of exchange bought [dɔkjuˈmentəri bil ɔv, iksˈtʃeindʒ bɔːt] 押汇
41. down payment [daun ˈpeimənt] 分期付款之定金
42. dealer [ˈdiːlə] 经销商
43. European Common Market [juərəˈpiːən ˈkɔmən ˈmɑːkit] 欧洲共同市场
44. embargo [emˈbɑːgəu] 禁止出口
45. exclusive distributor [iksˈkluːsiv disˈtribjutə] 总代理
46. excess of export [ikˈses, ɔv, eksˈpɔːt,] 出超
47. excess of import [ikˈses, ɔv, imˈpɔːt] 入超

48. exchange position [iks'tʃeindʒ pə'ziʃən] 外汇头寸
49. exchange table [iks'tʃeindʒ 'teibəl] 汇兑换算表
50. exchange settlement certificate for L/C [iks'tʃeindʒ 'setlmənt sə'tifikit fɔ:ˌ el/si:] 结汇证实书
51. export declaration [eks'pɔ:tˌ deklə'reiʃən] 出口申请书
52. export trade [eks'pɔ:tˌ treid] 出口贸易
53. exports [eks'pɔ:tˌz] 出口签证
54. export [eks'pɔ:tˌ] 出口
55. exporter [iks'pɔ:tə] 出口商
56. foreign exchange ['fɔrin iks'tʃeindʒ] 外汇
57. facsimiles of authorized signature [fæk'simili ɔvˌ 'ɔ:θəraizd 'signitʃə] 有权签字人签字样本
58. favorable trade balance ['feiərəbl treid 'bæləns] 顺差
59. general imports ['dʒenərəl im'pɔ:ts] 普通进口签证
60. inward remittance ['inwəd ri'mitəns] 汇入汇票
61. import permit [im'pɔ:t pə:'mitˌ] 输入许可证
62. import trade [im'pɔ:t treid] 入口贸易
63. import [im'pɔ:t] 进口
64. importer [im'pɔ:tə] 进口商
65. imports amendment [im'pɔ:ts ə'mendmənt] 进口签证更改
66. imports without exchange settlement [im'pɔ:ts wið'aut iks'tʃeindʒ 'setlmənt] 不结汇进口签证
67. indirect trade [indi'rekt treid] 间接贸易
68. inquiry sheet [in'kwaiəri ʃi:t] 询价单
69. installment [in'stɔ:lmənt] 分期付款
70. international market [intə(:)'næʃənl 'mɑ:kit] 国际市场
71. international trade [intə(:)'næʃənl treid] 国际贸易
72. invitation [invi'teiʃən] 招标单
73. list of bid [list ɔvˌ bid] 决标单
74. L/C issued [el/si: 'iʃu:d] 结汇
75. letter of commitment ['letə ɔvˌ kə'mitmənt] 公证委托书
76. letter of consent ['letə ɔvˌ kən'sent] 同意书
77. letter of hypothecation ['letə ɔvˌ haipɔθi'keiʃ(ə)n] 质押借款书
78. letter of indemnity ['letə ɔvˌ in'demniti] 赔偿保证书
79. multilateral trade ['mʌlti'lætərəl] 多边贸易
80. manufacturer [mænju'fæktʃərə] 制造商
81. negotiate purchase [ni'gəuʃieit 'pə:tʃəs] 议价
82. notary ['nəutəri] 公证人
83. notarize ['nəutəraiz] 公证
84. outward remittance ['ɑutwəd ri'mitəns] 汇出汇款
85. offer ['ɔfə] 报价（发盘）
86. official rate [ə'fiʃəl reit] 官价
87. open tender ['əupən 'tendə] 开标
88. purveyor [pə'veiə(r)] 买办
89. penalty ['penlti] 违约金
90. performance invoice [pə'fɔ:m 'invɔis] 估价单
91. performance bond [pə'fɔ:məns bɔnd] 履约保证金
92. prevailing price [pri'veiliŋ prɑis] 牌价
93. prevailing rate of interest [pri'veiliŋ reit ɔvˌ 'intrist] 现行利率
94. procurement authorization [prə'kjuəmənt ˌɔ:θərai'zeiʃən] 采购授权书
95. protest [prə'testˌ] 拒绝证书
96. public biding ['pʌblik bidiŋ] 公开投标
97. quota ['kwəutə] 配额
98. retailer [ri:'teilə] 零售商
99. ration export ['ræʃən eks'pɔ:tˌ] 限额输出

100. restricted tender [ri'striktid 'tendə] 比价
101. supplier [sə'plaiə] 供应商
102. selling rate ['seliŋ reit] 卖价
103. shipper ['ʃipə] 货主
104. sole agent [səul 'eidʒənt] 独家代理
105. special imports ['speʃəl im'pɔːts] 专案进口签证
106. specification [ˌspesifi'keiʃən] 规格
107. successful bidder [sək'sesful 'bidə] 得标商
108. superintendent [ˌsjuːpərin'tendənt] 公证商
109. tracer ['treisə] 追询书
110. tabulation ['tæbjuləiʃən] 算价
111. tenderer ['tendərə(r)] 投标商
112. unsuccessful bidder ['ʌnsək'sesful 'bidə] 未得标商
113. unilateral trade ['juːni'lætərəl treid] 单边贸易
114. unfavorable trade balance ['ʌn'feivərəbl treid 'bæləns] 逆差
115. unified foreign exchange rate ['juːnifaid 'fɔrin iks'tʃeindʒ reit] 单一汇率
116. vendor ['vendɔː] 卖主，供应商
117. verifications [ˌverifi'keiʃən] 核对批准案件
118. validity [və'liditi] 有效期
119. waiver ['weivə] 弃权证
120. whole agent [həul 'eidʒənt] 总代理

# Appendix Ⅷ : Abbreviation of Terms 附录Ⅷ：缩略语

| | | |
|---|---|---|
| 1. ABS | area bounded staple fabric | 面黏合非织造布 |
| 2. ADL | acceptable defect level | 允许疵点标准 |
| 3. AQL | acceptable quality level | 验收合格标准 |
| 4. AH | armhole | 袖窿，夹圈 |
| 5. AUD | audit | 稽查 |
| 6. ACC | accept | 接受 |
| 7. ATTN | attention | 注意 |
| 8. ACT | actual | 实际的 |
| 9. ACC or A/C | account | 账目 |
| 10. ADV | advice | 通知 |
| 11. BL | back length | 背长，后身长 |
| 12. BR | back rise | 后裤裆 |
| 13. BNL | back neckline | 后领圈线 |
| 14. BNP | back neck point（BNPT） | 后领点 |
| 15. BSP | back shoulder point | 后肩颈点 |
| 16. BT | back tackler | 固缝机 |
| 17. BWP | back waistline point | 腰围线后中点 |
| 18. BW | back width | 后背宽 |
| 19. BMT | basic motion time | 基本动作时间 |
| 20. BD | battle dress | 战地军服 |
| 21. BW | before washing | 洗水前 |
| 22. BTN | button | 纽扣 |
| 23. B/E | bill of exchange | 汇票 |
| 24. B/L | bill of landing | 提货单 |
| 25. BK | black | 黑色 |
| 26. BL | bust line/level, breast line | 胸围线 |
| 27. BG | blue green | 青绿色 |
| 28. BS | body shirt | 紧身衬衫 |
| 29. BS | body suit | 紧身套装 |
| 30. BP | bust point | 胸高点 |
| 31. BH | button-hole | 扣眼 |
| 32. BTM | bottom | 衣脚，下摆 |

| | | |
|---|---|---|
| 33. CAD | computer aided design | 计算机辅助设计 |
| 34. CAM | computer aided manufacture | 计算机助制造 |
| 35. CAE | computer aided engineering | 计算机辅助工程 |
| 36. CAL | computer aided layout | 计算机辅助排料 |
| 37. CAP | computer aided pattern | 计算机辅助画样 |
| 38. CIF | cost, insurance & freight | 到岸价 |
| 39. CB or C/B | center back | 后中 |
| 40. CTN | cotton | 棉 |
| 41. CBL | center back line | 后中线 |
| 42. CBN-W | center back neck point to waist | 后颈点至腰 |
| 43. CO LTD | company limited | 有限公司 |
| 44. CF or C/F | center front | 前中 |
| 45. CFL | center front fold | 前中对折 |
| 46. CI | corporate identity | 企业标识 |
| 47. CMT | cutting, making, trimming | 来料加工 |
| 48. CS | commercial standards | 商业标准 |
| 49. CVC | chief value of cotton | 棉为主的混纺物 |
| 50. CLR. | color | 颜色 |
| 51. CTN NO. | carton No. | 纸箱编号 |
| 52. C&F | cost and freight | 离岸加运费价格 |
| 53. D | denier | 旦尼尔 |
| 54. DEPT | department | 部门 |
| 55. D/Y | delivery | 交付，出货 |
| 56. 2D | two-dimension | 二维 |
| 57. 3D | three-dimension | 三维 |
| 58. DK | dark | 深色 |
| 59. DB | double-breasted | 双襟 |
| 60. D&K | damaged & kept | 染厂对疵布的认赔 |
| 61. DOZ | dozen | 打 |
| 62. DBL NDL | double needle | 双针 |
| 63. EL | elbow line | 手肘线 |
| 64. Ne | english yarn count | 英制纱线支数 |
| 65. XL | extra large | 特大号 |

| | | |
|---|---|---|
| 66. XXL | extra extra large | 超特大号 |
| 67. E.G. | example gratia / for example | 例如 |
| 68. EXP | export | 出口 |
| 69. EMB | embroidery | 刺绣 |
| 70. etc. | et cetera=and so forth | 等等 |
| 71. FM | from | 从…… |
| 72. FAB | fabric | 布料 |
| 73. FAQ | fair average quality | 中等品 |
| 74. FAQ | free at quay | 码头交货 |
| 75. FIT | fashion institute of technology | 时装工艺学院 |
| 76. FB | freight bill | 装货清单 |
| 77. FC | franchisee chain | 连锁加盟店 |
| 78. FCFS | first come-first served rule | 先到先服务规则 |
| 79. F/F | fully fashioned | 全成形服装 |
| 80. FQC | field quality control | 现场质量控制 |
| 81. FIFO | first in first out | 先进先出法 |
| 82. FNP | front neck point | 前颈点 |
| 83. FOA | free on aircraft | 飞机上交货价 |
| 84. FOB | free on board | 离岸价 |
| 85. FWP | front waist point | 腰围线前中点 |
| 86. F/X | foreign exchange | 外汇 |
| 87. GM | gram | 克 |
| 88. GW | gross weight（GRS. WT.） | 毛重 |
| 89. G. | green | 绿色 |
| 90. GL | grain line | 布纹，经向标志 |
| 91. HV | hand value | 织物风格值 |
| 92. HT | height | 高度 |
| 93. H | hips | 坐围，臀围 |
| 94. HL | hips line | 臀围线 |
| 95. HH | horse-hair | 马尾衬，马鬃 |
| 96. ISO | International Organization for Standardization | 国际标准化组织 |
| 97. ITO | International Trade Organization | 国际贸易组织 |
| 98. IWS | International Wool Secretariat | 国际羊毛事务局 |

| | | |
|---|---|---|
| 99. ID Card | identity card | 证明，身份证 |
| 100. IN | inch | 英寸 |
| 101. JKT | jacket | 夹克 |
| 102. KG | kilogram | 千克，公斤 |
| 103. L | large | 大号 |
| 104. LB | pound | 磅 |
| 105. LHD | left-hand side | 左手边 |
| 106. L | line | 莱尼（纽扣规格） |
| 107. LG | length grain | 经向，直布纹 |
| 108. LOA | length over all | 全长，总长 |
| 109. L/C | letter of credit | 信用证 |
| 110. L/G | letter of guarantee | 保证书，担保 |
| 111. LS | long stick | 长码尺 |
| 112. Max | maximum | 最大限度 |
| 113. MKT | market | 市场 |
| 114. MIN | minimum | 最低限度 |
| 115. M | medium | 中码 |
| 116. MFTR | manufacturer | 制造商 |
| 117. Nm | metric number | 公制支数 |
| 118. MHL | middle hips line | 中臀围线 |
| 119. MTF | metal fibre | 金属纤维 |
| 120. M/C | machine | 机械 |
| 121. MAT | material | 物料，材料 |
| 122. Meas. | measurement | 尺寸 |
| 123. NDL | needle | 缝针 |
| 124. NK | neck | 领圈 |
| 125. NIL | nothing | 无，没有 |
| 126. NP | neck point | 肩颈点 |
| 127. NW | net weight（NT. WT.） | 净重 |
| 128. NS | new style | 时髦式样 |
| 129. OVRLK | overlock | 包缝 |
| 130. O/N | order No. | 订单编号 |
| 131. OZ | ounce | 盎司 |

| | | |
|---|---|---|
| 132. OJT | on-the-job training | 在职培训 |
| 133. OS | over-size | 超大型 |
| 134. PLS | please | 请 |
| 135. POS | position | 位置 |
| 136. POB | post-office-box | 邮箱 |
| 137. P | purple | 紫色 |
| 138. PA | polyamide | 聚酰胺 |
| 139. PKG | package | 包装 |
| 140. PAP | posterior armpit point | 腋窝后点 |
| 141. PB | private brand | 个人商标 |
| 142. PC | price | 价格 |
| 143. P/C | polyester/ cotton | 涤棉混纺物 |
| 144. PSI | per square inch | 每平方英寸 |
| 145. PP | poly propylene | 聚丙烯 |
| 146. P-O-R | product-on-rail system | 吊挂系统 |
| 147. PV | polyvinyl fibre | 聚乙烯纤维 |
| 148. PVC | polyvinyl chloride | 聚氯乙烯 |
| 149. PCS | pieces | 每件，每个 |
| 150. PKT | pocket | 口袋 |
| 151. PO NO. | production order No. | 生产制造单编号 |
| 152. PO NO. | purchase order No. | 采购订单编号 |
| 153. QC | quality control | 质量控制 |
| 154. QTY | quantity | 数量 |
| 155. QPL | qualified products list | 合格产品目录 |
| 156. QLY | quality | 质量 |
| 157. RS | right side | 正面 |
| 158. REF | reference | 参考，参照 |
| 159. RTW | ready to wear | 成衣 |
| 160. RM | room | 场所，部门 |
| 161. RN | rayon | 黏胶丝 |
| 162. REJ | reject | 拒绝 |
| 163. SZ | size | 尺码 |
| 164. SPEC | specification | 细则，规格 |

| | | |
|---|---|---|
| 165. SMPL | sample | 样衣 |
| 166. S | small | 小码，小号 |
| 167. SB | single breasted | 单排纽扣，单襟 |
| 168. SC | shopping center | 购物中心 |
| 169. SD | service dress | 军便服 |
| 170. SD | sports dress | 运动服 |
| 171. SP | shoulder point | 肩端点 |
| 172. SNP | side neck point | 颈侧点 |
| 173. SPM | stitch per minute | 每分钟线迹数 |
| 174. SPI | stitch per inch | 每英寸针数 |
| 175. SQ FT | square feet | 平方英尺 |
| 176. SA | seam allowance | 缝份，缝头，止口 |
| 177. SLV | sleeve | 袖子 |
| 178. SGL NDL | single needle | 单针 |
| 179. STY | style | 款式 |
| 180. TQM | total quality management | 全面质量管理 |
| 181. T-S | T-shirt | T 恤衫 |
| 182. TEL NO. | telephone No. | 电话号码 |
| 183. T/C | terylene/ cotton | 涤 / 棉织物 |
| 184. UBL | under bust-line | 下胸围线 |
| 185. WT | weight | 重量 |
| 186. WB | waist-band | 裤腰 |
| 187. WTO | World Trading Organization | 世界贸易组织 |
| 188. W | waist or width | 腰围，宽度 |
| 189. WL | waist line/ level | 腰围线 |
| 190. WS | wrong side | 反面 |
| 191. WX | women's extra large size | 女式特大号 |
| 192. XL | king size | 特大号 |
| 193. YD | yardage | 码数 |